Science on American Television

Illustration from the pamphlet *What Educational TV Offers You*, published by the Public Affairs Committee, 1954. Courtesy of Smithsonian Institution Archives.

Science on American Television

A History

MARCEL
CHOTKOWSKI
LAFOLLETTE

The University of Chicago Press

CHICAGO AND LONDON

MARCEL CHOTKOWSKI LAFOLLETTE is an independent historian based in Washington, DC. She is the author of several books, including *Science on the Air: Popularizers and Personalities on Radio and Early Television* and *Making Science Our Own: Public Images of Science, 1910–1955*, both published by the University of Chicago Press.

The University of Chicago Press, Chicago 60637
The University of Chicago Press, Ltd., London
© 2013 by The University of Chicago
All rights reserved. Published 2013.
Printed in the United States of America

22 21 20 19 18 17 16 15 14 13   1 2 3 4 5

ISBN-13: 978-0-226-92199-0 (cloth)
ISBN-13: 978-0-226-92201-0 (e-book)
ISBN-10: 0-226-92199-9 (cloth)
ISBN-10: 0-226-92201-4 (e-book)

Library of Congress Cataloging-in-Publication Data

LaFollette, Marcel C. (Marcel Chotkowski)
Science on American television : a history / Marcel Chotkowski LaFollette.
page ; cm
Includes bibliographical references and index.
ISBN-13: 978-0-226-92199-0 (cloth : alkaline paper)
ISBN-10: 0-226-92199-9 (cloth : alkaline paper)
ISBN-13: 978-0-226-92201-0 (e-book)
ISBN-10: 0-226-92201-4 (e-book) 1. Science television programs—History. 2. Television in science education—History. I. Title.
PN1992.8.S33L34 2013
791.45'66—dc23
2012003668

♾ This paper meets the requirements of ANSI/NISO Z39.48-1992 (Permanence of Paper).

I have gone so far as to contemplate, rather facetiously, the production of a newspaper story [on science] . . . in which you mix with the material that you want to get across such words as gold, cancer, war, certain aspects of psychological sex . . . perhaps it is the price you pay to get to your audience.

WATSON DAVIS, 1938

Contents

List of Illustrations • ix

CHAPTER ONE • I
Inventions and Dreams

CHAPTER TWO • 9
Experimenting with Illusion

CHAPTER THREE • 25
Elementary Education, Basic Economics

CHAPTER FOUR • 43
Dramatizing Science

CHAPTER FIVE • 59
Taking the Audience's Pulse

CHAPTER SIX • 71
Saving Planet Earth: Fictions and Facts

CHAPTER SEVEN • 85
Adjusting the Lens: Documentaries

CHAPTER EIGHT • 101
Monsters and Diamonds: The Price of Exclusive Access

CHAPTER NINE • 121
In Splendid Isolation: The Public's Television

CHAPTER TEN • 137
Defining What's New(s) about Science

CHAPTER ELEVEN • 155
Entrepreneurial Popularization

CHAPTER TWELVE • 171

Warning: Children in the Audience

CHAPTER THIRTEEN • 185

Rarae Aves: Television's Female Scientists

CHAPTER FOURTEEN • 197

The Smithsonian's World: Exclusivity and Power

CHAPTER FIFTEEN • 217

All Science, All the Time

Acknowledgments • 231
Notes • 233
Manuscript Sources • 275
Selected Bibliography • 277
Illustration Credits • 289
Index • 291

Illustrations

Frontispiece. Science via television

1a. David Sarnoff being televised at the World's Fair, April 1939   2

1b. Image of David Sarnoff seen by local television viewers, April 1939   2

2. Milton Cross testing plastic dishes on *What's New*, 1948   22

3. Demonstration of closed-circuit science education, 1965   26

4. Harvey Elliott White on *Continental Classroom*, 1958   33

5. Charles Greeley Abbot, Leonard Carmichael, Jennings Randolph, Bill Jenkins, and Paul Garber, 1953   38

6. How educators perceived their relationship to television, 1954   41

7a. *Our Mr. Sun* publicity brochure, 1956   53

7b. *Our Mr. Sun* animated characters, 1956   53

8. Bruno Rossi in *The Strange Case of the Cosmic Rays*, 1957   55

9a. Television simulation of *Friendship 7* flight, 1962   79

9b. Map of *Friendship 7* flight, 1962   79

9c. "Glenn's voice" on *Friendship 7* flight, 1962   80

10a. Television coverage of *Apollo 11* lunar landing, 1969   82

10b. Television coverage of *Apollo 11* lunar landing, 1969   82

11. S. Dillon Ripley speaking to television cameras, circa 1970s   91

12. *Apollo 7* astronaut Walter Schirra filming *Smithsonian Adventure*, 1971   103

13. David L. Wolper and the Hope Diamond, 1974 113

14. Press conference about the Three Mile Island nuclear accident, March 1979 149

15. Carl Sagan, 1978 157

16. Educational booklet for *Mr. Wizard's World*, 1983 176

17. Adrian Malone and Thomas H. Wolf, 1987 209

CHAPTER ONE

Inventions and Dreams

Our inventions are wont to be pretty toys, which distract our attention
from serious things. They are but improved means to an unimproved end.

HENRY DAVID THOREAU, 1854[1]

[Television] is an art which shines like a torch of hope in a troubled world.

DAVID SARNOFF, 1939[2]

We can reason out to a certain extent what the men and women of tomorrow will be free to do, but we cannot guess what they will decide to do.

H. G. WELLS, 1939[3]

"We are going to have television in a real big way the moment peace is
declared," wrote inventor Allen B. Du Mont in early 1945.[4] What emerged
from the laboratory and eventually claimed a preeminent place in American lives was not, however, the ideal predicted by scientists and engineers
like Du Mont. Television evolved into more of an "improved means to an
unimproved end" than a dynamic tool for civic education.[5] That outcome—
its trajectory determined primarily by economic forces—had major consequences for how the medium would present, assimilate, and transform
popular science in the United States.

Imagined television had enlivened novels and Hollywood movies throughout the early twentieth century, validated by news reports about the inventions of C. Francis Jenkins and Philo T. Farnsworth.[6] Once the Federal
Radio Commission authorized experimental broadcasts, Radio Corporation of America (RCA) and other companies arranged to promote *consumer*
television at the 1939 New York World's Fair.[7] As the fair designers incorporated the technology into art and exhibits about the "World of Tomorrow," they visualized television's emerging role within American culture
as a window on that and other worlds.[8]

During the NBC telecast of the fair's opening ceremonies, cameras
panned slowly across the iconic Trylon and Perisphere, and showed the
crowds and parade. At the time, only a few hundred receivers were operating in the New York City area; a thousand people, at most, watched

FIGURE 1A. What the television camera broadcast. RCA president David Sarnoff is shown testing television transmission systems for the World's Fair opening, April 1939. Courtesy of Smithsonian Institution Archives.

FIGURE 1B. What the audience saw. The image of RCA president David Sarnoff as seen by local television viewers, April 1939. Courtesy of Smithsonian Institution Archives.

President Franklin D. Roosevelt and other dignitaries via television.[9] Within a few weeks, though, RCA was regularly broadcasting from its exhibition hall—jugglers, visitor interviews, historical reenactments, Fred Waring's orchestra, Walt Disney cartoons, beauty contests, and "actualities" (live broadcasts of events as they were occurring), all the content that would later become modern television's bread and butter.[10] Broadcasters tried out production techniques; engineers calibrated lighting and sound standards; producers learned how to beguile viewers; and directors discovered that, without close-ups, events could appear "stripped of action and interest"—"Every ounce of interest and variety that can be packed into the frame is necessary to attract and hold the attention of an unseen audience."[11]

Science's latest "brainchild," newspapers explained, could create magic from a distance: "You press a button, turn a dial, and watch a drama being acted miles away. . . . You are there; the illusion is complete."[12] By exploiting that illusion, television rewrote the relationship between audiences and what they saw. The technology enabled and encouraged viewing from afar: "The camera with the telephoto lenses will poke closer than [World's Fair] visitors can hope to push, and . . . television, with its reputation for intimacy, will reveal what advantages, if any, are promised for those nongregarious souls who sit comfortably at home looking through space with a length of antenna wire as a new sort of telescope."[13]

Although national broadcasting and the manufacturing of sets were suspended during World War II, engineering research and business planning continued.[14] Because early television sets emitted an unappealing greenish glow, engineers developed a rounded cathode-ray tube with a clearer, black-and-white image viewable from wider angles.[15] CBS, NBC, and DuMont Laboratories began preparing for intercity connection via cable, and for remote telecasts of proven audience winners like football games and wrestling matches.[16] The Federal Communications Commission (FCC) and the industry resolved questions about engineering standards, broadcasting limitations, and spectrum assignments.

The networks' plans copied radio's model for success: the schedule would be centered on entertainment. News, information, inspiration, and education would be interspersed among comedy, drama, sports, and anything else likely to sustain a viewer's attention to commercial advertisements. Television would offer "the complete form of radio entertainment," combining "sight and sound," news "embellished" with pictures, merchandise demonstrated, and public figures on view.[17] For those who dreamed of a medium dedicated to higher social purposes like informing citizens about science, television's ties to radio ("its lineal predecessor") hindered attempts to reach such goals.[18] The same business interests that had shaped radio would own the television stations and networks; the federal regulations devised for radio during the 1920s and 1930s would be adapted to the new medium; and because radio income in 1945 alone topped $300,000,000, broadcasters had deep pockets and well-established business relationships with advertisers who were eager to buy television time.[19]

Jack Gould, one of the first newspaper journalists to review television, observed in 1946 that little critical thinking was being applied to how this technology would be used for the public good.[20] In contrast to the energetic political debate concerning control and application of atomic energy, no similar public discussions involved the future of television. Instead, television's "idealistic potential" was being confused with its "practical probability," and the "bulk of thinking . . . [was] preoccupied with the marvel of sending pictures through the air and not with what kind of pictures would be sent."[21] The networks, rather than public officials or citizens, were determining the direction this powerful technology would take. Because broadcasters had been convinced that more listeners would twist a radio dial to be entertained than to be educated or enlightened, they made identical assumptions about how television viewers would behave. And because television production was far more expensive than radio, opting for something other than entertainment contradicted both accepted wisdom and prudent business practice.

DANCING ILLUSIONS OF SCIENCE

When historians of technology write about the "Automobile Age," they refer not just to automotive engineering, assembly-line production, stoplights, and the Interstate Highway System but also to how automobiles affected society, culture, land-use patterns, and many other aspects of modern life. Historians of science likewise trace the "Atomic Age" through cultural and political change as well as increasing megatonnage and missile range. Within a decade after the World's Fair, Americans were experiencing the "Television Era." Television was shaping American culture and being shaped by it, ironically becoming simultaneously a "global village" (whenever tens of millions of people watched the same telecast) and a "vast wasteland" (whenever millions of viewers found nothing satisfying and switched channels). Television invited royalty and rioters, presidents and poets, to appear in front of the cameras, selectively revealing the backsides and backstreets, bloopers and banality. Television combined sight and sound, forced spontaneity to follow a script, and became the ultimate agent of illusion. Television gave viewers laughter, suspense, vicarious adventure, and useful information, interspersed with commercials. Via remote broadcasts from street corners, legislatures, state funerals, and eventually outer space, television brought the parade of life into view, all the while giving each viewer, even if she were sitting alone in her living room, a sense of shared context with others in the television audience. In 1945, the word *television* connoted a technology or perhaps a new piece of furniture; within a decade, the term encompassed context, culture, industry, content, experience, and validation of importance ("Have you been on television?").

For science, the visualization of researchers and their laboratories at first seemed a promising possibility. A number of scientists experimented with using television to lecture or teach, and the dynamic accomplishments of fields like physics, biology, astronomy, and chemistry attracted producers and dramatists in search of inspiration and content. Very quickly, universities, scientific associations, and other nonprofit organizations like the Smithsonian Institution began facing important choices: should they stick with traditional means for public outreach (such as print or public lectures), invest in producing ephemeral content for educational television channels where only a few knowledge seekers might turn, or cooperate with commercial broadcasters? Given that the second option required substantial financial and institutional resources, many organizations contemplated the third. Television executives, though, demanded concessions and claimed that the type of programming acceptable to scientists (such as a roundtable discussion among brilliant physicists) would not draw sufficient viewers to compete against comedies and game shows.

Despite such tensions (or possibly because of them), American television arranged throughout its first half century a varied but paradoxical banquet of factual and fictional images and information related to science. Mr. Wizard performed magical experiments with milk bottles, and Dr. Frank Baxter bantered with a sneering Mr. Sun. Television took everyone to the Moon and back. Carl Sagan journeyed through the *Cosmos*; Marlin Perkins and Jane Goodall traveled to the jungle; *NOVA* offered weekly tastes and tidbits of science mixed with engineering. *Star Trek* exploited fantastic scientific devices to peer into the future, just as *CSI: Crime Scene Investigation* later used DNA analysis to solve past crimes. Electrons and embryos, blood testing and blasting caps, chimpanzees and chemistry, fictional Frankensteins and charismatic Nobel laureates—television used them all to offer windows onto the world of science and beyond. The multi-billion-dollar television industry transmuted science into entertainment, incorporated it into drama, dramatized science-related events, and muscled educational programs toward the edges of the daily schedule.

The science that eventually appeared on television resulted from contentious negotiations among scientific experts, institutional administrators, television executives and directors, and the foundations and corporations willing to underwrite production, with the television audience weighing in via remote control. Viewers gravitated toward science programs that were entertaining as well as relevant, preferring relevance within fictional entertainment and entertaining approaches to educational subjects. Television eventually established a level of expectation among viewers for *which* aspects of science really mattered. By the 1930s, radio had begun to privilege scientists over science, personalities over facts. Television slowly squeezed theories, processes, explanations, and conclusions into sound bites and accentuated the social problems, moral dilemmas, ethical challenges, and controversies related to science. The contrast between *The Johns Hopkins Science Review* in 1948 and *NOVA* in 1988 transcended technical innovations such as the shift from black-and-white to color transmission or the use of multiple cameras and special effects. Over those forty years, the focus of American television's science shifted toward politics and morality, toward the illusory and visually dynamic, toward social context and scientific celebrities. Fortunately, dozens of committed popularizers also persisted throughout the years, hosting, writing, and producing creative programs that were outstanding examples of how to bring technical content into the studio and make it sing.

The dance of popularization joins two partners, the scientific expert and the interpreter, who perform for audiences who either watch in fascination or change the channel. For centuries, lecture halls and print (magazines, newspapers, books) had offered comfortable venues allowing willing scientists to engage the public with minimal economic investment. The advent of mass broadcasting created bustling electronic marketplaces of the air, where the emphasis on profitability favored sophisticated, expensive entertainment over low-key education. It would be a mistake, as Robert T. Bower cautioned many years ago, to explain the impact of television in terms of the technology alone, without regard for adjacent influences, because social, economic, and historical forces (including those set in motion by the advances of science) affected public reception of the images displayed on the screen.[22] Nevertheless, in the television era, by reconstructing science's own messages, the medium assembled a new creation suitable to the small screen—science at a distance, science for the millions, science transformed into an entertaining illusion.

CHAPTER TWO

Experimenting with Illusion

Play it honestly and factually and give the viewer credit for intelligence.

LYNN POOLE, 1952[1]

During World War II, the General Electric Company television station in Schenectady, New York, telecast a mix of experimental content to the few hundred receivers in the area. The didactic nature of the occasional science shows—such as describing the uses of synthetic materials—undoubtedly accounted for the fact that, between 1939 and 1944, "light opera music" drew twice as many viewers as science.[2] The first regular science-related network series was not much more scintillating. *Serving through Science*, sponsored by U.S. Rubber Company on the DuMont Network from June 1946 to May 1947, featured a half hour of *Encyclopaedia Britannica* films and moderated discussions.[3]

The creative force behind *Serving through Science*, broadcasting executive Miller McClintock, had begun his career not as a scientist but as a student of literature and public policy. After finishing a Ph.D. at Harvard in 1924, McClintock helped establish a bureau of street traffic research to tackle Boston's notoriously hazardous roads. An advocate of such innovations as one-way streets and elevated expressways, he is credited with convincing local Chicago officials to ban horse-drawn vehicles from city streets in 1926. As he worked on various public safety issues, McClintock became intrigued with the role that advertising plays in civic education. From 1933 to 1942, he worked for the Advertising Research Foundation and the Advertising Council Inc. and then, between 1943 and 1944, served as president of the Mutual Broadcasting Company. By the time McClintock hosted *Serving through Science*, he was a board member and consultant to Encyclopaedia Britannica Films, and was attempting to persuade the New

York Film Council that television offered marketing opportunities for educational films.[4] The first science series on television thus sprang from a confluence of circumstances that would mark similar successes for the next half century: a clever entrepreneur with the ability to draw from expert knowledge in a wide range of fields, in and out of science; awareness of advertising and public relations techniques; and the coordination of content with corporate sponsors or underwriters.

Serving through Science's relatively unsophisticated content probably seemed novel enough to the 100,000 or so people who then owned sets. In 1947, watching television represented an "exclusive hobby"—or something to experience with others in a New York bar. Within a year, the number of receivers in homes doubled; manufacturers began shipping thousands more to showrooms. Dozens of stations began operation throughout the country, and coast-to-coast broadcasting seemed possible. In major urban areas, programs could attract around 750,000 viewers, with a few special telecasts, like baseball's World Series, drawing a million or more.[5] Because sets were still relatively rare, neighbors often watched with those who had already made the consumer leap. Americans, Philip Wylie observed, were spending "half [their] time swooning in the gloom and the rest sweating to pay installments on the sponsored merchandise."[6] This dynamic growth created markets for better content as stations recognized that novelty (not reruns) drew the masses of viewers needed to sell advertising time. The sponsors themselves, though, were no more concerned with the public good or with why people enjoyed their programs than with "why people drive on the highway beside which [the same company] erects billboards."[7] If a chemistry demonstration interested viewers, great. If something else interested more people, then all the better.

Station managers scoured their communities for potential programs and invited scientists and educators to develop programs for the same reasons that radio station managers had invited experts to give lectures on the air during the 1920s. In March 1948, WMAL-TV in Washington, D.C., pioneered real-time science programs that filtered views of Mars, Saturn, and the Moon through the U.S. Naval Observatory's forty-inch telescope.[8] The following year, the station's cameras peered through a microscope (a television first) so viewers could watch blood coursing through the veins of a mouse and, a few days later, the station broadcast images of a lunar eclipse caught by the observatory's telescope.[9] Although the technical quality rarely matched that of network programming, local television welcomed experimentation. Three of the most important pioneering science personalities—Roy K. Marshall, Lynn Poole, and Earl S. Herald—began their careers at local stations and directed their programs at adult audiences.

THE NATURE OF THINGS

Philadelphia station WPTZ-TV first invited astronomer Roy K. Marshall, director of the Franklin Institute's Fels Planetarium, to appear and promote a new exhibition called "A Trip to the Moon," and he soon began hosting remote telecasts from the museum. On February 5, 1948, Marshall premiered in *The Nature of Things*, a general science series that moved to the NBC network in December 1948, where it ran in prime time through September 1950 and as a summer replacement show and weekend afternoon feature until the fall of 1953.[10]

Marshall had had some experience in radio, so he understood the importance of precise timing and rehearsals, and how to pitch technical explanations for general audiences. The middle-aged Marshall was dark-haired with piercing eyes and a small moustache, personable and comfortable in front of the camera. After completing a Ph.D. at the University of Michigan in 1932, he had taught astronomy and mathematics at several universities and then joined the Fels Planetarium. In 1949, soon after the show became a hit, he became head of the University of North Carolina's astronomy department and director of the new Morehead Planetarium, and, according to the newspapers, "one of the best-paid astronomy professors in the country."[11] In addition to hosting *The Nature of Things*, Marshall appeared as paid pitchman in five-minute commercials for the Ford Motor Company, explaining how carburetors and speedometers work. He resigned the North Carolina job in 1951 to concentrate full-time on broadcasting, becoming the educational director for commercial radio and television stations owned by the *Philadelphia Inquirer*.[12] As one of the first "celebrity scientists" of the television era, Marshall made regular appearances on summer variety programs, narrated the live broadcast of a solar eclipse for *Howdy Doody*, and hosted special telecasts from American Medical Association meetings.[13]

Marshall's goal on *The Nature of Things*, critic John Crosby wrote at the time, was "to convince as many people as possible that science is not black magic, that it isn't even very hard to understand, and that it can be a good deal of fun."[14] Television's "show business," Marshall argued, allowed scientists to promote their work to the public and to share the "fun" inherent in research.[15] Although most episodes of the "pleasant, undidactic" series focused on astronomy or astrophysics, Marshall also explored practical, earthly topics like firefighting and steam generation and was willing to experiment with "actuality." In March 1948, a few days after the WMAL-TV telecast, Marshall's studio technicians attached a camera to a large telescope and transmitted their own live glimpse of the lunar landscape.[16]

In television's early days, however, that pretense of realism could occasionally alarm the audience. After one show, viewers complained that Marshall had revealed the "secret" of the atomic bomb. The Atomic Energy Commission (AEC) dutifully reviewed the script and then announced that, not only was there no secrecy violation, but Marshall had also performed a public service by helping to educate citizens about this important topic.[17]

HOPKINS GOES ON THE AIR

Television's development as a national communications medium coincided with a time during which scientists and their professional organizations sought to exploit their wartime accomplishments in a campaign for public funding of research.[18] Freed from the restraints of military secrecy, chemists and biologists stepped triumphantly into the spotlight to explain the wonders of DDT and new antibiotics like penicillin; physicists explained the basic principles of atomic energy.[19] *The Johns Hopkins Science Review* took full advantage of this enthusiasm for public communication when it premiered in 1948, thanks to the creative efforts of Lynn Poole, the university's public relations man. Poole had studied art, worked at the Cleveland Museum of Art, and then directed the Walters Art Gallery education department in Baltimore from 1938 to 1942.[20] During World War II, he served as an army public relations officer and, in 1946, became the first director of public relations at Johns Hopkins University.[21] When Baltimore station WMAR-TV began soliciting program ideas, Poole suggested a series that would revolve around the university's scientific and engineering faculty.

Poole had planned to remain in the background as writer and producer, but when he appeared on camera to soothe a nervous guest, he became an instant star, and he remained on the air in various science-related series until 1960.[22] His on-screen persona matched television's emerging archetype—cool, reserved, at ease—yet he always conveyed deep concern that every viewer understand what the experts were saying.

From the opening montage of the Hopkins campus to the closing assurance that next week's program would be just as fascinating, the series projected a carefully crafted image of academic science as a trusted authority worthy of public support. Ample institutional underwriting proved key to the series' survival, and in return for subsidizing all production costs (slightly under $30,000 for fifty-two programs a year), the university gained invaluable publicity.[23] When newspapers wrote that the program demonstrated that "TV Really Is Educational" or reported that Lynn Poole competed well with Milton Berle, such praise validated the investment.[24] Commentators consistently cited *Johns Hopkins Science Review* as a television bright spot during the early 1950s, its "thoroughly worthwhile" programs faithfully translating "the complexities of the scientific world into language readily comprehended by mass audiences."[25]

To Poole, television's greatest potential lay in the ability to combine sound with motion pictures and send those images directly to the home. Exploiting visuality would allow informational programs to compete in an entertainment marketplace.[26] Demonstrations, of course, had to conform to the medium's physical constraints, creating imaginative video illusions while avoiding scientific topics simply too abstruse for easy visualization. For a program on volcanoes, Poole and his geologist guest presented maps, cross sections of the earth, samples of lava, films of volcanic eruptions, and a model of Vesuvius. To demonstrate the flow of electrons, Poole filled a glass jar with puffed wheat cereal and pumped in air. Popcorn was used to represent molecules; one hundred mousetraps were baited with sugar cubes to simulate an atomic chain reaction.

Despite the difficulties of persuading scientists to appear—overcoming busy schedules and a disdain for the medium—Poole resisted replacing experts with actors. What scientists "lacked in thespian talents," he explained, "they made up in their enthusiasm and authority."[27] Viewers want to see real experts, not imitators, he argued, so he taught guests how to stand within camera range, pick up objects slowly and steadily, and refrain from weaving "in the cobra-like fashion used in the classroom."[28]

The show also pioneered the practice of having experts "appear" via phone or remote telecast. While one doctor in the Baltimore studio used a new type of X-ray machine to reveal steel splinters embedded in a patient's back, two other physicians, watching television monitors in New York and Chicago, offered comments via a telephone connection. Even though his scripts generally mimicked the formality of university lectures, Poole sometimes cracked little jokes or used models in bathing suits to hold his audience's interest, but he drew the line at "making a monkey out of a famous scientist or turning his presentation into a circus."[29] The cardinal rule was "Play it honestly and factually and give the viewer credit for intelligence."[30]

On Poole's first program, March 9, 1948, "All about the Atom," physicist Franco Rossetti described recent advances in nuclear physics. Over the next eight years, Poole interviewed several hundred scientists and engineers. Viewers learned about rockets, houseflies, influenza, schistosomiasis, light polarization, and forensic anthropology, and watched a lobotomy. Although many experts were "wary of submitting themselves to the new public outlet," producer Leo Geier explained, the series was able to attract eminent guests because it had "a reputation for presenting material in its true light [and] guests were not asked to do ridiculous things."[31] On "The World Is an Atom," Donald H. Andrews demonstrated how knowledge of atoms and molecules would assist development of new technologies. In "Don't Drink That Water," Abel Wolman, John Charles Geyer, and Cornelius Kruse enlivened the history of water purification and the science of sanitary engineering with simple animations, film of microscopic organisms, and models of filtration and sewage treatment plants. "The Magnificent Microscope" featured a new television microscope developed by Princeton University and RCA Corporation Laboratories. In "A Closer Look at Mars," astrophysicist John Strong, explaining new information about the "red planet," seemed a stereotypical academic scientist—tweed jacket, pipe, and wire-rimmed glasses.

Scientists appearing on these programs must have found the experience sufficiently novel to justify the effort, although the amount of preparation time eventually became a rationalization for declining to appear. Two curators in the Smithsonian Institution's division of physical anthropology, T. Dale Stewart and Marshall T. Newman, appeared in an episode called "Skeletons in the Closet." After a three-hour meeting with the curators, Poole and his staff prepared a draft script, requesting corrections and additional details, and then the "television men" returned later to the Smithsonian to review potential props. The curators spent several more hours mounting two skeletons from the collections, and an assistant accompanied the specimens to Baltimore to ensure safe transit and storage. On the day of the broadcast, Stewart and Newman arrived at the studio at 9:30 a.m. and participated in rehearsals. After the evening's live broadcast, they repacked the specimens for the return journey. By then, twenty-two stations around the country were carrying *Johns Hopkins Science Review*. "Considering what this means in terms of audience," Stewart assured his supervisor, "it seems to me that the Smithsonian got some good publicity at a very low cost."[32]

By 1950, Poole had published the first book explaining how to communicate "science via television," and skeptical scientists had difficulty dismissing such efforts as "mere" television.[33] The university's president, biophysicist Detlev W. Bronk, even admitted that research institutions "would be remiss" in their duty to inform the public if they did not exploit the medium to the fullest, a notable shift from his predecessor Joseph Sweetman Ames, who, in the 1930s, had characterized the science news in the *New York Times* as deplorable sensationalism.[34]

SCIENCE AS ACTION

Another groundbreaking series aimed at broad audiences was *Science in Action*, a regional project launched in 1950 by the California Academy of Sciences with underwriting from a local bank. *Science in Action* remained on the air in California (and was carried on a few other educational channels elsewhere via kinescope), with over thirty episodes a year, through 1966.[35] Although the staff approached the venture with apprehension, *Science in Action* proved to be a hit, winning awards and attracting visitors to the academy's museum, herbarium, and Steinhart Aquarium. Live demonstrations and an "Animal of the Week" feature were especially popular. Academy director Robert C. Miller and program producer Benjamin C. Draper wanted *Science in Action* to live up to its name: "The rule the Academy follows is this: show the thing itself, if that is not possible, show a film, if that is not available, use a photograph or drawing."[36]

Zoologist Tom Groody, the first host, and his successor Earl S. Herald, the academy's curator of aquatic biology, adopted on-screen roles similar to those of Marshall and Poole, posing questions to the guests and assisting in "demonstrations which the viewer very likely has in his own mind at the same moment."[37] On a set designed to look like a laboratory, Herald would use pointers, still photographs, and innovative camera techniques to create what Poole had begun calling the "atmosphere of here and now," that is, the illusion that viewers were standing next to the demonstrators. Both the Johns Hopkins and California shows enhanced that perception by confining action to an "uncluttered, small space" seemingly part of a larger laboratory.[38] Attired in a white lab coat, Herald would shake his guests' hands, welcome them, and thank them for coming, thereby creating a sense of familiarity and camaraderie.[39] His slight drawl made him sound a bit like a television cowboy, and brought a warmth to the host's role that differed from Poole's studied East Coast sophistication. Guests ranged from Nobel laureates (Melvin Calvin explained photosynthesis) and a young boy with a pet gopher (part of the "Animal of the Week" segment) to an air force brigadier general describing radar defense systems. Episodes focused on paleontology, the history of coffee cultivation, birdsong identification, and protozoology. Biologists described raptors and termites; chemists raved about new plastics; a professor of entomology presented "unusual insects" from around the world; a fisheries biologist recounted the habits and habitats of the striped bass; and a physicist explained how a Geiger counter works.

By 1953, *Science in Action* was attracting praise from television critics because it was "not a dull, blackboard talk, filled with scientific terms," and drawing applause from the scientific establishment because scientists were presented "realistically, truthfully, and entertainingly": "The show . . . has scripts approved by the scientists who appear. . . . There is no sensationalism, no talking down to the audience, and no glossing over of unsolved problems."[40] It was underwriting from a commercial sponsor, however, that ensured survival of *Science in Action*. The name of American Trust Company appeared on every page of the show's publicity material, and at one point the bank even hired advertising agency McCann Erickson to promote the series.

ACTUALIZATION

Early in television's development, Gilbert Seldes had predicted that its "prime fact" would be "instantaneous and complete transmission of actuality," the "artificially created illusion" reflecting real events.[41] Whenever science programs projected demonstrations into the viewers' living rooms or simulated experiments, tests, and expressions of amazement, they transformed the television set into a window on the laboratory. Networks also filmed a few science-related actualities along with political conventions, athletic games, and beauty contests. On June 30, 1954, a partial solar eclipse offered a perfect opportunity to showcase the medium's technical advantage. Humans cannot look directly at the Sun without damaging their eyes. Given the timing of the eclipse along its North American path, television offered safe, enhanced convenience for armchair amateur astronomers. As the *New York Times* explained, people could go outside with special viewing glasses at 6:06 a.m. (although the day turned out to be cloudy in New York) or remain in their pajamas and "turn on the television set and see the whole celestial display from Minneapolis," where the American Astronomical Society had arranged a telecast.[42] *Today* carried the Minneapolis feed, and some stations around the country broadcast kinescopes of the entire eclipse. During the 1930s, radio personalities had accompanied famous eclipse expeditions; there were broadcasts from the jungle, dirigibles, and planes. Now, television viewers could *watch* eclipses in comfort while scientists provided narration from the studio.

Atomic testing offered even more spectacular actualities, with those broadcasts presaging how networks would later treat space missions. Reporters had been allowed to observe the 1946 Operation Crossroads tests in the Pacific, but in 1952, the U.S. government permitted live television broadcasts from the AEC facility in Yucca Flat, Nevada. Ostensibly a scientific experiment, the Operation Tumbler-Snapper test on April 22 was scheduled at a convenient time for maximum network coverage, and local newspapers ridiculed it as "Operation Big Shot" because of the hundreds of journalists and broadcast personalities who attended.[43] That morning, the reporters "perched like animals in a zoo on hillside boulders," and thanks to a relay tower on a nearby mountain peak, which sent signals to Los Angeles and out to the networks, millions of other people could watch via television.[44] When the atomic bomb was dropped from an airplane, "hell burst from the skies."[45] Technical quality of this first broadcast was poor, however. Many viewers saw only "a tiny white spot in the middle of a large black oval" because the initial blast overloaded the camera lens; others complained that the event took place so fast that it seemed "a little anticlimactic."[46] By the next spring, the networks and AEC, with the assistance of the Advertising Council, were better prepared. A Sunday-afternoon program offered a drama-building peek at the Operation Doorstep test site. Television reporters like Walter Cronkite and Chet Huntley provided color commentary for the explosion on the morning of Tuesday, March 18, and a follow-up program, "The Effects of an Atomic Explosion," displayed the structural damage to test houses at ground zero.

Television critics reviewed these broadcasts as if the tests had been prime-time variety shows or football games. Jack Gould commended the networks for "fine public service" and said the telecasts "caught the sense of drama," but he faulted commentators for inadequate preparation and bland discussion.[47] Sonia Stein complained that "the blast was a bust on television," with "nothing frightening, dramatic, illuminating, clarifying, or even very interesting" about seeing the event "through television's eyes."[48] "Anything that you can put in your front parlor loses its terror potential," she wrote, and without Huntley's interpretation, the audience would have had "no sense at all" of what had occurred. Other critics suggested that competing broadcasts not only attracted greater audiences but had also demonstrated that "immediacy isn't everything" and that "well-edited highlights" of future tests might be more useful.[49] Around eight million people tuned in to see the explosion (one-third of the number of viewers who had watched the Oscar award ceremony the previous week), but Sidney Lohman actually recommended that New Yorkers watch a different "spectacle" that morning—the tip-to-tail coverage of the St. Patrick's Day parade—because it would have "more enjoyable connotations."[50]

The networks soon learned how to exploit similar illusions of actuality within their regular newscasts. Radio had already expanded the audiences for fast-breaking news, improved field reporting techniques, and introduced dramatic flourishes.[51] On television, the producers of NBC's *Camel News Caravan* (starring John Cameron Swayze) and CBS's *Douglas Edwards with the News*, which both premiered in 1948, combined movie newsreel film clips with texts read by on-camera anchors, creating "cocktail style" news. Factual reports and films were mixed with "a brew of other ingredients" like live interviews, still photographs, charts, graphs, and animation.[52] Visualization transformed the news from "names on paper" into "faces, vivid and realistic, recognizable," which could prompt new "judgments and prejudices," reinforce old attitudes, or merely fortify "the illusion of knowledge."[53] When the FCC ruled that radio and, by extension, television should "be as free as the press" and allowed editorializing as long as the platform included a diversity of opinion, that decision paved the way for a change with important implications for science news.[54] Scientific facts would be increasingly colored by tints of political controversy.

As Charles Siepmann and Sidney Reisberg pointed out in 1948, news is never "all we need in order to be intelligently informed."[55] During the 1940s, radio documentaries like *You and the Atom* (1946) provided additional insights by mixing factual and dramatic techniques to probe modern political and social issues.[56] Television developed a similar approach to documentaries tackling controversial topics such as medical research policy (230,000 *Will Die*, 1954), the promise of atomic power (*Three Two One Zero*, 1954), and whether atomic testing posed a health hazard to civilians (*Is Atomic Testing Endangering Your Life?*, 1957). One of the more inventive presentations, *1960? Jiminy Cricket!*, produced by ABC in 1947, animated the forecasts of twenty scientists to examine what life would be like in the future, with fictional characters Donald Duck, Jiminy Cricket, and the Seven Dwarfs outlining America's likely "needs and resources."[57]

CONVERSATIONS AND PERSONALITIES

Thanks to television, Americans could now see the scientists they had been reading about or hearing on the radio. Interview shows, panel discussions, and scripted debates favored experts who were glib, witty, concise, and willing to perform on camera, often for social or political reasons. In a 1950 appearance on *Today with Mrs. Roosevelt*, hosted by the former First Lady, Albert Einstein made one of his first public statements about the hydrogen bomb, declaring that development of that weapon would raise the potential of "radioactive poisoning of the atmosphere and hence annihilation of any life on earth."[58] On ABC's *Horizons*, which ran live on Sunday evenings from 1951 to 1955, Columbia University professor Erik Barnouw interviewed such guests as anthropologist Margaret Mead, who discussed "the future of the family."[59]

The top public affairs shows—NBC's *Meet the Press* (1947–) and CBS's *Face the Nation* (1954–)—paid only occasional attention to science. Television was already favoring celebrities and "personalities." Only 15 of the 188 episodes of Edward R. Murrow's popular *See It Now* (CBS, 1951–1958), for example, featured interviews with scientists.[60] The series' fifth episode, about neuroscience research, included a remote broadcast from MIT. Although other 1952 and 1953 episodes covered bomb testing in Nevada and construction of the Savannah River hydrogen-weapons plant, medical research attracted far more attention than nuclear physics. Murrow discussed heart disease and cancer with three different scientists in 1953 and interviewed polio researcher Jonas Salk twice during 1955 (before and after the Salk group announced development of a vaccine). Murrow's celebrity-oriented program, *Person to Person* (CBS, 1953–1961), where guests were interviewed in their homes and offices, featured only three scientists and engineers through the years, not the least because, as James L. Baughman observes, "fame" rather than accomplishment determined who was invited.[61]

Television ripped aside anonymity, and when the camera turned on scientists, it sought to transform them into friendly personalities.[62] One of Murrow's most memorable *See It Now* programs was "A Conversation with Dr. J. Robert Oppenheimer." By January 1955, Oppenheimer was such a controversial figure that CBS refused to promote the show to advertisers (government hearings on the physicist's security clearance had taken place seven months before), so coproducers Murrow and Fred Friendly underwrote the promotion themselves.[63] The "avalanche" of adverse viewer and critical responses never occurred, however, an outcome that critic Jack Gould attributed to Murrow's "generalized, low-keyed and non-controversial" questioning.[64] At a time when Oppenheimer was being demonized, Murrow had elicited a different side of the man, focusing their conversation on philosophy, human freedom, and the future of science. The interview's "cameo of a mind at work" also showed that the medium could allow scientists and other intellectuals to describe their research and ideas in depth and in ways that viewers might find interesting.[65]

Television producers preferred livelier encounters, of course, such as when physicist Edward Teller debated chemist Linus Pauling about the merits of the proposed 1958 nuclear test ban. In a program televised live on San Francisco station KQED-TV and kinescoped for rebroadcast elsewhere, Pauling urged an immediate halt to testing, and Teller argued that the risks were overrated and the tests essential.[66] Their exchanges alternated between the personal and the technical. Pauling accused Teller of making "misleading" and "untrue statements," and Teller asked Pauling to treat him with more "magnanimity." Their public squabbling continued into other television appearances, much to the chagrin of many leading scientists at the time.

INCREASING THE FUN AND GAMES

Such serious debate about science occupied only a small part of television content. The first comprehensive surveys conducted by the National Association of Broadcasters showed that, in 1951, 25.4 percent and, in 1952, 35.7 percent of New York City television consisted of drama.[67] When variety shows, quiz shows, and sports were added to the tally of entertainment content, the percentages (in both years) totaled 56 percent, while science shows accounted for only 0.3 percent. The disparity between television's capacity for visual education and how it was willing to treat science had widened. Few educational programs, science or otherwise, appeared in prime time, and most of those seemed "amateurish" in comparison to the sophisticated network drama and comedy productions.[68] The national television audience in the United States, which by the early 1950s exceeded forty million viewers, offered an irresistible market for entertainment.

Some of the dramas did incorporate science, but not always accurately. *Actor's Studio* mounted live productions in both 1949 and 1950 of "Joe

McSween's Atomic Machine," about an amateur inventor who builds a device that neighbors erroneously assume to be an "atomic machine." A far more grim 1954 drama about a hydrogen bomb dropped on New York City (*Atomic Attack*) offered, one critic wrote, a valuable lesson on "the importance and meaning of civil defense" yet ignored the consequences of surviving such an attack.[69] From 1952 through 1957, the DuPont Company's prime-time *Cavalcade of America* offered less sensational and more accurate science-related dramas with themes that echoed the company's earlier radio series.[70] "What Hath God Wrought" was a biography of Samuel Morse, inventor of the telegraph. "Mr. Peale's Dinosaur" followed Charles Willson Peale and the founding of America's first natural history museum. Other programs celebrated advances in medicine, such as Joseph Goldberger's research on pellagra ("G for Goldberger") and George Minot's work on pernicious anemia ("The Gift of Dr. Minot").

Live, seemingly unscripted academic competitions, such as the panel shows popular during the 1950s, occasionally incorporated science.[71] Rutgers University dean Mason Gross refereed *Think Fast*, and announcer Robert Trout used quotations from current news stories to challenge panelists like John Cameron Swayze on *Who Said That?* Panel shows were easy and relatively inexpensive to produce, and audiences enjoyed the participants' facial contortions and bemused expressions as they grappled with brainteasers. One of the most successful was *What in the World?*, hosted by Froelich Rainey, director of the University of Pennsylvania Museum. That series began as a local broadcast in the Philadelphia area in 1951 and became a CBS network broadcast, remaining on the air nationally until 1955, and then locally (with rebroadcast on educational stations in Boston and Chicago) until 1965. On each program, prominent experts like physical anthropologist Carleton S. Coon, archaeologist Alfred Kidder, and museum director Perry Rathbone were challenged to identify archaeological specimens and cultural artifacts, describing the origins, age, intended uses, and circumstances of discovery, while engaging in witty banter.[72] Viewers (informed "secretly" of the identity of the object) could follow along while experts asked questions and engaged in debate.[73] The artificial suspense of identification mimicked the supposedly spontaneous drama of commercial game shows—but here the prize was intellectual supremacy rather than money. The *New Yorker* declared (somewhat prematurely) that the series proved "that television can become a useful art without losing its liveliness."[74]

FIGURE 2. Surprises kept viewers entertained. Broadcast announcer Milton Cross and his assistant compare the durability of china and Melmac plastic dishes on *What's New*, WFIL-TV, Philadelphia, 1948. Courtesy of Division of Medicine and Science, National Museum of American History, Smithsonian Institution.

AN ARTIFICIAL PLACE FOR SCIENCE

Liveliness, clever banter, and entertainment soon dominated the schedule because no powerful entity—government or private—pushed for anything different. The airwaves may be a common good, but the Communications Act of 1934 had established a legal context, or what historian James L. Baughman calls a "regulatory expectation," which left little room for adjustment once television became a reality.[75] Congress had consistently restrained government control of radio broadcasting, endorsing the premise that "maximum private enterprise competition, rather than detailed regulation, would best promote the public interest by providing a wide diversification of cultural, educational, and entertainment program sources; a market place for the interplay of ideas and opinions; and a means for community self-expression."[76] U.S. law treated the spectrum like a natural resource that could, with permission and appropriate license, be exploited for commercial gain, much as had happened with forests and minerals on public lands. In the intervening years, the FCC and Congress weighed in on such issues as false advertising and offensive speech, but networks and station owners enjoyed a lax regulatory environment that encouraged growth and paid only token attention to educational and public service programming.[77]

Television developed first as a national medium, with cross-border or global broadcasting only possible after development of communications satellites.[78] The American system did not, of course, represent the only possible choice for how to structure and regulate television.[79] In Europe, although damage to the economies and communication infrastructures had delayed resumption of television broadcasting after World War II, over twenty thousand television receivers were operating around London by 1947.[80] The first postwar television programming from the government-controlled British Broadcasting Corporation (BBC) mirrored BBC Radio's tiered, class-conscious approach (Home Service, Light Program, and Third Program); offerings included science and natural history productions and tended to demand "intelligent attention" from the audience.[81] The BBC was soon making science available to about four million viewers via *Science Review*, *Matters of Medicine*, and programs starring scientists such as Julian Huxley and Jacob Bronowski, although BBC producers admitted in 1953 that even *their* educational programs were being increasingly assessed by a yardstick of "amusement or entertainment value" and that they were being pressured to adopt the American commercial advertising model because it seemed so successful.[82]

"Institutional jealousies, ideological beliefs, and economic interests," Robert J. Williams emphasizes, "have all played their part in determining the nature and effectiveness of regulation" of American broadcasting, as power shifted from equipment manufacturers to advertisers and then to entertainment conglomerates.[83] Even though private gain is supposed to be "secondary to the 'public interest, convenience or necessity,'" federal regulatory bodies allowed commercial broadcasters considerable laxity, a circumstance that served the public interest both well and ill, cultivating creativity while hobbling educational use, with long-lasting consequences for communication about science.

During that first decade of national broadcasting, television's cultural, political, and social effects expanded hand in hand with its economic and technical development (numbers of sets, numbers of stations, increasingly sophisticated lighting and sound). A 1945 government pamphlet, *What Is the Future of Television?*, predicted likely changes in family social structure, home décor, and purchasing patterns, but geographer Paul C. Adams has observed that television's impact eventually extended far beyond the physical (the living room) or the economic (investment by individual consumers and by networks) to the social fabric of American life. Television became a "center of cultural meaning," a "gathering place" that transcended national borders and ignored economic and class barriers even while it subtly reinforced them.[84]

Critic Robert Lewis Shayon inadvertently predicted how science would become transformed within this electronic space when, in 1951, he criticized a *Johns Hopkins Science Review* episode about volcanoes for sticking to the "facts" of geology and geophysics. Shayon wrote that he would have preferred to have learned "what the Johns Hopkins departments of literature, history, perhaps art and music—yes, even philosophy—have to say about volcanoes," and he added that "[t]o argue that this is not 'science' in review is merely to accent the artificial compartmentalization of the program."[85] Shayon was, in fact, calling for the very type of presentation of science in its cultural and social context that would become television's standard fare, a type favored by millions of viewers, although not without protest from educators.

CHAPTER THREE

Elementary Education, Basic Economics

A good teacher is likely to be a born ham.

FRANK BAXTER, 1955[1]

Attempts to teach science via commercial television began in earnest during the mid-1940s, with local programs such as *Your World Tomorrow*, an NBC physical science series created by high school instructor Joseph Mindel and the New York City Board of Education.[2] As a means of disseminating pedagogical content to millions of children and adults, television initially seemed an exciting alternative to public lectures and books.

Closed-circuit television programming—transmitted within a school system or single building—offered a popular short-term solution for teacher shortages and expanding pupil populations during the baby boom of the 1950s.[3] Open-circuit educational programs—disseminated via commercial or public stations and therefore accessible to all viewers within range—proved more problematic. Commercial channels with the best transmission capability balked at donating lucrative airtime to instructional programs that might not sustain viewers for whatever followed.[4] As a result, educational programs were scheduled in early-morning hours (not always the most convenient time for working adults) or on public channels with weaker signals. Local commercial stations might have taken advantage of universities willing to share their "vast storehouses of knowledge," and some stations did establish effective relationships with secondary schools, but conflicts between the television business and public education's social goals continually inhibited such projects.[5] By 1951, less than 5 percent of the programs shown on all New York City stations were categorized as instructional. In 1952, only 0.3 percent of television broadcasts in New York City focused on some type of science information, and most shows that might have appealed to adult audiences were scheduled in late-night hours.[6] By 1953, many major American cities had only one educational station. "Since programs in these classifications seldom produce revenue," critic Jack Gould observed, "they are regarded as of secondary importance, and often as a nuisance. . . . Television looks to the advertiser, not the audience, for its most important cues as to what to do next."[7] Economic considerations hobbled good intentions from the outset.

In 1951, Telford Taylor, counsel to the Joint Committee on Educational Television, emphasized that citizens, policymakers, and school systems must begin to consider "to what uses and values" this "marvelous new instrument for education" would be dedicated and whether educators would be able to "display the energy, and be granted the opportunity, to take advantage of it."[8] The FCC was being asked to reserve a portion of the limited number of broadcasting frequencies for educational use. Commercial operators resisted the proposed set-aside because in 1951 no more

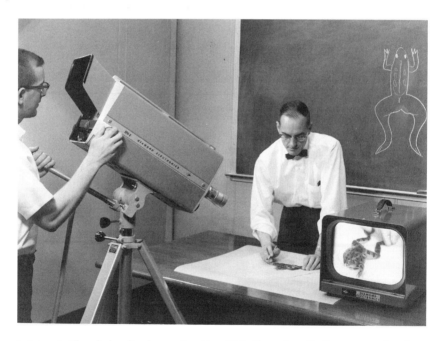

FIGURE 3. Closed-circuit science education. This dissection of a frog was used to demonstrate the reproduction quality of a $11,000 closed-circuit television system (two cameras, projector, monitors) being marketed to U.S. schools in 1965. Courtesy of Photographic History Collection, Division of Culture and the Arts, National Museum of American History, Smithsonian Institution.

than seven VHF signals could be broadcast in any one locality. There was considerable pressure to push educational content onto the UHF channels (which had weaker signals, and were not yet receivable on all home sets). A few policymakers urged that civic education be given priority, arguing that whether the content be election coverage or a geology lecture, the public would be well served. When decisions were left to commercial interests, Taylor pointed out, they chose to offer only a minimal number of cultural programs "as a showcase to which they can point with pride and a light-ning rod to draw off public and official criticism."[9]

Citizens involved in promoting educational television during the 1950s were reminiscent of earlier temperance-movement supporters—affluent businessmen, professors, and society matrons—and they tended to envis-age the technology as a force for reforming, enlightening, or educating the masses. Their dreams were worlds away from Milton Berle, laugh tracks, and quiz shows, from a medium dedicated to democratization through mass culture. And, as had happened with radio, science played only a minor role in educators' plans. Proposed programming tended to focus on dramatic arts, dance, music, literature, vocational classes, or craft demonstrations, not chemistry or physics experiments.

When educational programs did succeed, it was invariably because of a persevering entrepreneur, a "born ham" willing to learn how to exploit the technology creatively rather than merely replicate classroom lectures in front of a camera. These were rare individuals. Educational broadcast-ers lost the battle, R. D. Heldenfels later concluded, because, without the people, coordinated will, resources, and ability to create outstanding pro-grams different from what was already available on commercial televi-sion, the arguments for education could be easily undermined.[10] Science popularization was once again caught between the educators' biases and the media owners' control of the airwaves.

INTERNAL BARRIERS

In 1948, local Washington, D.C., stations gave extensive coverage to the cen-tennial meeting of the country's largest scientific organization, the Ameri-can Association for the Advancement of Science (AAAS), telecasting talks by prominent experts like astronomer Harlow Shapley and psychologist Arnold Gesell.[11] Given the medium's potential for public outreach and given that both the Poole and Marshall series were already garnering positive responses, one might have expected AAAS to have encouraged more such activities. A 1951 AAAS retreat on public outreach, convened by mathematician Warren Weaver, the association's president-elect, gave little attention to television, however, despite ongoing national initiatives to "make the home a classroom."[12] As historian Bruce V. Lewenstein points out, retreat participants construed popularization in moral terms. Even though they agreed that the power of scientific knowledge conferred upon scientists some obligation to communicate to society and even though their consensus statement urged members to consider how to "increase public understanding and appreciation of the importance and promise of the methods of science in human progress," the group's leaders had little inclination to move boldly into using new media.[13] The recommendations were never translated into significant action, either via television or other means. Cold War politics, political bickering, personality conflicts, and unwillingness to turn communication tasks over to media professionals reduced the effectiveness of any pilot efforts—a pattern repeated throughout science and academe at the time.[14]

When Dael Wolfle became the association's chief executive officer in 1954, he encouraged several projects that might have engaged with television, but resistance from within led AAAS to focus instead on solidifying relations with the press corps and on improving primary and secondary science education.[15] The practice of allowing citizens only to glimpse science from afar, via managed coverage of formal talks, remained standard for AAAS and most scientific organizations throughout the 1950s. After *Sputnik* prompted renewed attention to public education, AAAS established a committee on public understanding of science, but decades passed before that committee made significant efforts to promote science via television.

Popularization, Weaver had long asserted, should emphasize scientific values rather than "scientific accomplishment" or "spectacular achievements."[16] This approach exemplified what sociologist W. F. Ogburn characterized as *evangelistic popularization*, that is, communication directed at extolling the value of science as a way of life and knowing, as opposed to *didactic popularization* intended to explain scientific discoveries or fundamental knowledge.[17] The contrast between those two strategies lay at the heart of midcentury disagreements over science popularization via television.

CULTURAL ALCHEMY

"Popularized science"—and, especially, attention to science within broadcast media—embraces a mélange of ideas and images, factual and fictional, accurate and distorted, from which audiences choose at random. Academic suspicion of popularization had, somewhat ironically, coincided with the maturing of the social sciences during the 1930s. Scholars like Paul F. Lazarsfeld suggested that analysis of public opinion and attitude formation toward complex topics like science or politics must account for mass culture's overlapping content and overlapping audiences. Because people consumed many types of media at a time, the variability and range of choices affected the ability to predict or control impact.[18] Americans flipped through magazines while listening to the radio; the same person who purchased serious books explaining relativity would laugh at cartoons about absent-minded scientists. Such complexity of reception meant that an individual's "knowledge" about chemistry, for example, might be shaped less by her formal education than by the entertainment and information she consumed during her lifetime.[19] Each individual would react differently to what she absorbed from the media; each would rely on different sets of trusted filters or interpreters; each would select different items from the mass media's daily menu. Understanding how audiences interpreted the science they received thus required comprehensive attention to the whole as well as analysis of how different media outlets interacted and shared content. To reach mass audiences with specific messages, one might be forced to consider alternatives to more traditional, restrained, and controllable venues like print or formal lectures.

For scientists and other intellectuals who found the majority of mass media content vacuous or abysmal, adopting such a wide-angle perspective implied an unacceptable endorsement of popular culture. The terms "low" and "high" became "fighting words" in the debate.[20] Pandering to huge audiences might dilute appreciation for fine arts and classical music, for philosophy and poetry, and therefore for science.[21] "No art form, no body of knowledge, no system of ethics is strong enough to withstand vulgarization," critics argued, because a "kind of cultural alchemy transforms them all into soft currency."[22] T. W. Adorno's lament that "radio has made of Beethoven's Fifth Symphony a hit tune which is easy to whistle" and Hannah Arendt's observation that popular culture makes the "classics" into "something to be consumed rather than understood" epitomized the academic scorn.[23] The same criticisms of popular culture circulating during the 1930s reemerged two decades later: television degraded and sensationalized every subject; television detracted from more substantive and serious activities; television encouraged escapism, undermined moral standards, and distorted reality; and, through mass advertising, television encouraged people to buy things they did not need.[24]

Broadcasting's owners and decision makers tended to dismiss these critiques with the arrogance of monarchs. Go on and hurl your snide re-marks and brickbats! *We* control the studios! With representative disdain, CBS president Frank Stanton declared that academics could not bear to have their ideas diluted and that they feared popularization as a review of their own worth.[25] "[H]ostility on the part of the intellectuals toward the mass media is inevitable," he explained, because they are "not really rec-onciled to some basic features of democratic life."[26] Undisguised hostility for entertainment certainly did reflect experts' fears that their own author-ity and prestige might be challenged in popular forums, but, Neal Gabler suggests, the attitudes also sprang from deep-seated distrust of the pub-lic's tastes and sensibilities—and from indifference to what society might need.[27] Mass entertainment celebrated irrational "triumph of the senses over the mind, of emotion over reason."[28] Some academics even constructed mass audiences as victims, and demonized the media as corporate tools, refusing to admit that the content might well reflect "genuine tastes and values."[29] The distinctions between "high" and "low," Ivan Karp observes, became "intimately associated with notions of power and control, with ideas about who should be entitled to have a voice and who should be si-lent."[30] Whatever the root explanation, commentaries about media popu-larization tended to have a paternalistic tone and, noted Patrick Hazard, to read "like a coroner's report."[31]

Cooperation with producers of mass culture represented an unaccept-able option for scientists convinced that "processing and re-packaging scientific knowledge so that it can be understood by non-specialists" invariably degraded that knowledge into something "distorted and less true."[32] The fear was that popularization, even when aimed at educating adults, was becoming something *done to science* rather than *done by scien-tists*, initiated by those who neither respected science nor had its best inter-ests at heart. In the "wrong" hands, knowledge might be misinterpreted. Scientists themselves should control the flow of information, lending im-primatur only to works judged to be accurate and authoritative, an attitude reflecting age-old efforts to reinforce the cultural boundaries supporting the prestige and authority of technical expertise.[33] The problem with writ-ing, Anthony Smith once wryly suggested, is that "it travels . . . and gets 'into the hands of people who have no business with it.'"[34] Television was perceived as having similar dangerous potential.

CLASSROOMS ACROSS THE CONTINENT

In addition to those who, like Roy K. Marshall, became national celebrities, other scientists and academics began to step into the studios of local stations and even participate in game and talk shows. Northwestern University professor Bergen Evans, frequently invited to appear on panel shows, told *Time* magazine that an academic who engages with television "has the feeling of being out in the wide, wide world."[35] Anthropologist M. F. Ashley Montagu's appearances on the New York State Board of Education's *Camera Three*, discussing topics like evolution, led to invitations to discuss family relations and parenting skills on the weekday *Home* series and to parry jokes with the host of *The Tonight Show*.[36]

A less exciting but more effective forum for adventurous professors turned out to be college courses offered via television. These projects were experiments in democratic communication. They brought science classes to hundreds of thousands of viewers who cared to watch and, as with early radio, reached people who, for economic reasons or because of racial segregation or other bias, lacked access to public lectures or adequate formal education. Television allowed anyone to learn from established, accomplished scientists and scholars. *University of the Air* began in 1951 in the Pennsylvania area as a cooperative venture involving Triangle Television (publisher of *TV Guide*) and twenty-two Pennsylvania academic institutions, and soon expanded to Triangle-owned stations nationwide.[37] Course topics ranged from literature and economics to chemistry and natural resources conservation. Ecology professor Francis J. Trembley of Lehigh University summarized "The Imprint of a Million Years of Human Affairs." Twenty professors from Bryn Mawr, Swarthmore, and Haverford colleges collaborated on "Our Ties with Other Cultures," interweaving lectures on art history, music, sociology, geology, zoology, and engineering. Bryn Mawr professor Marguerite Lehr told fellow mathematicians that, while "the opportunity and risk were both great," she had enjoyed most the ad-libbing and viewer call-in sessions during her "experiment with television."[38] In 1959, 84 percent of the students registered for credit with *University of the Air* were female; most viewers were age thirty-six to fifty-five.[39]

Two network projects, *Continental Classroom* (NBC) and *Sunrise Semester* (CBS), relied on the charisma of individual instructors. After finishing his Ph.D. at Cornell University, physicist Harvey Elliott White had joined the University of California at Berkeley faculty in 1930 and become well known for his work in atomic spectroscopy. White was the middle-aged author of several major textbooks when he received a Ford Foundation grant and took leave to design a physics course for Pittsburgh station WQED-TV. The success of the WQED project led to development, with more Ford Foundation underwriting and the cooperation of the American Association of Colleges for Teacher Education, of *Continental Classroom* and the first physics course to be offered nationwide for college credit. Two months into the fall 1958 series, about five thousand students had enrolled for credit at $45 apiece, 250 colleges were offering credit, and 150 stations were carrying the broadcasts. "Physics for the Atomic Age" attracted over a quarter million regular viewers, both credit and "auditors," including high school students and their parents, blue-collar factory workers, military personnel, nuns, and five hundred inmates at San Quentin Prison.[40] Such success helped to validate the costs—over $1 million initially given by corporations and foundations for planning and production, and broadcast time donated by NBC affiliates.[41]

Although the NBC project was spurred by *Sputnik*-era goals of improving science education and was originally intended for high-school science teachers, White is credited with expanding the audience. Episodes were kinescoped in New York for use around the country, yet White managed to convey spontaneity and an early-morning camaraderie with viewers routinely "sacrificing sleep for science." In many locations, *Continental Classroom* ran between 6:30 and 7:00 a.m. The program would have been "a nucleonic turkey without its M.C.," *Time* wrote: "In the cold dark before winter dawn, by the TV screen's eerie blue glare, the show's rumpled star looks like an insomniac alchemist. With spectacles sliding down his nose, he brews electrons, protons, and mesons while evoking Newton, Faraday, Planck, Einstein, and Heisenberg."[42] A "lanky, friendly, precise talker," White made physics fascinating through preparation rather than showmanship. *Time* described him as droning "like a farm agent exhaling a market report," but he was always thoroughly prepared, spending many hours on script development and rehearsal. He also invited an array of scientific celebrities to make guest appearances, including seven Nobel laureates. Subsequent *Continental Classroom* courses focused on chemistry, mathematics, and American government, with their instructors chosen in cooperation with the relevant professional organizations.[43]

The CBS competitor, *Sunrise Semester*, had begun in 1957 as a local show under the auspices of New York University and, by the following year, had been expanded to six hours per week, to include Saturday broadcasts. As a low-budget, unsponsored operation, *Sunrise Semester* subsisted on donated

FIGURE 4. Some scientists became television celebrities. University of California physicist Harvey Elliott White taught the first course offered on NBC's *Continental Classroom*, 1958. Courtesy of Emilio Segrè Visual Archives, American Institute of Physics.

airtime and willing volunteers. Around the country, other stations and universities engaged in similar outreach, such as anthropology courses sponsored by the University of Utah and University of Michigan, general science lectures by MIT professors, and mathematics courses from the University of Alabama.[44] The University of Chicago program *Science '58* ran live every weekday morning at 6:30, providing early risers with accounts of the latest scientific events and accomplishments.[45] To coordinate publicity for its affiliates' educational ventures, NBC established a *Science Calling* project during the late 1950s. New programs at its San Francisco station, for example, were developed in conjunction with Stanford University, and those at Los Angeles, with the California Institute of Technology.

Despite good reviews and enthusiastic pupils, these educational programs had little effect on television overall. In some markets, the courses were the only science-related programs on the air, reflecting an academic preference to tie cooperation with television "closer to the classroom than the living room."[46] Even without budget pressures (and *Continental Classroom* demonstrated that underwriting could be obtained for high-quality series), the attitudes and enthusiasm of academic and educational groups made all the difference. One critic suggested that educators should step back, provide the material and overall direction, and let professionals create the presentations: "Until there is some such fusion of talents the educational program is not going to be very satisfactory to the educator or the broadcaster—or to the audience."[47]

COST ESTIMATES

Money proved a persistent barrier. Science popularizers who had labored without compensation to produce sustaining programs for radio were understandably leery of donating more time and effort to television. Science Service director Watson Davis explained in 1949 that television was then in the same period of development as radio had been two decades before, although far more expensive: "If we are to produce television programs, they should be self-supporting."[48] Without charitable subsidy, a series needed commercial sponsorship, which nonprofit groups like Science Service or AAAS perceived as incompatible with their missions.

For *Johns Hopkins Science Review*, the university accepted small grants and donations from the DuMont network (annual gifts that increased from $10,000 to over $50,000), and the local station donated airtime and endowed an academic fellowship.[49] Johns Hopkins agreed briefly to sponsorship by a chemical company, but soon thereafter prohibited commercial advertising.[50] Nevertheless, Lynn Poole did accept equipment, props, and assistance from government agencies and companies (for example, General Electric and RCA) in exchange for on-air acknowledgment and publicity.[51] Direct expenditures in 1955 totaled around $30,000 for fifty-two programs, an amount that did not include the salaries of Poole and his staff or the donated time of guests or actors (recruited from family or university to dramatize historical episodes and demonstrate scientific principles).

Without major grants, universities establishing their own educational stations faced even tougher challenges. In 1952, the estimates for a television studio and kinescope equipment at the University of Wisconsin were almost $200,000.[52] To broadcast throughout the state and produce content competitive with the commercial networks' "wrestling, westerns, and deep décolletages" would have required millions.[53]

Lack of enthusiasm for television within the larger academic community did not help. When the FCC held hearings on educational uses during the early 1950s, few representatives of top universities bothered to testify. "They are letting go by default the greatest educational medium ever devised [just] as they let radio go by default," John Crosby complained.[54] Meanwhile, the revenues of the commercial television industry skyrocketed. In 1955, the four national networks, and the stations they owned, generated revenues over $374 million. That year, the DuMont Network folded; ABC, CBS, and NBC could pick and choose among independent productions and already had a financial interest in over half of prime-time content.[55]

Although the FCC allowed nonprofit groups to acquire television licenses, Congress had made no provision for funding production facilities or transmission towers. Educational institutions had to secure the money from private donors or state legislatures; only in the mid-1970s did substantial federal funding for educational (public) television become available. In the interim, the commercial networks shaped audience expectations, and as R. D. Heldenfels has noted, that situation "blurred" any coherent vision of what educational television might have been.[56] When the FCC lifted the licensing freeze in 1952, over two hundred channels were available for educational use. Two years later, many remained unclaimed. Some universities eventually relinquished their licenses to commercial operators because they could not afford (or did not want) to support a station.

POLITICAL PRESSURE AND PSYCHOLOGICAL BARRIERS

The Smithsonian Institution, a multilayered, quasifederal organization, carries out research and education activities with a mission to "increase and diffuse knowledge." The Smithsonian might therefore have been expected to embrace new communication technologies with enthusiasm. Instead, the reaction resembled that of other elite scientific and cultural institutions: creating committees to study the medium, entertaining proposals from external producers, and then balking. Biologist Alexander Wetmore and psychologist Leonard Carmichael, the two men who led the Smithsonian during television's infancy, would continually express mild interest, contemplate the choices, and then hesitate. In June 1948, for example, the director of radio and television operations at a nearby university suggested a series that would "be even better" than the Franklin Institute's *Nature of Things*.[57] After a month of deliberations, Secretary Wetmore explained that the Smithsonian expected "to enter into television as a public service later on" but would not pursue this particular idea. Wetmore cited reasons that he and his successors would repeat for decades, such as the reluctance to transport valuable and fragile objects to a studio, the additional work for Smithsonian staff ("Would curators welcome this additional demand on their time?"), and a conviction that the medium was perhaps not all that important (Wetmore referred to "the relatively small scope of television broadcasting . . . as compared to world-wide radio").[58] Even once television's vitality and "scope" were proven, the potential burden on staff remained a convenient excuse to decline requests.[59]

When CBS News proposed a series about the Wright brothers' plane and similar historic artifacts, Smithsonian officials established a committee to study the suggestion.[60] Its members applauded the prospect of reaching substantial audiences "at comparatively slight cost to the Institution," acknowledged television's potential for cultural and scientific expression, found the proposal compatible with the Smithsonian's mission, and even concluded that television might offer useful publicity.[61] A successful program "would take the Smithsonian into the very homes of the people" who represented its constituency. Initially, the committee expressed approval of suggested arrangements for script review and filming, but soon focused on three questions: who would pay for the production, was there a demand for such content, and, if so, could that demand be met by exploiting other media? Wetmore eventually declined the CBS offer, explaining that the institution would only become involved "at a high level" and that the estimated internal costs of $40,000 to $50,000 (plus staff time) exceeded available resources.[62]

From time to time, individual Smithsonian scientists or curators did agree to appear on local or network programs. In spring 1953, there were several broadcasts in conjunction with the fiftieth anniversary of powered flight, such as a segment of *The Dave Garroway Show*. On April 8, 1953, a local program about Charles Lindbergh's *Spirit of St. Louis* involved appearances by Carmichael, former Smithsonian Secretary Charles Greeley Abbot, and Paul Garber, the head curator at the National Air Museum. That same spring, the institution came under fire from Senator Francis Case of South Dakota, who argued that, instead of requesting additional funds for exhibitions and buildings, the Smithsonian should develop a television project.[63] Case explained to reporters that television would allow people everywhere in the country to view the contents of the Smithsonian.[64] With prompting from Case's office, various television, advertising, and public relations companies began to contact Carmichael, proposing either to set up production units or develop series.[65] By the middle of 1953, the institution was engaged in discussions with both NBC and CBS.

These network negotiations were bedeviled by two aspects that would trouble similar Smithsonian ventures for decades—dramatization and money. Television required a narrative thread to connect its visualizations, and drama offered a more creative device to tell the stories behind museum research and objects than would, say, filmed tours through galleries. One producer explained that "things which are sheerly educational . . . do not get large audiences."[66] When NBC pushed the "interesting and dramatic," Smithsonian officials expressed concern that dramatization might erode the "genuineness" and "originality" of the research and collections.[67] The CBS idea, sweetened with per-program payments and generous division of subsidiary rights, stumbled over similar issues.[68] For that series, scientists and curators who appeared on camera would have been required to join labor unions with which CBS had contracts, and would have been prohibited from appearing on any other radio or television program during the duration of the contract. Moreover, CBS wanted final approval of broadcasts and the right to sell them to sponsors. After many months of stalled negotiations, both NBC and CBS withdrew their offers.

Other producers attempting to broker similar projects encountered the same obstacles. In early 1954, the Smithsonian regents (who serve as the institution's trustees) pressured Carmichael to keep trying.[69] And by then at least one applicable model existed. In 1952, the Ford Foundation had funded Robert Saudek's experimental TV/Radio Workshop, which resulted in creation of *Omnibus*, a variety show aimed at elite, well-educated viewers.[70] A single hour might include a scientific demonstration, poetry reading, musical or dance performance, and a scene from a Shakespeare play. Prompted by Carmichael's meetings with network executives, Saudek approached the Smithsonian with a typical *Omnibus* idea—making high

FIGURE 5. Television "show and tell," April 8, 1953. During this Washington, D.C., broadcast, former and current Smithsonian Institution Secretaries Charles Greeley Abbot and Leonard Carmichael, Capitol Airlines executive Jennings Randolph, WTOP-TV announcer Bill Jenkins, and National Air Museum curator Paul Garber discussed Charles Lindbergh's famous transatlantic flight. Using a map and the aviator's original flying suit (on wall at back), Garber explained how Lindbergh's plane *Spirit of St. Louis* made one final flight to Washington and was now on permanent display at the Smithsonian. Courtesy of Smithsonian Institution Archives.

culture into "fun." Comedian Fred Allen, accompanied by curators, might tour "certain interesting areas" within the museums during a live telecast. Allen's repartee would be "of a humorous nature," and Smithsonian curators would play the straight men. Saudek suggested that Allen might, for example, joke about the "unstuffed skin of General Pershing's horse" (an artifact recently pictured in *Life* magazine).[71] When Carmichael responded that he imagined a quite different project, Saudek explained that, even for "a feature of a definitely educational sort," compromise was essential: "[I]t is especially important to make the feature attractive, in the sense that Walt Disney has treated nature to pleasures of dance music and sometimes accented the eccentricities of birds and animals. . . . Our philosophy of programming . . . does try to utilize techniques which will attack the special psychological blocks which most people have to most serious subjects."[72] Allen's humor would not dominate the segment. Laughter would merely serve as "the bait" to attract an audience "somewhat in the way that the full-page picture in *Life* of General Pershing's unstuffed horse must have been calculated to attract readers to read the more serious text about the Smithsonian," Saudek wrote. Fred Allen would be much more dignified than, say, "a hundred other comedians from Jackie Gleason to Arthur Godfrey."

Admitting that, as a psychologist, he was "sympathetic" to the need to adapt communication techniques in order to sustain viewer attention, Carmichael nevertheless expressed a typical academic reaction to television. He was reluctant to approve an approach that would be "primarily humorous" for fear of appearing undignified: "You may recall that Mark Twain spoke of his sorrow at not being able to make serious comments because everything he said was supposed to be funny."[73]

Carmichael commented at the end of 1954 that he had had "many informal discussions with the members of the Board of Regents concerning the matter and, with reluctance, these discussions always end with the same idea, that we should not formalize our relationship with the television networks."[74] Nevertheless, Smithsonian officials continued to negotiate, deliberate, and decline.[75] The reasons for refusal would vary: transporting a precious or historically significant artifact to a studio was impossible; filming on site could disrupt research or inhibit visitor access to exhibitions; the cost of producing programs of acceptable quality exceeded the institution's resources; and no producer or network could be granted exclusive access. The "insurmountable" barrier, however, remained the attitudes of the scientists and curators.

SEARCHING FOR AUDIENCES

After *Johns Hopkins Science Review* went off the air, most new science series for adults during the 1950s were "all-business-no-laughs." *The Search* premiered on CBS in 1954 and continued, as a summer replacement show, until 1958. Hosted by journalists Charles Romine and Eric Sevareid, the series showcased academic research on such topics as stuttering (University of Iowa), auto safety (Cornell University), child development (Yale University), and race relations (Fisk University). For those who championed civic education via television, such humorless, didactic approaches represented failures of vision. Instead of shaping television, the educators and community leaders remained on the sidelines, carping and kibitzing but refusing to initiate productions.[76] When radio began, critic Jack Gould explained, educators either "did not understand it and resigned without a fight to commercial interests coveting the few available wave lengths" or else brought their ivory-tower attitudes into the studio and bored the audiences. Now, that pattern was being repeated. Professors who predicted that television would demote cultural standards did little to produce competitive high-quality programming.[77] Gould refused to let the academics off the hook. Merely expressing "articulate despair" over commercial television was insufficient and irresponsible.[78]

The networks, too, shared in the failure. Throughout the radio age, advertisers had shaped content by paying for productions or vetoing potential topics or approaches.[79] CBS began producing its own radio shows in 1946, and that practice carried over into television. Advertiser influence, while indirect, remained substantial. Sponsors demanded high ratings. Whenever broadcasters interpreted past audience selections as expressing preference for conservative, imitative plot lines, that assumption pushed programming ever farther from more challenging content. National and international syndication, which greatly increased a series' profitability, made generalized, homogenized approaches even more attractive.[80]

Harvey Elliott White and dozens of other dedicated, enthusiastic science educators had offered positive, accurate science lessons to supplement the more entertaining presentations of Lynn Poole and Roy K. Marshall. The rest of television—sadly, the majority of television—promoted a different reality. Its economic and cultural power derived from being able to hold viewers' attention with drama and stereotypes. While real science teachers were lecturing on local stations, a competing image was being projected by actor Wally Cox.[81] Broadcast live from July 1952 to June 1955, *Mr. Peepers* was the first television series to feature a fictional science teacher. Cox's character Robinson J. Peepers was lovable but ineffectual, a "mild-mannered, dry, owlish wisp," impractical and lost in theory. Bewildered and "detached from the world at large," peering "dreamy-eyed" at students, colleagues, and blustering school officials through his thick eyeglasses, Peepers offered viewers a comforting stereotype who was worlds away from the self-assured, well-organized experts lecturing on early-morning television.[82]

FIGURE 6. Cover illustration for the pamphlet *What Educational TV Offers You*, published by the Public Affairs Committee, 1954. Although some college and high school teachers became engaged in using television to reach beyond the classroom, other American educators adopted more aloof attitudes to the medium. Courtesy of Smithsonian Institution Archives.

A "scientist" could thus be materialized in any form that television desired. Walt Disney's 1945 article "Mickey as Professor" in *Public Opinion Quarterly* should have signaled to educators that the battle for control of the small screen required strategic engagement.[83] Disney claimed to embrace "attitudes that are fundamentally educational although expressed in the manner of entertainment" through animation and dramatization.[84] To be successful, he argued, products must accommodate to the filmmaker's craft, the educator's goals, and the constraints of finance and promotion. Television could only supplement the classroom, not substitute for it, he insisted.

Scientists' televised lectures could not compete with an amiable animated mouse. Entertainment programs incorporating natural history, physics, medicine, and space, including many produced by or influenced by Disney, soon became television's principal vehicle for communicating about science, for better or for worse.

CHAPTER FOUR

Dramatizing Science

In television it will be natural to emphasize types of program material where the addition of visibility will enhance the emotional effect—such as drama, news, or sporting events.

DAVID SARNOFF, 1939[1]

Given that science's public face tends to de-emphasize emotionality, drama might seem an incongruous device for communicating about the subject. Laughter and humor thread through the rhythm of laboratory culture (and ambition and competition can bedevil professional interactions), but researchers' public presentations have traditionally eschewed any suggestion of passion, fear, lust, anger, or sorrow. Journal articles are scrubbed of references to personality; sentences are constructed into the passive voice.

Drama offered a solution to the television popularizer's dilemma, just as storytelling had long been used to enliven print and lecture presentations about science.[2] Hollywood director Irving Lerner once complained that in science "everything that happens is inside the machine" and therefore difficult (if not impossible) to visualize on a television screen.[3] Point a camera at a cyclotron, he said, and viewers see only a gleaming metal shell. Dramatization helped to overcome such obstacles. Dialogue between characters could be used to explain the physics, for example. Moreover, by midcentury, drama had become a familiar mode of communication. As Raymond Williams observed, television emerged during a time when drama was no longer confined to the stage.[4] Sporting events and political conventions were routinely crafted to infuse every moment with "drama." Television scripts could distill human emotions to elicit sympathy, invite admiration, trigger laughter, and provoke disgust; scriptwriters adapted classical dramatic arcs to achieve tragedy or comedy, parody or pathos, among siblings, spouses, soldiers, and scientists.[5] Such techniques exploited the medium's intimate style and contributed to what Williams called the "drama of the box," that "small enclosed room, in which a few characters lived out their private experiences."[6] To compete in this context, science *had* to be dramatized. In television's artificial world, where every word and movement was carefully scripted, the spontaneity was wrung out of science for the sake of dramatic effect. Criminals caught, game shows won, scientific experiments completed—on to the next diversion.

Early television fiction did occasionally feature characters with scientific expertise—jungle adventurers, space travelers, heroic physicians, and stereotypical "mad scientists." In 1940, NBC's *Magnolia Floating Theater* included an adaptation of Robert Louis Stevenson's *Dr. Jekyll and Mr. Hyde*, the first of dozens of television versions of the tale of a scientist who self-experiments in order to banish his subconscious, evil side. Television also explored science through "docudramas," in which fictionalizations blended with factual narratives. As Roger Silverstone points out, film *documentaries* had been traditionally distinguished from film *drama* by the former's "claim for authenticity and truth, for factual accuracy and accurate reproduction of certain aspects of unfilmed reality."[7] In a docudrama, the lives and work of real scientists could be tweaked to add suspense ("Will the experiment prove successful?"). Directors combined staged events, dramatizations, and animation with original film and thereby stretched the boundaries of exposition.[8] It became acceptable to edit, add, and rearrange situations, to invent characters, to embellish events with sound effects, music, and narration, even for science programs.[9] Some docudramas stuck close to the facts; others were unapologetic hybrids that recreated what *might* have happened had cameras been rolling in Marie Curie's laboratory or when archaeologist Howard Carter uncovered Tutankhamen's tomb.[10]

For natural history documentaries, the use of dramatic techniques could enhance the glorious mysteries hidden within nature's kingdom.[11] A camera would pan slowly over a darkening forest, or track brilliantly colored fish darting ahead of a shark, or linger on eggs hatching in a nest. Beginning with *Meet Me at the Zoo* and *Zoo Parade*, television's natural history series intentionally mixed facts gleaned from contemporary research with "acceptable fictionalizations" of animal "life stories." Storytelling overwhelmed the science. Scientists were stereotyped into observers of nature, trackers of animals, protectors of habitats, or political advocates for conservation while their research work (measurement, analysis, and careful but unexciting accumulation of data) went unacknowledged in the narrative. The drama intended to heighten viewers' interest in science thereby inadvertently served to obscure it.

VISUALIZING NATURE'S KINGDOM

From museum "cabinets of curiosities" to lantern-slide shows, the use of natural history images for entertainment predated television. Early filmmakers like Martin and Osa Johnson attracted huge audiences to their African safari travelogues, many developed in cooperation with the American Museum of Natural History.[12] The Johnsons and their competitors in the 1920s and 1930s exploited connections with museums and zoos to enhance the illusion of authenticity, even while blithely using re-creation, re-construction, and artful re-arrangement of facts; Hollywood then marketed the films as "educational," "informative," and suitable for family audiences.[13]

Because dioramas, exhibits, and exotic animals contain rich visual potential, early television reached out to museums and zoos, and the institutions usually welcomed the publicity. In 1948, NBC's *Museum of Science and Industry* was broadcast live from New York.[14] ABC's *Sunday at the Bronx Zoo* premiered in 1950, cohosted by zoo curator and writer William Bridges and comedian-announcer Durward Kirby; telecasts from the New York Zoological Park included "conversations" about the animals. Other stations persuaded zookeepers to bring their menageries to the studio. *Animal Clinic*, hosted by Chicago veterinarian Wesley A. Young in 1950 and 1951, established a third formula that persists on television today: a congenial, telegenic veterinarian conveying basic information about animal behavior and biology while confidently handling energetic but cooperative dogs, cats, birds, reptiles, and rodents.

Director William M. Mann allowed a live broadcast from the Smithsonian's National Zoological Park in late 1947, to mixed reactions from staff members, who characterized the result as pleasing but mediocre. One zookeeper observed that the Smithsonian facility contained extensive possibilities for entertainment mixed with education: "There are many interesting facts and comments that might be made regarding almost every animal in the zoo such as its name, range, habits in the wild, economic importance, structural peculiarities, how and why it does some of the things it does, individual peculiarities that that particular animal has developed in the zoo, what it's fed, how it reacts to its keeper and others."[15] Although the local television crew had made up (or mixed up) names for animals (showing the white-tailed marmoset while the announcer described the "titi monkey"), viewers probably did not know the difference (or necessarily care). To the experts, however, such errors undermined the value of the communication.

By 1949, Smithsonian zookeepers and natural history curators were making regular appearances on local stations. Invited to bring "a small animal" (most television people did not seem to care which species) for appearance on the Junior League of Washington's community show *Playtime*, the zoo's assistant director selected a Colombian cuzumbie. Other staff carried reptiles to the *Ranger Hal* studio several times through the years.

Most zoos received no compensation for such cooperation even though staff members invested considerable time and effort. *Meet Me at the Zoo*, broadcast live on Saturday afternoons during 1953, featured regular appearances by Freeman M. Shelly, director of the Philadelphia Zoo. Shelly bragged to Mann about the program's success ("the publicity medium is good") yet admitted that "it is one super headache and causes all sorts of staff problems. . . . I am led to wonder many times if it is all worth the time and energy expended."[16] One proposal for an evening show, to be broadcast four times a week, would have featured the Smithsonian's zoo animals about one-third of the time. While Smithsonian Secretary Leonard Carmichael had no objection to the show's host (a local beauty queen), he disapproved of the commercial sponsorship.[17]

Invitations increased throughout the 1950s, and eventually the burdens of cooperation began to outweigh any perceived rewards. In 1960, for an appearance on the local *Jungle Jim* show, Smithsonian zookeepers spent twelve hours getting the animals ready, crated, transported, and safely returned. Staff member Mario DePrato even used his own automobile to take a ball python to a station.[18] That same year, Mann's successor, Theodore Reed, called a halt. He explained that, although "[t]he National Zoological Park enjoyed participation in the 'Jungle Jim' movie [*sic*] on Saturday afternoons by bringing live animals to the program for the past three weeks, . . . circumstances at the Zoo necessitate . . . that we will have to discontinue sending animals to your program on a regular basis. On occasion, when we can spare a keeper from the park, we will be glad to cooperate."[19] The Smithsonian's zoo had less time-consuming ways to publicize its mission, and other more telegenic zookeepers were attracting national attention.

THE LONGEST KINGDOM

As a young man, Marlin Perkins had dropped out of the University of Missouri to pursue his wildlife interests.[20] He began working in 1926 as a sweeper at the St. Louis Zoo, and two years later became curator of reptiles. By 1938, he had sufficient experience to run the Buffalo Zoological

Gardens and then, in 1944, to direct Lincoln Park Zoo. When Perkins was invited to bring small mammals to a Chicago station and "talk about them" on camera, he "jumped at the chance," he later wrote, because "publicity and promotion" were a zoo's "lifeblood."[21] The short-lived experimental show delighted local critics: "Star performers, joining in the rehearsal yesterday, were an alligator, which Perkins intends to mesmerize tomorrow; an owl, a crow, a snapping turtle, a chameleon, a California king snake and a Wisconsin bull snake, both harmless."[22]

In 1949, the local NBC affiliate asked Perkins to broadcast live from the park in *Visit to Lincoln Park Zoo*. An agile host with an "engaging personality," and described as "more handsome than Hollywood's leading men," Perkins was also a shrewd negotiator.[23] When he learned that the station had sold the program (renamed *Zoo Parade*) to a Midwest food chain and would be going commercial, he requested compensation for himself and the zoo. That series (sponsored by a dog-food manufacturer and eventually by such companies as Quaker Oats and Mutual of Omaha Insurance Company) premiered nationwide on twenty-eight NBC stations. Within a year, it was carried in forty-one cities, and by 1952, its weekly audience averaged eleven million, making Perkins a national celebrity. In 1954, *Zoo Parade* had the highest audience ratings of all science-related shows then on the air nationally (its Nielsen ratings[24] averaged almost 21.5, reaching about half of the potential audiences for fifty stations nationwide), but was canceled three years later.

Perkins returned to the St. Louis Zoo as director in 1962. Within a year, he was developing another television series, solely sponsored by Mutual of Omaha, and negotiating a lucrative deal for the zoo, eventually adding a succession of sidekicks to share the camera time (including Jim Fowler and Stan Brock). *Wild Kingdom* ran on NBC from 1963 to 1971 and in regular syndication until 1988 (long after Perkins's retirement from the show and even after his death in 1986).

Popular natural history films, like children's nature books, had frequently anthropomorphized animals, giving them names and imagining intentionality.[25] Television now extended this practice of establishing emotional connections with viewers and purporting to tell "the animals' side of the story."[26] When narration implied that a film crew (or host) was the first to observe a species or particular animal behavior, viewers were encouraged to "experience" the wonder (and dramatic moment) of discovery along with the camera. Scripts emphasized animal survival and sentimentalized daily struggles with predators. Achieving these dramatic moments, however, required either trained, tame animals or deliberate reconstructions. Film crews would routinely chase an animal, film the pursuit, catch and tranquilize the animal, and then "capture" it again for the host's pretended rescue, establishing practices that continue in natural history filmmaking today.[27]

As several historians have observed, even when programs refrained from explicit sentimentality, they conveyed messages about universal moral principles, reaffirmed essential values of home and family, and provided sanitized versions of the natural world.[28] And yet, *Wild Kingdom* was only mildly red in tooth and claw. "Fight or flight" scenes of violence were invariably followed by sermons about environmental responsibility.

DISNEY OPENS A NEW WINDOW

Walt Disney Studios was already a successful producer of natural history dramas when the television era began, having turned first-rate nature photography into "narrative entertainment" in such box-office hits as *Seal Island* (1948), *The Living Desert* (1953), and *The Vanishing Prairie* (1954).[29] Those films then became the backbone of a television series that Gregg Mitman has characterized as "sugar-coated education."[30]

The first *Disneyland* program in 1954 featured a salute to Mickey Mouse. That season's third program offered excerpts from Disney's latest full-length nature films combined with "behind-the-scenes" documentaries about the filmmaking techniques. Even though Disney claimed that "nature wrote the screenplay," its cinematographers continually assisted that composition by building camera windows into the sides of anthills or beehives.[31] Nature never really appeared "on its own terms" in a Disney program, notes historian Steven Watts, and never appeared without a moral.[32] "True-Life Adventures" began with dawn and birth and ended with sunset and rebirth and featured "courtship, nest-building, and parenting" story lines, "transparent allegories of progress," or "utopian fantasies" with visions of pristine, unexploited nature.[33] They starred anthropomorphized characters like Ida, "the Off-Beat Eagle." Edited to heighten the drama and draw connections between animal actions and motivations, Disney's fictionalizations of nature constructed a tidy world devoid of mess, chance, and human error, and upheld 1950s-style environmental stewardship, which emphasized domination and control, punctuated by sentimentalized wonder, as the appropriate relationship between humans and other species.[34]

The Disney approach also promoted an ambiguous image of science. Scientific authority was used to impart authenticity, yet scientists' real contributions to knowledge about the natural world were often ignored. Sonorous narration might describe an animal's diet, mating practices, and survival techniques in detail, conveying conclusions based on years of methodical fieldwork, while scientists rarely appeared in the films themselves. When scientific knowledge was referenced (for example, when a script described diet or migration patterns), facts were offered almost as folk knowledge, as if simply part of what humans had always known, rather than the result of systematic research.

To enhance their dramatic effect, programs interwove animation and live action.[35] Film of wild lions or elephants might be followed by excerpts from animated fantasies, segments on space travel followed by talking mice. The cartoon character Professor Ludwig von Drake was used to explain mathematical principles. Science was celebrated as part of life, but no more valuable than other products of human culture. During the first twenty-one seasons, about one-fifth of the Disney Studio television programs dealt with science, nature, space, or technology, mostly in "Adventureland" or "Tomorrowland" segments and, later, wildlife films.[36]

Natural history programs dispensed unthreatening content to fill the hours outside prime time, and the genre found lucrative markets in worldwide syndication. *Wild Cargo* (syndicated in 1963) described the capturing of wild animals for zoos; *Zoorama* (syndicated in 1965) was broadcast from the San Diego Zoo. Sometimes a successful series would involve an expert rather than an actor, such as NBC's *Animal Secrets*, hosted by anthropologist Loren Eiseley, but more often it was local educational programs, produced or underwritten by government agencies and aimed at special constituencies like bird-watchers or hikers, that filled the unsold time in late night or early morning and offered only token competition to Disney's sophisticated natural magic.[37]

NATURE AND CULTURE

In 1954, the cultural series *Omnibus* included the first Jacques-Yves Cousteau documentary shown on American television ("Undersea Archeology"), and its programs often featured natural history films along with segments on music, dance, and literature. Similar interspersing of science with the arts and humanities characterized a series that CBS launched in 1953 with the American Museum of Natural History.

The network initially claimed that "staff scientists" would appear to tell the "stories behind some of the museum's $40,000,000 collection of exhibits," and anthropologist Margaret Mead and other curators did appear on *Adventure*'s first season.[38] The producers promised that "scientific accuracy would be blended with . . . new electronic devices for creating illusions."[39] Critics praised the series ("This is not a hopped up Captain Video show," one wrote), and episodes attracted about two million viewers every week, yet no commercial sponsors bought time.[40] CBS eventually absorbed over $750,000 in costs during the first season. The next season, audience shares ran between 20 and 30 percent in major East Coast markets, helping to sustain viewers for *Omnibus*, which followed in the Sunday afternoon lineup, while advertisers remained leery.

CBS news and public affairs director Sig Mickelson, who oversaw production, said that the network knew that, to "provide the most useful educational service," the series must balance accuracy with appeal to a mass audience.[41] Because the museum had rejected the idea of "flavoring fact with fiction" and insisted upon using curators and staff scientists, the director attempted instead to layer the experts' "stories" with suspense: "The provocative question, the unpredicted result—in science as well as fiction—is always more exciting than a review of an accomplishment or an exploration of the past in which one pictures the result as a *fait accompli*. By the injection of this factor—television's priceless ingredient of immediacy—we hoped to capture and hold the imagination of the viewer by making him part of an experience."[42]

That desire to inject "immediacy" and include cultural topics may help to explain *Adventure*'s devolution. About one-third to one-half of the first season's shows dealt with anthropology (one of the museum's main areas of expertise) or topics like "a trip through space" via the museum's Hayden Planetarium (a segment one critic found "richer in promise than accomplishment").[43] The focus then shifted to include more cultural and social topics, such as discussions of religion and atomic weapons testing. Comedian Henry Morgan hosted several programs on the "human body and its functions." The second and third seasons featured more use of television "personalities" as hosts, explainers, and guides and therefore less on-camera time for scientific experts.

CARTOONS, PATTER, AND GIMMICKS:
THE BELL SERIES

In addition to influencing nature programming on television, *Disneyland*'s combination of live actors, film segments, and animation inspired other educationally oriented shows. Early in the 1950s, the Bell Telephone System (AT&T, its research group Bell Laboratories, and its regional subsidiaries) began developing an ambitious group of films designed to premier on prime-time television and then be distributed to schools "for use long beyond their brief hours on the air."[44] Aimed at "public education through entertainment," four of the films were made by Hollywood director Frank Capra (*Our Mr. Sun*, 1956; *Hemo the Magnificent*, 1957; *The Strange Case of the Cosmic Rays*, 1957; and *The Unchained Goddess*, 1958); four by Warner Brothers Studios (*Gateways to the Mind: The Story of the Human Senses*, 1958; *The Alphabet Conspiracy*, 1959; *The Thread of Life*, 1960; and *About Time*, 1962); and a final, shorter film, *The Restless Sea* (1964), by Disney Studios.[45] *Our Mr. Sun* attracted about twenty-four million viewers in Canada and the United States—almost one-third of the television audience likely to watch any program that night—and was praised by critics and the industry. The remaining films were similarly well received, although viewer excitement dwindled over the years.

In addition to adapting Disney's blend of animation and documentary film, the Bell project enthroned a new celebrity, a "face" who served as the audience's trusted surrogate, who translated complexity and guided viewers through the unfamiliar territory of the latest science. Like Lynn Poole, this host was not a scientist; instead, he played the role of a science advocate and explainer. Frank C. Baxter, a respected University of Southern California professor of literature educated at Trinity College, Cambridge, had already appeared on award-winning television programs when he was persuaded to play Dr. Research.[46] Viewers accepted the sixty-year-old Baxter as the epitome of a wise science teacher—rotund, jolly, smiling but formal, stiff in posture, slightly rumpled.[47] After his first appearances, Baxter's family joked about descriptions of him as a "pudgy, tweedy, twinkling, pink, bald bunch of enthusiasm," but he grew to enjoy both the celebrity and the additional income.[48] In the first few programs, Baxter shared the screen with professional actors playing the role of a wisecracking, skeptical, anti-intellectual writer. For the four Warner productions, Baxter served as sole narrator and star.

Although the visual style of the four Capra films differed from those produced by Warner Brothers, all expressed similar themes, probably reflecting the continuity of oversight from a prestigious advisory board cochaired by engineer Ralph Bown (recently retired from Bell Laboratories) and Warren Weaver, vice president of the Rockefeller Foundation.[49] Capra, a graduate of California Institute of Technology who maintained a lifelong interest in science and engineering, insisted that one "need only put a plank" across the gap between scientists and the public and "the people will rush over." Science, he explained, is full of "adventure stories" that can "grip the heart and mind."[50] To heighten the drama, Capra set many scenes in a crowded studio-laboratory, with a "magic screen" on the wall (mirroring the smaller screens on which viewers watched at home). Beginning with the fifth film, Warner Brothers removed the clutter from the sets and brushed emotion into the corners, but these differences were only superficial. All programs in the project incorporated similar Cold War concerns about national security and scientific secrecy, referenced the development (and uses) of nuclear energy, and celebrated an image of a science that cohabitated rather than quarreled with religion, even to the extent of suggesting religion as an alternative answer to nature's puzzles.

Capra and Warner Brothers strove to appeal to the largest possible audiences. No matter how many scientists criticized the productions as "the type of corn" that "entertainment people" push, the high ratings and positive critical reaction vindicated the livelier approach.[51] *Our Mr. Sun*—which critics called both "fascinatingly informative" and "irritatingly precious and condescending"—opened with choral music, a colorful sunrise, and a Bible verse.[52] A cartoon character then dispensed brief scientific explanations ("You know me as the Sun but of course what I really am is a star"), yet his solar history tended to be more cultural than technical. The Mr. Sun character decried the shift from worship to analysis ("instead of temples, rituals, and hymns to the Sun . . . it's domes, gadgets, charts, and numbers. . . . I'm demoted to a specimen").

Hemo the Magnificent's "short course" on the circulatory system alternated frivolity ("Professor Anatomy") with brief tutorials. Here, too, the central cartoon character (Hemo) complained about science's deconstruction of nature—humans cannot really tell Hemo's story because they do not "understand" him—and described blood as the "song of the lark, the spring in the lamb . . . the sacred wine in the silver chalice" whose mystery might be stripped away through scientific analysis. Baxter consoled viewers with the importance of science, explaining that acquiring knowledge requires hard work, extensive investment, self-sacrifice, and "burning the

FIGURE 7A. Cover of publicity brochure distributed to promote *Our Mr. Sun*, the first Bell Telephone System science special, 1956. The brochure emphasized the involvement of scientific advisers. Courtesy of Smithsonian Institution Archives.

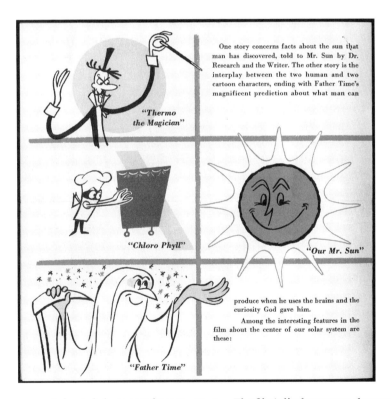

One story concerns facts about the sun that man has discovered, told to Mr. Sun by Dr. Research and the Writer. The other story is the interplay between the two human and two cartoon characters, ending with Father Time's magnificent prediction about what man can

"Thermo the Magician"

"Chloro Phyll"

"Our Mr. Sun"

"Father Time"

produce when he uses the brains and the curiosity God gave him.

Among the interesting features in the film about the center of our solar system are these:

FIGURE 7B. Animated characters from *Our Mr. Sun*. The film's lively cartoon characters signaled its entertainment goals. Courtesy of Smithsonian Institution Archives.

midnight oil . . . to find out why, when, and what." Nevertheless, such serious statements about researchers' dedication and creativity alternated with kaleidoscopic and satirical animations. As critic John Crosby explained, "Television itself has taught the crowd that if it hasn't got a gimmick, then to hell with it."[53]

The most visually complex of the Capra films, *The Strange Case of the Cosmic Rays*, featured an "out-of-this-world true detective story."[54] Marionettes (created by Bil and Cora Baird) representing Edgar Allen Poe, Charles Dickens, and Fyodor Dostoyevsky, serving on an "Academy of Detection Arts and Sciences" committee, are shown debating the "best detective story of the first half of the twentieth century" when the Writer interrupts and suggests, instead, "the strange case of the cosmic rays." The problem of how to detect and measure cosmic rays has as much mystery and adventure as any fictional story, he argues, because "science dicks" and "cap-and-gown private eyes" are currently engaging in "global legwork" to understand the universe's basic forces and discover new particles: "Think of it. As late as 1932, science had the universe all neatly wrapped up in three basic packages—electrons, protons, neutrons. Just twenty-five years later, in 1957, the number of separate and distinct particles of matter had jumped to at least twenty." When the marionettes ask the meaning of this "veritable deluge" of data "for common people, who work, love, hate," Baxter replies that "it's possible that how people work, love, hate has been, and is even now being affected by cosmic rays; it's possible that . . . these cosmic space bullets could be partly . . . responsible for evolution by causing mutations as they smash into our reproductive genes."

Private screenings at college campuses, sponsored by local telephone companies, were held after each television premiere. By 1958, when Capra's *The Unchained Goddess* was released, over six million students in the United States had seen the previous three films. *Goddess* contained hints of sex—the animated character Meteora, a "petulant, tempestuous, and irrational" queen of weather, flirted boldly with Baxter—and was less technical than earlier programs, with a gimmicky, arch style.[55] Nevertheless, *Goddess* included spectacular films of weather phenomena (hurricanes, snowstorms, torrential floods), praise for meteorologists at laboratories like the National Hurricane Research Project ("what a Buck Rogers device they've developed!"), and promises that research might someday reveal how to "steer" hurricanes to extinguish "oil fires on the ocean." *Goddess* also offered cautious warnings about climate change. Melting polar ice caps may indicate, Baxter warned, that humans are "unwittingly changing the weather through the waste products of civilization."

FIGURE 8. MIT physics professor Bruno Rossi, shown in a publicity still from the third Bell Telephone System science special, *The Strange Case of the Cosmic Rays*, 1957. In addition to appearing in the film, Rossi served as technical adviser. Courtesy of Smithsonian Institution Archives.

In the first Warner production, *Gateways to the Mind: The Story of the Human Senses*, a television sound stage served as metaphor for the connection between human brain and senses. Baxter and the actors played congenial, relaxed technical translators, while the real scientists and engineers shown in film clips were stilted, halting in speech, nervous, seemingly unable to look directly at the camera. Warner's *The Alphabet Conspiracy* attempted to explain linguistics by entwining science with magic and make-believe, and included not-so-subtle promotion for Bell's business interests. When a fictional character asserted that linguistics could not be a science because "[s]cience is test tubes, telescopes, microscopes, and chromium-plated machines," Baxter countered with examples like the telephone, phonograph, wireless, motion pictures, sound spectrographs, and the latest voice-recognition device allowing computers to "talk to each other over ordinary telephone circuits."

The Thread of Life, which premiered in late 1960, refurbished the professor's surroundings and recast Dr. Research as an industrialist. In a well-tailored gray suit, Baxter stood on an antiseptically spare set (no microscopes, glassware, or other scientific instruments visible) next to six television screens, explaining heredity and genetics ("why we are alike—in some ways—and yet why each person is different") to the *images* of the other actors, rather than face to face. This use of television as part of the plot placed emotional distance between the questioners (surrogates for the viewers) and science's representative, shielding both from embarrassment as they discussed human reproduction. The device also signaled that science remained distant from most viewers' experience. *The Thread of Life* ignored emotional considerations in human relations, and disconnected the "coming together" of a sperm and egg from anything vaguely hormonal. Baxter even used gambling devices as props to explain the odds of various genetic combinations. When a disembodied face complained that science apparently could not "really predict how things will come out," Baxter responded smugly: "No more than you can foretell the toss of a coin. Both are governed by that complexity of causes we call 'chance.'" *The Thread of Life* raised the ominous potential of terrible consequences from a nuclear age, yet stopped short of probing the topic in depth. When Baxter was asked whether all effects of radiation are bad, he merely responded that "most of them are" and "therefore we must be very careful when we use radiation." As critic Jack Gould observed, "An educational program that leaves unanswered the biggest question it raises can hardly be regarded as altogether successful."[56]

The final Warner Brothers production, *About Time*, opened with the declaration that "[i]t is written, to everything there is a season and a time to every purpose under heaven." Baxter grabbed an hourglass and suggested an "imaginary trip to . . . Planet Q," where people know little about time. His rote responses to questions from the planet's king, and dull examples from biology and navigation, dampened the drama, however, and the question-and-answer format introduced concepts that, in turn, just prompted more questions.[57] How, the king wonders, do we know that "Einstein guessed right, and that his theory is really true?" Well, Baxter suggests, "let's ask a physicist—Dr. Richard Feynman." Feynman stood in front of a blackboard covered in equations and appeared self-assured and engaging, but his judicious explanations fell flat. The king understandably complained, "I have more questions than ever," to which Baxter smugly replied, "Good . . . the more we learn, the more questions we have," reminding him that "[i]n the beginning, you had only one question: what time is it?" The greater the complexity, the greater the mystery.

Such production approaches either left viewers hungering for more information or changing the channel in search of livelier entertainment. Critics who had praised the Bell project's creativity in 1957 grew increasingly impatient with the later programs' vagueness. As John Crosby explained, "the whole point of an educational television show is lost if you try to mix too much entertainment in with it. People who want education seek education; people who want entertainment seek it. You can . . . wind up being neither entertained nor educated."[58] When scripts interspersed discussion of serious subjects like nuclear weapons with clever cartoons about "The Atom Bomb Boys," the science could easily be lost in the laughter.

CHAPTER FIVE

Taking the Audience's Pulse

The power of the media resides in its power to choose which ideas to present.
LESLIE GELB, 1991[1]

During radio's first decades, the American Medical Association (AMA) had cooperated enthusiastically with commercial broadcasters, sponsoring health talks and subsidizing production of medical dramas.[2] Well into the 1940s, AMA officials and prominent physicians continued to encourage use of broadcast media as a way to combat the "morbidity" of public "ignorance, indifference, and misinformation."[3] In 1942, New York City radio stations alone carried 283 broadcasts about health, many developed in cooperation with local medical groups. Nevertheless, a New York Academy of Medicine report dismissed much of that programming as ineffective: "the so-called drama is not dramatic, but consists merely of a motley of situations represented in excited chatter . . . almost entirely devoid of health education."[4]

Medicine's first use of television technology centered on professional training. During the late 1940s, medical schools began to install closed-circuit systems in surgical amphitheaters so that students might better observe operations. In the United States, the first remote televising of surgery took place in 1947 at Johns Hopkins University; later that same year, an appendectomy conducted at a Washington, D.C., hospital was observed by physicians attending a meeting elsewhere in the city.[5] In 1948, delegates to the AMA annual meeting watched a cesarean birth at a nearby hospital, and in 1951, a live birth was transmitted in color to AMA participants (reportedly causing several audience members to faint).[6] Within a few years, medical schools were offering closed-circuit refresher courses and long-distance demonstrations aimed at physicians in rural and remote practices, and health officials began planning how to exploit television for public instruction in first aid, hygiene, and accident prevention.[7]

Pharmaceutical manufacturers funded many of these early demonstrations. In 1951, with sponsorship by Smith, Kline and French Laboratories, the AMA conducted the first national live telecast of surgery, as physicians in Chicago and New York watched a heart operation being performed in Los Angeles.[8] In 1952, again with Smith, Kline underwriting, television star Roy K. Marshall narrated network broadcasts from an AMA meeting, allowing members of the public to see medical exhibits, demonstrations of new resuscitation techniques, and reports on new drugs.[9] With no advance notice, the June 1952 program included the first prime-time telecast of a live operation; viewers saw a few minutes of a complicated surgical procedure to remove a portion of a man's stomach.[10]

Close cooperation between the medical-pharmaceutical industry and professional associations also characterized the first general-interest programs. ABC's *Medical Horizons* focused on discoveries, research programs, and medical techniques, interspersing interviews of physicians and researchers with films of surgical operations or demonstrations of experimental equipment such as a new artificial kidney machine.[11] During the first seasons, episodes were hosted by past AMA presidents and, from 1955 to 1957, sponsored by CIBA Pharmaceuticals. Throughout 1952, many medical spectaculars were produced for national audiences in cooperative arrangements between the AMA and pharmaceutical companies, including remote pickups from hospitals, a living beating heart observable during a catheterization procedure, and an interview with a patient recovering from extensive jaw reconstruction.[12] Today's viewers have become inured to lurid news film from accident scenes or to simulated bloodletting on crime dramas, but such images did not represent standard fare on 1950s television. The medical professionals often seemed more enamored with the operations than did the critics, who complained about the overly "intimate details" of cesarean births and other procedures.[13]

Despite limited resources, many groups attempted to supplement the dearth of network public health programming. By 1954, about ninety county and state medical societies were producing television content on their own or cooperating with government agencies.[14] The locally produced series tended to be either symphonies of praise ("real doctors . . . men who today save lives once considered lost") or tedious instruction.[15] The AMA had to remind members that "three-dimensional montages" could increase "realism and impact," visual aids could help sustain viewer attention, and low-budget formats like roundtables could help preserve good relations with the station staff: "Equipment, floor-space, time and personnel are all commodities in such short supply that the local station manager is rapidly developing ulcers, or a coronary, or both, over implementing his own concerns, let alone ours. In addition, television requires tremendous outlays for scenery, costumes, artwork, props, special effects and lighting."[16] Whenever physicians requested local station assistance, the AMA cautioned, they should be mindful that they were "asking a great deal"—television time should be treated as a "gift."[17]

In exchange for sensitivity to broadcasters' constraints, the medical community apparently expected cooperation in the campaign to discourage television commercials making unsubstantiated health claims or offering inappropriate medical advice. "The use of stethoscopes and white coats" in advertisements, the *Journal of the American Medical Association* claimed, was "a cheap attempt to mislead audiences into believing that the claims had been proven medically."[18] In a formal complaint to the National Association of Radio and Television Broadcasters (NARTB), the AMA charged that commercials did not inform viewers that a man shown in a white coat "was only a hard-working trouper in greasepaint." Although the NARTB tightened its code on dramatized material involving statements by someone appearing to be a medical professional, the problem did not disappear.[19] Five years later, the New York State Medical Society again objected to pitchmen dressed in white coats, and called on federal regulators to control references to "doctors" or "your doctor" that implied that accredited physicians used or endorsed a product.[20]

In a news story about the first NARTB ruling, Jack Gould argued that the AMA's complaints were misplaced. The association should have taken on the overall absurdities of the claims made in medical commercials rather than quibbling over the "fate of its 'men in white.'"[21] The viewers, Gould explained, recognize that such dramatizations are fictional, whether or not the actors are labeled. He wondered (presciently) how the AMA might react if a "genuine doctor . . . stepped before the camera and begun reciting 'actual experiences' with a drug product that the sponsor was selling"—something that did occur on television fifty years later, much to the chagrin of many medical professionals.

MAKING IT ALL UP

Television dramas about hospitals, operations, diseases, and charismatic physicians brought medical information (both accurate and misleading) to far more viewers—and did so far more effectively—than most of the educational programs. Fictional series reinforced news reports about medical advances and, with the assistance of stereotypical characters, "played out various assumptions" regarding health care delivery and the efficacy of modern medical research, pharmaceuticals, and diagnostic procedures.[22] The plots of *City Hospital*, a live studio production that premiered in 1951, centered on a hospital director, Dr. Barton Crane, and a continuing character, Dr. Kate Morrow, one of television's first major heroines. The next season, episodes of another new series, *The Doctor*, were unified by a character called simply The Doctor (played by Warren Anderson), who introduced the opening scenes and returned at show's end to discuss the medical outcome.

These approaches were soon merged into a formula with sufficient drama and realism to satisfy viewers and please advertisers.[23] At the beginning of each episode of NBC's *Medic*, the narrator and central character Dr. Konrad Styner (played by Richard Boone) celebrated the physician as "guardian of birth, healer of the sick, and comforter of the aged."[24] Praised for its "seriousness and high-mindedness," *Medic* added medical information to the plots "in a taut and intense manner, without humorous relief."[25] During the second season, in 1955, *Medic* became the first fictional series to incorporate film of the birth of a real baby, prompting one columnist to observe that there was a "sound reason why doctors in the days before television did not invite every Tom, Dick and Harry . . . to sit in the operating room."[26]

To avoid offending the medical establishment and to reinforce an aura of authenticity, *Medic*'s producers sought validation from local and national medical associations, just as the popular crime series *Dragnet* closed each episode with a "seal of approval" from the Los Angeles County Sheriff.[27] Motion picture producers had sought similar assistance and endorsements for years. In 1937, Paramount Studios hired Hollywood Hospital's chief resident physician to read the script of the first "Dr. Kildare" movie (*Internes Can't Take Money*) and to check all sets and props for accuracy. Studio publicity then claimed that "[e]very article seen in any part of the picture pertaining to the medical profession was there by [the expert's] advice and with his approval."[28]

Medic's creator, James Moser, had been a principal writer for *Dragnet*, and he wanted the medical series to combine similar features: an institutional hero, puzzle-solving plots, and "rusty nail realism."[29] In order to film inside local hospitals and receive a "seal of approval," Moser courted the Los Angeles County Medical Association (LACMA) and allowed the association's advisory committee to review scripts.[30] In 1955, the AMA also established a Physicians Advisory Committee on Radio, Television, and Motion Pictures to address concerns about medical accuracy and "to protect the image of the practicing physician."[31] By 1960, that committee was fielding several hundred questions a week from writers and media producers, reviewing as many as 150 scripts a year, and recommending experts for programs other than medical dramas.[32]

At first, these review arrangements seemed mutually beneficial. Motivational researchers had advised the AMA that attaching a friendly fictional personality to the medical profession (much as General Mills Company was doing with Betty Crocker) could "establish an emotional connection with members of the public."[33] During *Medic*'s first seasons, association officials praised the program ("Here, as never before, television informs while it entertains").[34] Even if scripts occasionally veered into "doom-touched melodrama," the predominant images were positive.[35] Young actors like Dennis Hopper, Denver Pyle, Vera Miles, Michael Ansara, Charles Bronson, and Richard Crenna appeared in episodes about such sensitive topics as manic depression and postpartum psychosis. A 1955 episode about the destruction of a city with an H-bomb ("Flash of Darkness") featured an appearance by the U.S. civil defense administrator. *Medic* sometimes added practicing doctors and nurses to the cast, and Moser arranged for physicians to be present on the set to check such details as the correct use of medical instruments.[36] The networks then used this participation in its publicity (encouraging viewers to imagine they were watching real surgeons operating on real patients).[37] One of *Medic*'s first sponsors, Dow Chemical Company, ran advertisements that declared that the series made "no compromise with truth."[38]

From the outset, historian Robert Alley writes, LACMA functioned as both "censor and prod."[39] Physicians corrected errors and tinkered slightly with tone; in return, *Medic* received the validation of affiliation. The arrangement initially favored the doctors, Joseph Turow adds, because producers did not want to offend the medical establishment and (as with Frank Capra and the Bell series) the shows' creators admired and believed in the importance of technical expertise.[40] Although the medical community's support served as a defense when network executives or pressure groups criticized *Medic*'s scripts or pushed Moser to tone down realistic touches, LACMA's insistence on accuracy eventually played a role in the program's demise. In early 1956, the Roman Catholic Archdiocese of New York, concerned that *Medic* planned to include film of a cesarean birth, made a formal protest to the network.[41] In response, NBC removed the segment from the program, replacing it with stock footage of surgical operations. This censorship angered LACMA because the change altered medical aspects without their consultation, and the group withdrew its seal of approval for that episode, weakening the series' standing with advertisers.

Not everyone found the trend toward naturalism in medical dramas and documentaries appealing. In 1954, a Sloan-Kettering Institute special included graphic film of experiments on animals; an episode of the CBS series *The Search*, produced in cooperation with Yale University, showed a woman in childbirth and a newborn with the placenta still attached; another NBC special, produced in cooperation with the AMA, included closeups of an operation to remove a diseased aorta. "Medicine is not going to build confidence and faith in its service to mankind by turning the living room into an operating room," Jack Gould warned.[42] Under "the guise of education," he wrote, physicians were now sanctioning an "element of shock" more graphic than in horror movies.

The March of Medicine, a group of documentary specials produced by the Medicine Television Unit of Smith, Kline and French Laboratories, in cooperation with the AMA, further tested the line between educational realism and entertaining sensationalism.[43] In December 1952, a *March of Medicine* special originating from an AMA meeting included a live heart catheterization carried out with what one television reporter called "brute force, undue haste, and little sense of direction."[44] During 1954, the specials attracted almost fourteen million viewers per showing, and in 1955, *The March of Medicine* received the Albert Lasker Medical Journalism Award, the first television project to receive that award.[45] Nevertheless, the specials continued to draw mixed reactions. One viewer's informative "realism" was another's lurid "sensationalism."[46]

SENTIMENTALITY AND SOAP

Television's fascination with medicine continued throughout the 1950s and 1960s. Dramas based on the Sinclair Lewis novel *Arrowsmith* were shown on *Robert Montgomery Presents Your Lucky Strike Theatre* (1950) and *Kraft Television Theatre* (1954). "Bulletin 120" on *The Philco Television Playhouse* (1951) glamorized the world of medical researchers. NBC's *Producer's Showcase* production of "Yellow Jack"—a television version of Paul de Kruif's popular account of yellow fever research, which had already been adapted as a Broadway play, Hollywood movie, and radio drama—featured an all-star cast (Jackie Cooper, Wally Cox, E. G. Marshall, Raymond Massey, Broderick Crawford, Rod Steiger, Eva Marie Saint, and Lorne Greene). In the 1961–1962 season, 2.4 percent of network programming consisted of medical dramas; that amount doubled to 4.9 percent in the 1962–1963 season, and by spring 1963, 7.5 percent of programming on NBC alone consisted of medical dramas.[47]

Heroic physicians proved perennial favorites, in roles ranging from surgeons and psychiatrists to folksy country doctors and military medics.[48] The lead characters of the two most popular series, ABC's *Ben Casey* (1961–1966) and NBC's *Dr. Kildare* (1961–1966), possessed universal medical expertise as well as social consciences. American audiences already knew James Kildare's too-good-to-be-true character because of the Kildare novels, radio series, and ten MGM films. Ben Casey, however, was a new hero created by James Moser. Played by dark-haired, craggy-faced Vince Edwards, Ben Casey exemplified the 1960s angry man, a neurosurgeon with "a conceited, contemptuous smirk" who battled injustice and flouted authority.[49] Blond, dewy-cheeked Richard Chamberlain's Kildare offered a softer heartthrob, but his character, too, held no brief for officious administrators.

Both series projected positive images of physicians and, like *Medic*, received endorsements from medical associations. *Kildare*'s producers hired special technical consultants to vet scripts, both *Casey* and *Kildare* sought advice from the AMA advisory committee, and (in return for advance access to scripts) the association allowed statements on screen mentioning AMA "assistance."[50] In general, the relationships between studios and advisers worked well, but the AMA committee could neither change nor censor what they reviewed. Even when individual committee members went on record as being "appalled" by scripts, the association refrained from severing the arrangement because that would have eliminated any possibility of influence.[51] Publicity had also become linked to politics. As early as 1955, AMA officials had suggested that television's overwhelmingly positive images of medicine might help in fighting "socialized medicine" and "compulsory health insurance."[52] The medical establishment expressed similar opinions about the effectiveness of media attention through drama during the political debate over establishment of the federal health program that became known as Medicare.[53]

Despite these elaborate advisory structures, scripts included relatively few medical facts. Dramatic allusions to medicine became overwhelmingly cultural, social, and political rather than technical. Plots emphasized moral dilemmas and the righteousness of treating the afflicted; they rarely criticized inefficient care or medical error. Scripts also assumed a level of familiarity with experimental results that was probably exaggerated for most physicians outside of medical school faculties.

The success of *Casey* and *Kildare*, Erik Barnouw points out, triggered "a television stampede to the operating table."[54] *Medical Center* and the "New Doctors" segments of *The Bold Ones* focused on hospitals associated with research institutes. Two principal characters on *The Donna Reed Show*, broadcast from 1958 to 1966, were physicians—pediatrician Alex Stone and his next-door neighbor Dave Kelsey. An AMA president even appeared in a cameo role on that series.

This positive stereotype reached its apex on television in 1969 with the premiere of *Marcus Welby, M.D.* At the height of its popularity, that program received over five thousand letters a week from people seeking medical advice. Even though Welby's "perfect father" and "counselor and confessor" image was unrelentingly "over-sentimentalized," and the show received over thirty awards from medical associations, the show eventually provoked debate within the medical community over whether such exaggeratedly positive representations might actually ricochet back on the profession and its political goals.[55] The AMA's television advisory panels folded soon thereafter, and the association ceased providing formal endorsements.

Between 1954 and 1984, the vast majority of television's medical shows were produced by the same men who created *Medic, Casey, Kildare*, and *Welby* and who regarded the profession with great respect.[56] Dramas like *Medical Center*, which premiered in 1969, discussed errors, misdiagnoses, social controversies, and physicians' personal problems, while maintaining a respectful, admiring tone. After the 1970s, the AMA made fewer efforts to influence network coverage and focused on encouraging local programming; from 1971 to 1978, LACMA even produced its own nationally syndicated public affairs series, *Medix*.[57] Producers in search of advice began hiring individual consultants, much as Hollywood film studios had been doing for years.[58]

Television's afternoon "soap operas" offered one long-lasting context for positive images of medicine and health care. Many popular radio serials—such as *Young Dr. Malone*—had had physicians as their central characters, and these programs moved easily onto television during the 1950s. The melodramatic plots about human relationships frequently revolved around disease, disability, disaster, and death, with medicine exploited as "artistic prop."[59] *The Doctors*, which began as a prime-time anthology drama series in 1964, was the first daytime serial to win an Emmy award.[60] *General Hospital*, which premiered in 1963, was the top-rated television soap opera for decades.

PSYCHOLOGY 101

Not every advisory relationship proved advantageous for the professional associations. As the television industry experienced dynamic economic growth, securing expert imprimatur and approval became increasingly irrelevant. In the case of one drama centering on psychotherapy and psychology, the involvement of association committees actually exacerbated controversy.

At the beginning of 1962, building on the success of *Dr. Kildare*, NBC's publicity machine trumpeted the network's intention to focus forthcoming shows on the treatment of mental illness. An episode of *Dr. Kildare* served as a pilot for a series to be called *The Eleventh Hour*. Around the same time, the network hired a psychologist to evaluate its children's programming and promised that a suitable consultant would be appointed for the new project.

The Eleventh Hour revolved around a court-appointed psychiatrist (played by Wendell Corey) and a clinical psychologist. Most episodes exploited the relationship between clinical assessment of mental illness and the criminal justice system, which allowed scripts to explore a progression of sensational characters and crimes (such as a "sexy murderer feigning insanity").[61] When the producer boasted that dialogue was "realistic" because it included such words as "crazy," "nut," "loony," and "bughouse," the American Psychological Association (APA) objected.[62] The discussions of diagnoses and treatments were often inaccurate, and, the association statement explained, the characters' exaggerated personalities would only contribute to the social stigmatization of mental illness. Critic Jack Gould called the series "Disturbed Television." Television's occasional fine documentaries on the problems of treatment of mental illness did not "excuse use of insanity as a regular staple" in its fiction, Gould wrote; and "the sordid insistence on examining the madman again and again" in one particularly "dark" episode did nothing to advance "the ostensible theme of the drama."[63]

The psychologists' group, in apparent coordination with the American

Psychiatric Association, began a public relations ping-pong game with the network, sending a series of letters and issuing a formal press release in early December, all answered by the network with lengthy defenses.[64] Concerns centered on a credit line that thanked the AMA rather than the psychiatric or psychological associations. The APA complained that because the studio had ignored input from a psychologist provided to advise on the series (he was shown only one preliminary script), scripts were riddled with inaccuracies, yet the closing statement implied professional endorsement and, therefore, authenticity. The AMA committee, the APA pointed out, was composed of physicians rather than psychologists or specialists in the treatment of mental illness. By then, however, neither the networks nor the Hollywood studios really needed *any* adviser's approval. The blessing of experts had become a mere publicity bonus. NBC rather archly dismissed the imbroglio as related to the APA's "running controversy with the medical and psychiatric profession over the role of the nonmedical psychologist." The show's star even felt confident enough of his ratings to make public statements in which he dismissed psychiatry as "somewhere between tea leaf reading and snake oil remedies"—despite the fact that he was playing the role of a psychiatrist.[65] Case dismissed.

DIMINISHING RETURNS

During television's first decade, viewers were able to glean bits of information about contemporary research from nature programs and medical dramas, and from special broadcasts relating to nature and medicine. More nourishing servings of science became increasingly rare. From 1949 to 1951, educational programs (on all topics, including science) accounted for only a small percentage of available programming, and while the absolute number of nonfictional science programs increased on major networks up through 1955, that amount soon declined.[66] In 1954, New York City viewers could watch *Adventure, Johns Hopkins Science Review, The March of Medicine* specials, *Watch Mr. Wizard, Omnibus, What in the World?*, and *Zoo Parade*. Within a few years, only Mr. Wizard remained on the air. As networks sold more advertising and airtime became more precious, the unsponsored programs and anything else considered "educational" were pushed out of prime time.[67]

The creators of radio's first science programs had taken considerable interest in *who* was listening; they read and counted each listener letter. By the 1950s, the same broadcasting networks had transitioned to a business model that valued aggregated demographic data more than feedback from individuals. Television's evening schedules were planned as a whole, each successive program selected to support the ones that preceded and followed, with sensitivity to competition and with the goal of orchestrating attention so that viewers would not switch channels.[68] Medical dramas had attracted lucrative commercial audiences, but if science or medicine were not perceived as drawing large audiences, then networks had little interest in scheduling other science-related programs in prime time, whether dramas or documentaries.

CHAPTER SIX

Saving Planet Earth: Fictions and Facts

[C]ute gimmicks tend to obscure the point rather than spread entertainment.

JOHN CROSBY, 1958[1]

By the mid-1950s, *U.S. News & World Report* was calling television "the biggest of the new forces in American life," an insidious influence on consumer purchases, election campaigns, meal schedules, and bedtimes.[2] A significant proportion of Americans now regarded television as a more complete, accurate, interesting, and comprehensible source of news than their daily newspaper.[3] When viewers were asked which television programs provided their prime source of information about science, over half named one of three entertainment series: *Medic, Disneyland,* or *Science Fiction Theater.*[4] Slightly less than one-third mentioned educational series like *Science in Action, Medical Horizons,* the Bell Telephone specials, or *Watch Mr. Wizard,* and many reported that they routinely turned to fictional series, including afternoon soap operas, for health information.[5]

Television's skill in creating "realistic" dramas confounded this situation. A well-produced show introduced as "based on fact" might look just like one described as "resting on fantasy."[6] The skillful interleaving of reality and fantasy related to space, for example, built on decades of science fiction, from the novels of Jules Verne and H. G. Wells to *Astounding Science Fiction* magazine, Flash Gordon movie serials, and, in the 1950s, Hollywood films like *Rocket Ship X-M, Destination Moon, When Worlds Collide,* and *The Day the Earth Stood Still.*[7] The images created in these venues took on new meaning when human astronauts actually headed for the Moon.

SUPERHEROES AND SPACE

Building on the long-standing popularity of science fiction, early television quickly ventured into space, with fictional characters like Captain Video, Captain Midnight, Tom Corbett, Rod Brown, Buzz Corey, Rocky Jones, and Johnny Jupiter offering new stereotypes of galactic adventurers unintimidated by science. First telecast live from New York in 1949, DuMont's *Captain Video and His Video Rangers* was the most popular of these series.[8] Until moving into syndication in 1956, *Captain Video* was telecast four to five times a week, drawing audiences, young and old, of over 3.5 million.[9] Armed with pseudoscientific gadgets like "opticon scillometers," "thermal ejectors," atomic rifles, electronic straitjackets, and portable force fields, and combining courage with "scientific genius," Captain Video and his civic-minded Rangers battled the evil scientist Dr. Pauli and the Asteroidal Society.[10] Scientists themselves were rarely the primary heroes or villains of these shows, however. They were more apt to be characters on the periphery of action, calm observers applying cold rationality or the inventors of weapons. The devices used on *Space Patrol* ("paralyzer ray guns," "brainographs"), for example, had often been invented by friendly scientist Dr. Von Meter.

During the early 1950s, space heroes—*Buck Rogers, Rod Brown of the Rocket Rangers,* and *Commander Cody, Sky Marshall of the Universe*—captivated viewers by spotlighting "excitement" rather than science.[11] Plots centered on "morally didactic" struggles and individual challenges; references to contemporary science were used to lend verisimilitude to the illusions.[12] Nevertheless, the writers of *Tom Corbett, Space Cadet* did seek advice from the Hayden Planetarium and similar institutions, and hired rocket scientist Willy Ley to review scripts and help them fictionalize concepts like "variable gravity forces, asteroid belts, and anti-matter."[13]

The set designers probably had the most fun in creating stage tricks for a new medium. Painted basketballs became "meteors" battering a spaceship. On ABC's *Tales of Tomorrow*, an atomic explosion was simulated by mounting cameras on springs: "Balsa wood columns fall and furniture collapses. Giant 5,000 watt spotlights flash on and off, simulating cosmic lightning. Six sound effect records rotating together pour out the eerie sound of typhonic winds created by a bursting world. Add smoke pots liberally, and the televiewer holds on to his chair with both hands."[14]

While some plots drew inspiration from Cold War politics (the "Atom Squad" fought to "protect America's atomic secrets from her adversaries"), most programs just sprinkled occasional facts among the illusions. *Science Fiction Theater*'s opening set was filled with scientific equipment, yet the setting was intended, Patrick Luciano and Gary Colville point out, to signify "the study of a gentleman scientist" rather than the "refuge" of a mad scientist.[15] At the beginning, host and narrator Truman Bradley would deliver a factual commentary and demonstrate a scientific principle, implying that the drama that followed "was not so much fiction as extrapolation."[16] Producer Ivan Tors reportedly placed heavy emphasis "on the 'science' half of the science fiction equation," employing a full-time scientific adviser and a six-person research department to check script accuracy with university and government experts.[17] Trade advertisements for *Science Fiction Theater* then exploited these connections, declaring: "Out of scientific truth comes STIRRING TV DRAMA . . . AMAZING because it's science! COMPELLING because it's fiction!"[18]

SPACE REALITY

Rod Serling's paean to irony, *Twilight Zone*, which premiered in 1959, made no comparable promises about authenticity. Each opening scene proudly declared that the series searched a "fifth dimension as vast as space and as timeless as infinity . . . the middle ground . . . between science and superstition." Such expansiveness had become necessary because science fiction had always offered "extrapolations of *known* science into the prehistoric past or the possible future," describing television, radar, and atomic power long before realized in fact. Now, real life threatened to overtake fantasy. In August 1955, Willy Ley was being quizzed by journalists on *Face the Nation* when the Associated Press announced that the Soviet Union had probably launched "experimental rockets bearing live animals more than 300 miles into space."[19] In November 1955, a telecast from the Hayden Planetarium, part of the CBS *Adventure* series, displayed a model of a satellite then being built by the U.S. government. Two years later, on October 4, 1957, the Soviet Union launched *Sputnik*.[20] The subsequent space race and political debate over "missile gaps" helped to make all science seem relevant, as well as to entangle science with international politics.[21]

Television quickly appropriated the space programs and helped transform their efforts into another type of show business. Until 1958, the U.S. military's rocket tests, while not a secret, had been off-limits to the public. Then, in response to *Sputnik*, and much as had happened with atomic testing in 1951 and 1952, "the media were invited inside."[22] Establishment of the National Aeronautics and Space Administration (NASA) signaled the opening phase of an international race to the Moon.

Because imaginative fiction could pale in comparison to the drama of real rockets and astronauts poised on a launch pad, television responded by artfully blending fantasy and fact, incorporating fiction into its factual coverage and adding more touches of "realism" to space-themed entertainment. Fictional series like *The Man and the Challenge* were enlisted to help strengthen political support for space exploration. *Men into Space*, whose 1959 plots extrapolated from current science and engineering and imagined how humans might function on a permanent space station, was produced in cooperation with the Department of Defense, which reviewed scripts for accuracy, and with the U.S. Air Force, which arranged filming at military bases.

CONQUEST

Media attention to science also received support during the 1950s from an ambitious quest for knowledge about Earth. The International Geophysical Year (IGY), modeled on previous International Polar Year projects, comprised geophysical research conducted during 1957 and 1958. The American component of IGY, spearheaded by Joseph Kaplan and Hugh Odishaw, was coordinated through the National Academy of Sciences (NAS) in Washington. Two years in advance of the research phase, Odishaw's Office of Information began to orchestrate IGY publicity via film, print, and television.[23] Odishaw and other scientists were guests on such disparate series as *Home* and *Johns Hopkins Science Review*; the office arranged a *Disneyland* production about the challenges of Seabee construction in Antarctica, and coordinated publicity connected to Frank Capra's *Unchained Goddess* through the Bell Telephone Company public relations firm.[24]

Tension between the military and civilian components of IGY research complicated the media campaign. After *Antarctica—The Third World* (an NBC documentary about Operation Deepfreeze) attracted over five million viewers, the network sought permission to return to document the U.S. Naval Task Force project in Antarctica and gather newsreel footage and radio reports for NBC News.[25] IGY officials, seeking to project a civilian image of their project, sought reassurance from the network that the documentary would focus only on "civilian scientific contributions," even though the military partners controlled access to the research site. When NBC grew frustrated with restrictions being imposed on filming another documentary at government laboratories, the network simply switched to studio production and added more animations.[26] For one program, IGY officials even tried to keep narrators from using the word "missile" in connection with rocket launches at White Sands Test Facility.[27] And as a segment was being developed for Dave Garroway's *Wide Wide World*, the NAS public relations officer asked NBC for assurance "that the program as a whole will afford an appropriate framework in which to portray the civilian scientific and peaceful nature of the IGY program."[28] The resulting broadcast's "critical" tone and preoccupation with rocketry's military aspects displeased NAS officials considerably.[29] They soon realized that, once commercial productions were under way, they would have little control over content.

During 1957, the IGY project produced ten programs in cooperation with NBC, disseminating the series first through educational stations and later through network affiliates.[30] Although intended as an informational series, *IGY: A Small Planet Takes a Look at Herself* worked at keeping viewers interested. Producer Evans G. "Red" Valens suggested that Kaplan add "some personal leavening—by this I mean some short tales of personal experiences you have had in your own work with the IGY, humorous incidents definitely included" and asked the scientist to "speak off the cuff instead of being glued to a teleprompter."[31] Valens discouraged other guests from seeking to "impress" with their qualifications and pushed them to show something "significant and intriguing" instead—"the very modesty of such an attitude will reflect the high quality of the show far better than any boasting, however subtle."

Commercial support for IGY media efforts was not rejected, though, especially if underwriting might ensure better-quality content. In 1957, NAS, CBS, and the Monsanto Company, along with the AAAS, began development of *Conquest*, a magazine-format series highlighting "critical shortages . . . in America's scientific education and practitioners."[32] NAS agreed to provide technical assistance, facilitate connections with likely guests and advisers, and suggest ideas, in return for a modest fee and advance consultation on program ideas and scripts.[33] *Conquest*'s premiere described the ocean floor, interviewed biochemist Wendell Stanley, showed films from a record-breaking balloon ascent, and discussed Soviet scientific progress. The initial reaction at NAS appears to have been positive. The programs were regarded as providing "accurate glimpses of the scientific frontier."[34] Nevertheless, one NAS official cautioned, "The word 'glimpses' is perhaps the keynote," because the "average" person "without some background in science would probably not really appreciate what effort and thought and work have been necessary to come as close as we are to the 'edge of life.' . . . The thing missing is a little of the build-up of what went before."[35]

Subsequent episodes about chemotherapy, volcanoes, and IGY Antarctic research did not elicit great enthusiasm from NAS president Detlev W. Bronk, despite Monsanto funding for additional educational outreach.[36] The next year, explaining that the organization no longer considered *Conquest* a "pioneering experiment," NAS withdrew from cooperation in the production, and AAAS assumed the role of principal adviser during the second (and last) season. Programs were reduced to thirty minutes, most with specific single themes, and many simply edited versions of earlier shows.

Perhaps the most unusual IGY publicity proposal involved the comedy team of George Burns and Gracie Allen. The incident demonstrates how the scientific elite continued to regard popular culture with wary skepticism. Explaining that the offer represented a private conviction "of the need for a greater national awareness" of science, one of the couple's representatives sent a lengthy telegram to NAS, pitching a story line "in which Gracie experiences a complete reversal of her zany characterization and becomes a great brain."[37] In that role, Allen would then convey an appropriately dignified message about science and the public welfare (albeit in her character's quirky voice). The NAS public relations officer politely declined, adding how "touched" they were by the suggestion. They would be happy to cooperate if Gracie Allen wanted to make a statement about science before or after the "story portion" of her show, but were concerned about the proposed discussion of IGY within a comedy context.[38]

CREATING FANTASIES

Television's cartoon characters participated enthusiastically in science and space adventures and, early on, began taking imaginary trips to the Moon. After Walt Disney read a *Collier's Weekly* series describing the future of space travel, his studio hired Wernher von Braun and Willy Ley as advisers, and the rocket scientists appeared in three films (*Man in Space, Man and the Moon,* and *Tomorrow the Moon*) shown on the *Disneyland* show, which combined newsreel footage with animation and mock-ups of space installations. These and other mid-1950s "Tomorrowland" segments promoted the message that space exploration would be possible if only humans had the will to try, a message coincident with U.S. government policy.[39] In contrast to "mad scientists" in other Disney cartoons, these space segments celebrated science as a sensible force for good.

Disney programs never completely isolated real science from fantasy, of course, even when the topic was atomic energy. A 1957 "Tomorrowland" included both *Our Friend the Atom* (a film combining live-action footage and state-of-the-art animation) and an excerpt from the new Disney movie, 20,000 *Leagues under the Sea*, about a fictional nuclear submarine, *Nautilus*.[40] *Our Friend the Atom* formed part of a larger project, including a theme-park exhibit and children's book, which promoted atomic energy as a genie who cannot be forced back into the bottle but whose power might be channeled toward peaceful uses, in deliberate contrast to other popular culture attention to the terrors of the "military atom" and nuclear weapons.[41]

By then, nuclear fear had become a "shaping cultural force" throughout American life, and the associated images infused network television, providing counterpoints to official government communications.[42] As historian Paul S. Boyer has documented, postwar novels, movies, and television dramas frequently explored atomic energy's negative potential, while government and industry sources promoted the positive. Public attitudes coasted up and down: through 1940s respect for atomic physicists and "relief" at war's end, 1950s anxiety over the bomb and the rise of other nuclear powers, and then into what Boyer describes as subsequent unending cycles of activism, apathy, and anxiety.[43] Television series like *The Outer Limits*, Boyer writes, imagined new menaces without explicitly addressing conceivable threats like "radioactivity, genetic mutation, and atomic war," and programs like *Disneyland* offered soothing images of the future.[44] Sometimes television's science fiction heroes battled the evil within human nature—villains motivated by greed or ambition—but more often the demons had been enabled, assisted, or even created by the same science later used to defeat them.

JUMPING ON THE MOON

By 1960, reaching the Moon had become a political and scientific obsession, and television planned to go along for the ride, just as radio had followed Richard Byrd's Antarctic expeditions and William Beebe's ocean descents. Viewers could now watch rocket launches in real time. Live broadcasts of audio transmissions from Alan Shepard's 1961 flight proved so popular that the networks began pressuring for "unrestricted coverage" of subsequent missions.[45]

Television eagerly followed the expansion into space, building on the journeys taken by fictional dramas and exploiting many of the same stereotypes.[46] Astronaut "explorers" became television's newest celebrities and superheroes, prompting some scientists to criticize the "saccharine family interviews" and "breathless chronicling" of every mission and to bemoan the failure to explain technical terms and concepts.[47] Television producers ignored such complaints and stuck with what they believed the audience wanted: more personality and more drama. As critic Robert Lewis Shayon later explained, the model promoted by NASA and embraced by the networks guaranteed that all future explorers would be "accompanied by television cameras" and thus become "actors, making their nebulous exits and entrances for the benefit of multi-planetary audiences."[48] It became standard practice for astronauts to conduct press conferences from orbiting capsules, answering personal questions, with the live presentations scheduled to coincide with television's morning or evening news shows, just as Byrd and Beebe had timed their radio broadcasts to achieve maximum audiences. Half a century later, despite real accomplishments, education, and expertise, the space pilots and crew from all countries were still waving and joking for the cameras, transformed into performers for the magic box, siblings to the late-night comedians.

Although space exploration soon became sufficiently unexceptional to generate satire (such as *The Jetsons* or *My Favorite Martian*), most of television's space fiction exaggerated the dangers of exploration or the threats that might lurk outside Earth's atmosphere. "Science" served as a tool for both offense and defense whenever television's unworldly visitors used advanced technologies and hypothetical scientific advances in their attempts to conquer Earth on series like *The Outer Limits* and *The Invaders*.[49] That first decade of space dramas reinforced images of science's cultural value, social relevance, and authority, intertwined with mystery, spiritual explanation, and political intrigue, even as the plots raised questions about scientists' omniscience and consistency. In some fictional dramas, the "villain" could be technology (for example, an uncontrollable computer system or a genetically engineered virus imagined as a futuristic "Frankenstein's monster"); in others, technical expertise would guide heroic teams to victory against aliens who were also armed with science.[50] Science became a universal prop. Viewers increasingly encountered space-related science in contexts where fantasy, fact, and fiction swirled together like so much cosmic dust.

Reality had also begun to seem fantastic. In 1962, all over the United States, students trooped into school gymnasiums to watch on tiny television sets as John Glenn orbited the earth in *Friendship 7*, experiencing his adventure vicariously just as they had galloped in their imaginations alongside the Lone Ranger or flown next to Superman. Millions of people around the world heard Glenn describe three sunrises, three sunsets, and a dust storm over the Sahara. And the networks, not yet able to televise Glenn while in

FIGURE 9A. *Friendship 7* capsule simulation, February 20, 1962. For the first space flights, NASA provided only audio transmissions, so the television networks created visual simulations to keep viewers interested. Here an "astronaut" is looking at a monitor. *U.S. News & World Report* then photographed a television screen to obtain illustrations for its news coverage of the *Friendship 7* flight by astronaut John Glenn. Courtesy of Library of Congress.

FIGURE 9B. *Friendship 7* capsule simulation. A model of the capsule is being dragged across a map. Courtesy of Library of Congress.

FIGURE 9C. *Friendship 7* capsule simulation. As viewers heard John Glenn's account of the view from *Friendship 7*, they could watch "Glenn's voice." Courtesy of Library of Congress.

orbit, created simulations to show on the screen while viewers listened to his voice.

Liftoffs and landings, of course, offered irresistible visual opportunities. These actualities, unlike the atomic bomb tests, could stretch over days. Real-time events (athletic contests, inaugurations, state funerals, and space missions), edited to provide the best perspectives, increasingly mixed with faked actuality. At the very moment viewers reclined on their couches, the contestants on a game show were winning or losing—or so audiences were meant to assume. In each shared experience, viewers were encouraged (sometimes prompted by the reactions of skillfully manipulated studio audiences) to gasp, laugh, or cry with millions of other people. And so the Apollo project, too, became a television show, the astronauts clowning for the cameras and holding up messages to the people at home.[51] In 1961, CBS devoted slightly over 10 hours to covering the space program; in 1969, its coverage exceeded 109 hours.[52] By the end of the 1960s, NASA's space extravaganzas, culminating with choreographed landings on the Moon, had become sophisticated mixes of actuality and fantasy, news and entertainment.

The space project's main themes conformed well to how space technology and science were already being presented in popular culture. Fictional teams like the "government-funded" IMF (Impossible Missions Force) on *Mission: Impossible*, which premiered in 1966, lent credence to dreams that science and technology might solve the world's problems after all. *Star Trek* had also premiered that same year, its characters engaging in noble missions, confronting threats with ingenuity and courage, and depending upon teamwork marked by civility.[53] Thanks to sets and video techniques that made anything seem plausible, uninformed viewers could easily assume that the science fiction dramas mirrored reality. Science teachers, in fact, constantly wrote to the producers of *Mission: Impossible* to suggest devices for future episodes.[54]

During the Christmas holiday of 1968, the first manned spacecraft to leave Earth orbit provided television viewers with a new perspective on their home planet. While circling the Moon, the *Apollo 8* astronauts transmitted a live television image of Earth.[55] Since the mid-1940s, analysts had frequently characterized television as a "window" either *on* or *to* the world. After the 1968 broadcast, that metaphor, exploited in Dave Garroway's *Wide Wide World* and in *The Wonderful World of Disney*, carried a new implication. Television had enabled humans to look back on themselves and to glimpse the whole world all at once.

MOON PARTY

When the *Apollo 11* capsule landed on the Moon on July 20, 1969, the astronauts' actions provided familiar close-up views of humans walking, jumping, and planting flags, choreographed and carefully timed for evening broadcast. Breaking news, entertainment, reality, science, technology, and a sense of history in the making combined to create the ultimate "reality" show.

In Washington, the Smithsonian Institution held an all-night party, a television-linked Lunar Weekend in the Victorian-era Arts and Industries Building (where the Wright brothers' *Flyer* and Lindbergh's *Spirit of St. Louis* hung on display). NBC and CBS set up broadcasting operations under the historic planes, and installed large color monitors for visitors to watch (just as giant screens were set up in New York's Central Park and elsewhere).[56] Because the INTELSAT satellite network had been completed nineteen days earlier, the lunar landing became the first truly global broadcast event.[57] When Neil Armstrong stepped onto the Moon's surface, 720 million people around the world participated in his "giant leap."

Space missions supplied the ultimate television spectaculars—more thrilling than Hollywood's movies because viewers knew that the danger and risks were real. Throughout the duration of the Apollo program, a few prominent scientists complained that important opportunities for public education were being squandered: television spent too little time explaining the relevant scientific and technical aspects; no network

FIGURE 10A. *Apollo 11* Moon Party at the Smithsonian Institution, July 20, 1969. CBS and NBC broadcast from the Smithsonian Institution's Arts and Industries Building, interviewing experts and setting up monitors so museum visitors could watch the astronauts step onto the Moon. Charles Lindbergh's *Spirit of St. Louis* is visible at top right. Courtesy of Smithsonian Institution Archives.

FIGURE 10B. *Apollo 11* Moon Party at the Smithsonian Institution, July 20, 1969. The Wright brothers' *Flyer* is visible at top right. Courtesy of Smithsonian Institution Archives.

carried significant informational discussions about space apart from the NASA coverage; and news stories focused more on human-interest aspects (such as the astronauts' personalities and families) than on the scientific principles behind rocket propulsion or zero gravity. Such biases, while regrettable, had a firm foundation in television economics. Commercial networks assumed, probably correctly, that viewers hooked on drama and comedy would be uninterested in space seminars. During the long waiting times preceding launches, broadcasters occasionally inserted "minilessons" about rocket technology, but they never hesitated to mention Jules Verne to keep the audience entertained.

CHAPTER SEVEN

Adjusting the Lens: Documentaries

Give me a good commercial TV writer and a physicist-consultant with imagination and a sense of humor, and I'll teach ten million Americans more about the fundamentals of high-energy physics in half an hour than science writers and seminars can get across in the next 50 years.

JOHN K. MACKENZIE, 1965[1]

Print reporters had long visited places like disaster sites, mines, and medical laboratories where (for reasons like safety or practicality) not everyone can be admitted, with the understanding that the journalists would describe accurately what they observed.[2] With the development of television, its producers and cinematographers sought similar privileged access for filming news segments and documentaries. For science in particular, the television documentary offered exciting opportunities to accompany researchers to remote sites or show complex equipment in operation.

Documentaries did not only visualize and explain, however. Whenever they explored what scientists were discovering about degradation of the environment or about threats to public health, they engaged in the same type of investigations through which print media rattled the pillars of power. If the public good includes having a well-informed citizenry, then should not television pay close attention to the science related to current political debates? What responsibility did the television industry have to support and encourage science documentaries (even if they rarely attracted large audiences or commercial advertising)?

In the United States, trusteeship of a broadcasting frequency (that is, possession of a license to broadcast) involves not merely the right to provide entertainment and sell advertising time but also, as part of a free press, a legally prescribed duty to serve the public interest, to bring "to the people facts and opinions in order that they may learn to know the true from the false."[3] Commercial realignment of the media industry during the 1950s had complicated where that responsibility was being assigned.

Newspaper corporations had purchased radio and television stations, and film-production companies and movie studios had developed cooperative agreements with the broadcast networks. Each such arrangement had, by the 1960s, added to the entanglement of financial and organizational "interrelationships within interrelationships."[4] Communications syndicates were acquiring unprecedented power over the content available via television, with consequent ability to spotlight *or ignore* social and educational topics, as they wished.

Ultimately, two disparate forces controlled which science-related documentaries appeared on the air—illusion and money. Television's most distinctive quality remained an ability to replicate the world, in and out, up and down, mountaintop and mining shaft. Visualization, however, brought no assurance of accurate representation. As video artists were discovering during the 1960s and 1970s, television could be used to create both "illusory view(s) of reality" and "realistic portrayal(s) of illusion." Via stunning cinematography and animation, documentaries could "distort and sharpen" visions, could offer plausible *versions* of truth.[5] And without adequate subsidy or commercial sponsorship, the most creative documentary could not be produced or broadcast. Television may have given the documentary film "its greatest impetus," historian William Bluem notes, by providing "a vast audience and enough money to produce."[6] Ironically, through underwriting certain programs and not others, political organizations, corporations, government agencies, and other sponsors could shape what was made available to that audience.

American scientists and their organizations rarely exploited this potential by financially supporting documentary programming on television.[7] In the absence of scientific community engagement or pressure, it was thus the level of controversy surrounding a topic, the amount of money needed to make a competitive program, or an individual producer's interest in science that most often determined what appeared.

EARLY WINNERS

Television documentary programming on all subjects actually increased during the 1960s. In 1962 alone, ABC, CBS, and NBC together broadcast 387 documentaries—ten times the number shown in previous years, most of these as "specials" or multiple-part miniseries preempting regular programming.[8] Some scholars attribute this rise to the quiz show scandals, in that the networks used cultural specials as a public relations ploy to forestall government regulation.[9] Improvements in video-production techniques also played a role. And the availability of color television increased the attractiveness and appeal of nature documentaries.

Whatever the reason, viewers were the beneficiaries. Many of these programs offered important sites for "ideological work," exploiting well-researched scripts and expert commentary to decipher complex contemporary social and political issues.[10] Television documentaries played significant roles in galvanizing public opinion on workplace safety and family planning, for example. Fred Friendly's *Harvest of Shame* is credited with helping to improve conditions for migrant farmworkers, and *The Population Explosion* addressed access to contraception. Many 1960s documentary projects also served as training grounds for the professionals who would produce the next decade's science miniseries and apply the same narrative and editing techniques they had previously used to investigate poverty, hunger, and electoral politics.

Science was not ignored altogether in the documentaries of the 1960s. When Rachel Carson's first book, *The Sea around Us*, drew attention to the fragility of interfaces between humans and the marine environment, an award-winning film based on the book led to invitations for the biologist to become involved in various television projects.[11] Historian Mark Lytle describes how Carson agreed to develop a script about "clouds" for *Omnibus*. Even though she admitted she did not watch television and did not know the series, the prospect of huge audiences for her environmental message was, she said, "irresistible," and the producers had "offered her wide latitude" in determining script content.[12] Far more controversial attention to Carson's work appeared after publication of *Silent Spring*, when a *CBS Reports* documentary "The *Silent Spring* of Rachel Carson" (April 13, 1963) described industry attacks on the book. As host Eric Sevareid explained, although the decade had been dubbed the age of the atom and the "frozen TV dinner," Carson had now also revealed it as the age of "the calculated risk" because pesticides sprayed to produce perfect fruits and vegetables could have potentially negative impacts on all living things. When Carson's opponents pressured CBS to cancel the documentary, their campaign attracted even more media coverage and, the following month, prompted another CBS project, *The Verdict on the Silent Spring of Rachel Carson*, about a major scientific report that vindicated Carson's warnings.[13]

The overall number of science and medicine specials had increased slightly in the early 1950s, primarily because of the films subsidized by the Bell Telephone System and by Smith, Kline and French Laboratories.[14]

During the 1960s, other subsidized "specials" and miniseries (like those created by the National Geographic Society) provided some of the only nonfictional science available during prime time. Topics ranged from dire warnings about pollution to the artistic celebrations of natural beauty in films about the Sonoran desert region and the Grand Canyon narrated by writer Joseph Wood Krutch. Shorter science-related segments could also be found within television's public affairs programming. *CBS Reports*, which premiered in October 1959 as a direct outgrowth of Edward R. Murrow's *See It Now*, devoted over 5 percent of its episodes between 1959 and 1979 to science, space, medicine, or environment, with the latter two topics receiving the most attention.[15] On "magazine format" programs like *60 Minutes* (which premiered on CBS in 1968), the independently produced minidocumentaries included occasional interviews with scientists, interspersed among features about movie stars and criminals.[16]

CHIMPANZEES AND OCEANS

Even though the Disney and Bell projects had demonstrated color broadcasting's potential, enough consumers had to purchase new color sets to warrant a wholesale shift to color film for documentary and news programs. By 1967, however, color capability was being credited as a "principal exponent" in the popularity of "event" and "reality" programming ("television is recognizing its opportunity to break through the TV tube and really link hands with the viewer," one executive explained).[17] Predictably, sports (like Olympics coverage) drove most of this innovation, but the "true life" simulacrum of the natural world also proved a winner. Prearranged "nature moments," displayed in spectacular color images, could impart illusions of spontaneous contest between competing species or would-be mates.[18]

The specials produced by the National Geographic Society demonstrated well how color documentaries could establish a sense of vicarious adventure. Since 1888, the organization devoted to "increase and diffusion of geographical knowledge" had set a high standard for natural history popularization. The society's familiar yellow-bordered magazine described inhabitants, flora, and fauna of exotic locations, from Death Valley to Arctic tundra, and covered the breadth of science and engineering—astronomy, archaeology, physics, geology, rocketry—through crisp prose and stunning photography.[19] Television, especially color broadcasting, represented the logical next step. In 1961, after years of negotiations with other institutions, National Geographic established a television service, and contracted with Hollywood production groups to make and market programs. From the outset, the series strove for excellent cinematography, scientific accuracy, and real-life entertainment.

"Would you like to stand with Barry Bishop on the glittering top of Mount Everest? Or visit Jane Goodall as she camps among wild chimpanzees?" "Beginning next month," the *National Geographic Magazine* promised, "you can enjoy such adventures in your own living room."[20] The first specials aired September 1965. Filmmakers (and therefore viewers) walked behind ornithologists, geologists, and archaeologists on their field trips, or stood next to Jane Goodall in Africa. The specials seemed a natural extension of the society's commitment to diffusion of knowledge. Earlier explorers had brought back photographs; Richard Byrd's polar expeditions had made motion pictures; now films shot specifically for television exploited drama and intimacy to enhance each viewer's experience of "being there." On these "photo safaris," the photographers could sometimes be glimpsed next to the researchers. "Americans on Everest" sought to have every member of the audience imagine, at the end of the program, that he or she had also "climbed" the mountain.[21]

With the second special, "Miss Jane Goodall and the Wild Chimpanzees," the project solidified its reputation for reliable family entertainment. Scenes of an attractive young woman, carrying out exotic research in a remote location, invited viewers to "share discoveries about animal behavior that have startled the scientific world" and thereby to gain "rare insight" to the animals' "personal lives."[22] Goodall later said that she had been "appalled" by the "blatant anthropomorphizing" in the final cut and successfully persuaded the filmmakers to revise the program.[23] Other films that first season examined the Mojave Desert and the efforts to save grizzly bears from extinction, while the program "Dr. Leakey and the Dawn of Man" built on National Geographic's long-standing attention to anthropology.[24] Sustaining a high level of quality required substantial investment; to subsidize subsequent productions, the society agreed to full commercial sponsorship.[25]

National Geographic perceived television as a way to attract new members and financial support and fulfill its mission of popularization. French oceanographer and explorer Jacques-Yves Cousteau had more specific political motives. In 1951, he outfitted the research vessel *Calypso* for a four-year voyage around the world. Films about that trip, shown in movie houses, helped turn him into an international star and spread his message about marine conservation.[26] Although National Geographic had cosponsored oceanographic research with Cousteau since 1952, his first appearance on American television had been in the 1954 *Omnibus* segment. That program was followed by films created for the *Adventure* series in 1955, and feature-length documentaries like *The Silent World* (*Le Monde du Silence*, 1956).[27] Soon after starring in a National Geographic special in 1966 ("The World of Jacques-Yves Cousteau"), Cousteau agreed to develop his own specials for ABC. That project, *The Undersea World of Jacques Cousteau*, premiered in 1968, combining "adventure, entertainment and education."[28] Cousteau's first television series coincided with the birth of the environmental movement in the United States. In the first episode, the crew of the *Calypso* followed a school of sharks through the Red Sea, Indian Ocean, and Gulf of Aden, continually reminding viewers of the importance of marine creatures and the threats to their survival.

Until 1976, when the series was moved to cable, Cousteau starred in four new broadcast specials a year, weaving themes of ecological unity and biodiversity into dramatic narratives. His personality and passion dominated the screen: "When you think of Cousteau you don't think of a particular [television] program," environmentalist Bill McKibben observes, "just an endlessly repeated and demonstrated love for the ocean."[29] Television, Cousteau believed, could assist "diving for science," advance public understanding of the marine environment, and encourage financial support for research.[30] *Calypso* was not simply a vehicle to transport cinematographers but also a research home for marine biologists, zoologists, and ornithologists and, like the resulting series, a platform for oceanographic politics.

THE SMITHSONIAN'S ENIGMA

Even after the National Geographic Society had demonstrated that a scientific organization could use television for dignified public outreach, the Smithsonian Institution continued to wait on the sidelines. Leonard Carmichael set up vaguely defined roadblocks to every suggested television project, citing "the matter of sponsorship and some of the other problems that might stand in the way."[31] By the mid-1960s, the Smithsonian had a dynamic new leader—S. Dillon Ripley—but the institution still had no comprehensive policy for evaluating media proposals, whether a modest request to show a single object live on camera or a bold idea for a major network series. Instead, museum directors and curators continued to react on an ad hoc basis, cooperating to greater or lesser degrees depending on their moods and schedules, and usually mentioning time or resources as justification for saying no. In 1962, the head of Warner Brothers, Jack Warner, Jr. (who had been involved in the Bell Telephone series), attempted to persuade a Smithsonian official that "the modern marvel of television" could introduce the institution to millions of people unable to visit Washington, but he was rebuffed.[32]

Ripley confronted the mass media with far more enthusiasm than his predecessors, and he established new offices to interact with the press, handle public relations, make films for exhibitions, and work with radio and television producers.[33] Ripley recognized that the institution's inherent "enigma" would be tested by its response to television. Is the Smithsonian merely a collection of museums and objects, he asked in a 1965 speech, or is it (as the institution's original charter declares) a creator and diffuser of knowledge energized by "a company of scholars, brought together to

FIGURE 11. S. Dillon Ripley in front of television cameras, circa 1970s. Throughout his tenure as Secretary of the Smithsonian Institution, Ripley championed the use of new media to diffuse information about scientific research and collections. Courtesy of Photographic History Collection, Division of Culture and the Arts, National Museum of American History, Smithsonian Institution.

use and interpret objects, and [free] to follow wherever their inquiries may lead"?[34] If the Smithsonian is the latter, how might broadcast media assist that mission? As an ornithologist who had conducted research around the world and who, through his social connections and political (and military intelligence) activities, had become friends with powerful and accomplished people, Ripley represented science's cosmopolitan face. His vision for the behemoth organization perceived it not as a dusty "attic" preserving relics of the past but as a dynamic repository of "living" collections used to uncover "new correlations" between organisms and their physical and temporal environments. Dinosaur bones, giraffes, sculptures, photographs, stone axes, postage stamps, and china dolls should be made accessible to both researchers and the public. Television might help millions to appreciate the diverse collections in new ways. Ripley's "enigma" speech galvanized plans for a television series that would emphasize scientific experimentation and combine attention to systematic and environmental biology, anthropology, and astronomy with global concern for conservation of vanishing and endangered species.

For reasons tied to television economics, however, the Smithsonian's first national network series was produced not for adults but for preteens. The parameters apparently changed during the project, and there was miscommunication about the intended audience. Internal Smithsonian memos describe potential viewers as similar to those for the National Geographic programs (which were aimed at adults), while NBC publicity mentions "school age youngsters and information-minded adults," and the network scheduled the series for midday on Saturdays.[35]

Compared to other television ventures, the project was not lavish. The amount invested by NBC, about $20,000 per program, was less than the usual cost of prime-time documentaries, and the Smithsonian perceived the series as a low-cost public relations effort rather than a source of external revenue. NBC reimbursed the institution around $150 per day for curators' time and other associated costs, promised 15 percent of nonbroadcast (aftermarket) distribution income, and retained exclusive rights to the series during the term of contract.[36] Other parts of the arrangement, though, point toward conditions that would become more contentious with each successive project. Although the Smithsonian retained "final review for factual accuracy by the . . . curatorial staff" and the network agreed that "principles of good taste will be observed in offering the programs for sponsorship on a scale commensurate with the dignity of the institution," all final decisions on "program content and the course of the series" were reserved to NBC.[37] The contract required the Smithsonian to "cooperate with and assist" NBC in planning, which should have been some comfort, but curators apparently complained constantly—about the extra work required to assist script development and review, about process (such as scripts received at the last minute or not reviewed at all by the appropriate in-house experts), and about inaccuracies in the finished product.[38]

NBC News eventually produced seventeen half-hour specials, hosted by a network correspondent and telecast during the 1966–1967 season; twelve of these were rebroadcast the following summer on Sunday evenings, as a lead-in to *Walt Disney's Wonderful World of Color*. Ten of the programs had science-related themes. "The Sky Is Falling" discussed meteors, "Treasure under the Sea" looked at underwater archaeology, "Dem Dry Bones" explored osteology, and programs like "The World around Us" and "Our Vanishing Lands" incorporated conservation and ecology themes.[39]

Institutional expectations ran high. Ripley declared that the programs would "stress the excitement of the search for knowledge" and were "of such potential importance to the Smithsonian's educational mission that we will engage in no other major television programming until their conclusion."[40] The initial broadcasts attracted between four and six million viewers, with favorable critical reaction.[41] Aftermarket sales generated modest revenues, and the institution did, of course, receive valuable publicity. Very quickly after the broadcast, however, senior staff expressed disappointment with production quality. Any future projects, they urged, should have more "direct supervision of production to maintain SI [Smithsonian] standards and point of view."[42]

THE NEXT ROUND

Ripley remained convinced of the medium's potential value: "Television permits behind the scenes viewing in the laboratory and in the field, close-up examination of scientific experiments, and the step-by-step process involved in the pursuit of knowledge. Television can provide a dramatic view of a particular subject and analyze the inter-relationships between many different disciplines."[43] He was also aware of the potential of individual delivery systems like videocassettes, which were just starting to be sold for home use, and wanted the institution to be proactive in producing educational video.

Even though politicians and private foundations were encouraging use of television for public outreach, other scientists did not necessarily share Ripley's progressive enthusiasm. The Smithsonian had attempted, for example, to persuade AAAS to cooperate in convening a major conference to discuss "science on television," but neither AAAS officials nor physicist Walter Orr Roberts, chairman of the AAAS Committee on Public Understanding of Science, would agree.[44]

Early in 1967, ignoring internal criticism, Ripley directed his staff to resume negotiations with NBC. The request triggered intense debate over strategy. Philip Ritterbush, one of Ripley's senior assistants, argued that the Smithsonian should, in fact, reorient its television efforts away from "publicizing" the institution and toward "informing the public on broader matters, topics within our competence as interpreters."[45] The earlier series ideas had focused on collections and objects, offering "pretend" museum visits and trumpeting scientists' accomplishments. Ritterbush now advocated a shift emblematic of how television would eventually appropriate science and intellectual activities generally, incorporating scientific knowledge and ideas within general news and entertainment and presenting science as relevant to political and social debate on a wide range of issues. Such an approach severed (or at least weakened) the direct connection between knowledge creators and their knowledge. Rather than beginning with panoramic, validating views of a university campus or glimpses into a laboratory, programs would assimilate scientific information within discussion of other topics anthropology might illuminate discussion of nationalization, zoology inform descriptions of natural resources projects, satellite images illustrate the information explosion. The Smithsonian and similar scientific institutions would, in Ritterbush's phrasing, "figure in the story" but no longer be "the only character in the drama."

Because negotiations were already under way for the 1968-1969 season, Ritterbush's suggestion raised a logistical problem. NBC News had agreed to the other type of content, that is, "positive" statements about the Smithsonian conveyed through single-topic programs aimed at family audiences. Television production required long lead times. Accepting the pending NBC offer would assure some Smithsonian presence on network television the next year. Hesitating would delay all presentations for a full season. Of the nineteen topics under consideration, most involved scientific topics: primates, birds, elephants, the ecology of the demilitarized zone between North and South Korea, and "Life on Mars" (a program that would have featured astronomer Carl Sagan). Then, in the type of reversal characteristic of the industry, NBC producer Craig B. Fisher informed Ripley that the network had decided not to pursue the series after all. Instead, he proposed "two or perhaps three one-hour specials to be aired in prime time," which would allow longer production times and larger budgets.[46] Fisher wanted a program about expeditions and had apparently already spoken to a Smithsonian scientist about describing the "ecology of the Serengeti and the animal movement on the plain."[47]

Eventually, NBC chose two projects: the expedition film and *The Enormous Egg*, a children's drama about a boy who "hatches" a triceratops and takes it to the Smithsonian zoo.[48] *The Enormous Egg* was filmed at locations throughout Washington, including the zoo and the National Museum of Natural History. Sinclair Oil Company (a sponsor whose corporate logo featured a green dinosaur) donated a twenty-two-foot-long model used in the film (and built with advice from Smithsonian paleontologists) for display in front of the museum.

Marshall McLuhan had long contended that the *content* of modern mass communication mattered less than the *act* of communication itself. In many ways, the Smithsonian, National Geographic, and similar institutions were testing McLuhan's assertion. Their goal had become achieving "presence" on television—always of the right sort, appropriate and accurate as possible, but "presence" nonetheless. Analysts confidently forecast that society would be permanently changed by emerging technologies such as satellites and miniaturized devices for communicating, recording, editing, and storing visual content. Business magazines described the economic impacts, while sociologists predicted either more social cohesion or more alienation. For science, the breadth and the technical improvements in what could be shown through television seemed promising.

Arrangements for the NBC expedition documentary, *Man, Beast, and the Land*, resembled those made by natural history museums almost half a century earlier for their film safaris.[49] Budgeted at over $100,000 (more than four times the estimated production costs for *The Smithsonian* series programs and relatively large for single television documentaries at the time), the project included underwriting for the African expedition of scientists Lee M. and Marty Talbot. The potential entertainment value of vicariously observing an expedition's "adventures" thus helped finance its research. The documentary's premiere had a Nielsen rating of 33.5, outdrawing some of National Geographic's top-rated programs, and the program won several cinematography awards.[50] In addition to projecting a family-friendly image and strong conservation message, successful films like these could be revenue generators, creating funds for the production of exhibition or research films.

Such one-shot appearances still did not constitute a significant television "presence," though. As one Smithsonian official wrote, the negotiations with NBC indicated that "at some point in time we are going to have to balance [such proposals] very carefully in our minds along with other offers we are now taking under consideration" and to examine each proposal "quite coldly in terms of our commitments and aspirations," choosing according to "whichever path or paths appears to be best for us."[51]

After the success of *Man, Beast, and the Land*, NBC News proposed assigning a photographer to "film at the scene on Smithsonian projects" year-round, making a "long-term, carefully logged, record of a project through all its stages." We have learned, a NBC News vice president wrote, that "you cannot combine several weeks of full-dress, production filming with a scientific enterprise without both suffering."[52] In proposing to "embed" a photographer within a research project, NBC News anticipated battlefield techniques then being applied for medical documentaries about hospitals. The proposal left open the number of finished productions but suggested that the arrangement must be exclusive, thereby precluding the Smithsonian from developing programs with any other commercial network. The invitation probably seemed tempting, especially from the perspective of scientists who would obtain a record of their work for later use, but other producers were also making offers (including ones with more flexible approval clauses, flat payments of $10,000 per special, and 10 percent net profits from aftermarket sales), and that situation reinforced institutional indecision. The "predictable pressures of compulsively phrenetic [*sic*] producers" clashed with internal "inertia and mixed judgments."[53] Discussions with various networks and studios continued, with no firm decision until Ripley declined ideas suggested by both NBC News and MGM, and then directed his staff to undertake an ambitious study of television.

The Smithsonian's assistant secretary for public service, William W. Warner, concluded that most proposals depended "too much on Smithsonian scientists, who are not always as cooperative as the Talbots. . . . In other words there really aren't enough of our people who want to do this kind of thing to sustain a series."[54] Some staff scientists urged caution in agreeing to any deal that involved commercial sponsorship; others touted the long-term funding potential of establishing a relationship with a corporation like DuPont—if a company underwrote a documentary, then it might fund other activities. Within months, however, CBS was proposing to use the Smithsonian's Center for Short-Lived Phenomena in a special being coproduced with Time-Life Inc., CBS News had asked to include camera crews with as many as five center expeditions, and other producers had arranged to film at Smithsonian installations.[55] Ripley also began to meet with David L. Wolper, who, as the next chapter describes, would persuade the institution to leap back onto network television with high-profile specials.

In 1969, Craig B. Fisher and NBC News, rebuffed by the Smithsonian, successfully brokered a deal with the National Academy of Sciences for a group of one-hour specials about preservation of prairie grasslands. Broadcast in 1970, the show was one of the network's highest-rated documentaries (over thirty million viewers).[56] Based on that success, the consortium planned more specials on the Great Barrier Reef, Everglades, Alaska wilderness, and human migration across the Bering Strait, and its producers often sought advice and script review from Smithsonian scientists.

CLOSED DOORS

Histories of popularization tend to analyze those programs (or books or films) that were produced. Sometimes, the story of what was *not* produced can be equally instructive.

In 1967, in residence in the United States for a project on public understanding of science, British historians and philosophers June Goodfield and Stephen Toulmin were also in the midst of planning a television miniseries to be called "Man and His Environment."[57] To read the program descriptions decades later is to marvel at the authors' vision. Goodfield, Toulmin, and their team outlined environmental problems that have persisted into the twenty-first century; they proposed using video and narrative techniques that eventually became standard in television's environmental documentaries; and a formidable array of "special advisors" had agreed to be involved. The language in the prospectus soars, arcs, inspires.

As one reads through the documents, a profound sense of loss chills the skin. In hindsight, the themes seem so sensible and obvious. Would a single television miniseries have made a difference? Probably not. Nevertheless, the failure to produce it—despite an experienced creative team, exciting ideas, and prestigious endorsement—reflected the media landscape's economic whimsy and the cultural roadblocks hindering television investigations of how humans use and abuse natural resources.

Goodfield and Toulmin envisioned four episodes about distinct yet interdependent contemporary challenges. "Man and Nature" (advisers: René Dubos, Paul Weiss, S. Dillon Ripley, Ernst Mayr, and Evelyn Hutchinson) centered on a speech, "The Quality of Man's Environment," in which

Dubos had discussed the concept of "Spaceship Earth." There is a balance to nature. The natural world is interconnected and interdependent, and too often, humans upset several balances simultaneously, draining estuarine marshes to build towns and then decrying the resulting stagnant ponds or periodic flooding. Short-term demands for "use" of nature are privileged over long-term interests in preservation. The script called for better scientific information about the consequences of air and water pollution, destruction of natural landscapes and rare species, and effects of pesticides. Such a summary, read in an era when similar messages echo throughout popular culture (from *An Inconvenient Truth* to "environment" segments on the evening news) might seem unexceptional, but Goodfield and Toulmin were articulating these issues at a time when communities and nations heavily promoted progress through unrestrained development, despite mounting scientific evidence of the negative consequences of unrestricted, unlimited, thoughtless growth and destruction of natural resources.

Each film would have adjusted the lens differently. "The Environment of Man" (advisers: Catherine Bateson and Margaret Mead) took an anthropological approach. What cultural factors influence how humans interact with their environments? How do the practices of primitive tribes resemble the pleasure hiking of urban dwellers? "Man and the City" (adviser: Asa Briggs) drew from new work in urban studies to consider the consequences of unchecked creation of megalopolises. The fourth and final film, "Creating Man's Environment" (advisers: Kenneth Clark and Asa Briggs), would have looked forward to the "choices and decisions of unprecedented range and complexity" likely to confront humanity during the twenty-first century.[58] Here, Goodfield and Toulmin emphasized that the miniseries would move beyond specific decisions—where to build a road or whether to drain a bog—to the huge, intractable choices reflecting entrenched political approaches and social attitudes, such as systematically clearing old growth forests to build suburban single-family houses, or altering nature to create "totally contrived and synthetic environments."

Ripley threw his weight behind the project and did his best to broker partnerships with potential funders, assigning William W. Warner to secure funding for this "interesting but difficult" proposal. The Smithsonian was being asked to become heavily involved in production and to assist in arranging foundation underwriting.[59] Goodfield, Toulmin, and their producer Richard Crewdson had earlier been approached by Standard Oil of Great Britain about creating public service films to explore human influence on the environment. When the company "started making suggestions, which seemed to add up to potential interference," that project "dissolved."[60] Now, because of his commitment to conservation issues, Crewdson had even offered to produce the film for a flat fee and transfer all rights to the Smithsonian.

Warner began the search for money, starting in March 1969 with the most likely prospect, the Ford Foundation.[61] Over seven months later, foundation officials turned the filmmakers down, saying that "[t]he approach . . . seemed too academic and seemed to lack the essential component of an explicit action by the viewers. The feeling of our people is that what is needed now are films which show why we are having environmental problems and how individual citizens can be effective in alleviating such problems."[62] By then, enthusiasm had also begun to wane at the Smithsonian.

MARKET FORCES

Merely convincing citizens in the developed world to switch to energy-efficient lightbulbs and carry out their recyclables, we now know, will not be sufficient. Preventing the long-term, wholesale, and irreversible environmental destruction that the script proposal predicted, and that continued unchecked during the following decades, requires wholesale social and political change around the world, the type of change enabled by major shifts in public attitudes. The Goodfield-Toulmin project represented one of many worthy documentary projects that, for diverse reasons, never located sufficient funding to be produced. To capture the support of commercial networks (or foundation or corporate underwriting) demanded a more entertaining type of popular science, one that exploited the audience's increasing addiction to the spectacular and "the real."

Marketplace assumptions about viewer preferences continually shaped science and nature documentaries on television, rejecting *Zoo Parade* gentility and emphasizing violent encounters between predators and prey, narrated by television personalities rather than scientists. Many nature series were developed expressly for syndication: Bill Burrud's *Animal World* and *Safari to Adventure*; *Water World*, with actors Lloyd Bridges and James Franciscus; *Wild, Wild World of Animals*, hosted by actor William Conrad; *Secrets of the Deep*, hosted by astronaut Scott Carpenter; and *Lorne Greene's Last of the Wild*, featuring the star of *Bonanza*. Instead of serious documentaries such as that proposed by Goodfield and Toulmin, where politics and philosophy would have provided the framework, 1970s commercial television increasingly explored nature through splendid visual illusions or through violence, taking on the coloration of its context.

CHAPTER EIGHT

Monsters and Diamonds: The Price of Exclusive Access

We are talking about a technology that has already arrived; perhaps our very certainty that it has arrived underscores our uncertainty as to how it will be put to use.

University of Chicago Committee on the Role and
Opportunities in Broadcasting, 1972[1]

No matter how persuasive the arguments for increasing the amount of science on television, the commercial mass media responded first to market forces, not to educators' or scientists' social agendas. Moreover, as projects were being planned, the need to be competitive drove producers to suggest sensationalistic approaches and to demand exclusive access to research and collections. Scientists unfamiliar with the television business often reacted brusquely to such proposals, inadvertently fueling conflicts between organizations like the Smithsonian and the very professionals they needed to create interesting, successful public outreach. Protection of intellectual territory conflicted with the desire for publicity. Time after time, projects would flower, falter, and then collapse, so much so that Senator William B. Benton (voicing attitudes prevalent within the telecommunications policy community) blamed educational broadcasting's failure on *everyone* involved—"the networks, the foundations, and educational institutions themselves."[2]

LOVE AND ADVENTURE

After several more ineffectual attempts to develop a Smithsonian television series, pressure from the institution's regents pushed S. Dillon Ripley to jump-start arrangements with CBS News for a group of specials.[3] Even though, in exchange for greater content control and the freedom to develop single-shot programs with other producers, the Smithsonian had already agreed to forgo royalties, CBS demanded exclusive access to staff and research collections, all rights to the productions, final word on content, and permission to incorporate the films into an existing series.[4] Smithsonian specials on such topics as archaeology, anthropology, conservation biology, and the Ceylon elephant project would have a separate narrator (former astronaut Walter Schirra) but would otherwise resemble regular CBS documentaries, that is, be slightly sensationalistic and "firmly oriented to entertainment value."[5]

The CBS proposal represented exactly what many scientists found repellent. Senior assistant William W. Warner told Ripley that staff at the natural history museum had been especially critical: "a minority of the department heads think that the Smithsonian should never get into the arena of broad public appeal television in the first place, believing as they do that the networks will demand concessions to popular taste that will prove unacceptable." In a recent meeting, staff members had asked, "Are we all to become Jacques Cousteaus?" and suggested that the institution "[l]eave that sort of stuff to the National Geographic."[6] Despite such internal opposition, an agreement with CBS News was signed, with the understanding that the network would produce four specials a year.

The first "Smithsonian Adventure" appeared in 1971, sponsored by a high-end automobile manufacturer. *Search for the Goddess of Love* starred a charismatic young archaeologist, Iris Love, whose claim to have identified a "lost" nude statue of Aphrodite had drawn considerable controversy.[7] *Life* magazine coverage of Love's work provided favorable advance publicity, but internal curatorial reaction was fierce and negative. Archaeology department chairman Clifford Evans complained about inaccuracies ("Maybe dynamite is a common tool for the archaeology of Dr. Love, but it is not a common tool for scientific archaeologists that I know in the world") and called the show "a disgrace," yet decided that the institution's scientific reputation remained untouched because Love was not actually a staff member.[8]

Within the Smithsonian hierarchy, concern centered less on curatorial critiques than on whether the network would produce the anticipated number of specials. At least two more were allegedly in the pipeline—one reenacting John Wesley Powell's explorations of the Colorado River and another about human evolution. Then, in July 1971, CBS suspended all production on the project.[9] The second program, eventually completed under the series title *Smithsonian Adventure with Walter Schirra*, was rescheduled for March 1972, was publicized by the Smithsonian, and then was abruptly preempted. By then, no more programs were in the planning stages, so the Smithsonian moved to cancel the contract.[10] After more pressure, and after CBS located a sponsor, the Powell documentary ("99 Days to Survival") did air in October 1972, although network publicity ignored both the expedition's scientific significance and the Smithsonian's intellectual contribution to understanding the expedition.

The CBS project offered an important lesson in television economics. The costs of color production, viewers' rising expectations of quality, and the expense of filming on location challenged the resources of even major organizations. National Geographic Society productions, both specials and syndicated films, were reportedly several million dollars "in the red" during the early 1970s because of declining advertising revenues and increased expenses (the society paid up front for production and then sold each film to a network).[11] The Smithsonian (like the National Academy of Sciences and most scientific associations) neither wanted nor could afford similar financial and logistical burdens. In that context, then, Ripley ordered more analysis of how the institution might approach television. Still hoping for a commercial venue, Ripley and his staff also explored cooperative arrangements with independent production companies who

FIGURE 12. Former astronaut Walter Schirra, filming a segment of *Smithsonian Adventure with Walter Schirra*, 1971. In 1968, the *Apollo 7* mission made the first live-television broadcast to the public from inside a manned spacecraft, and Schirra received an Emmy Award for his role. Courtesy of Smithsonian Institution Archives.

might broker national broadcast via syndication. After briefly discussing a multiprogram partnership with National Geographic, the Smithsonian began to evaluate proposals from commercial groups, including that of independent producer David L. Wolper.[12]

PARAMETERS

Technological innovation reshaped the mass media landscape during the early 1970s. Until then, the cable industry had been most active in bringing network television into rural and remote areas, relaying signals via landlines yet constrained by marketplace rivalries and government regulation. In 1971, only 8 percent of American homes were served by cable (CATV). When new FCC regulations allowed more competition and required CATV businesses to include free channels for education, municipal governments, and public access, the number of cable systems increased and content improved. Broadcast television began to worry that expansion of outlets would siphon off and subdivide the potential audience, and that cable might kill off "free TV" altogether.[13] Affordable home recording devices now enabled time-shifted content consumption. Over the next decades, the broadcast and cable television industries eventually joined forces, became intertwined, interrelated, and multiplexed.

The role that educational organizations might play in this newly configured system was not obvious. As a result, many groups pulled back from active participation in television production for mass audiences. In Chicago, the University Broadcasting Council had been using radio and television for decades to promote local universities as places where "thoughtful" people devoted their careers to understanding science and similarly important topics.[14] In 1972, the University of Chicago withdrew from the cooperative, citing restrictions imposed by the network partner, a lack of autonomy in content development, and the stations' increasingly frosty attitudes toward educational programming.

Involvement in public communication required scholarly and scientific institutions to contemplate uncomfortable compromises. If an institution took its responsibility for public outreach via television seriously, the University of Chicago Committee on the Role and Opportunities in Broadcasting determined, then it must consider (1) the increasing "fragmentation of assumed audiences" by age, gender, occupation, and location; (2) the disassembling of "the 'conventions' of conventional broadcasting," especially time constraints, resulting from cable expansion and individual recording capability; and (3) the audience's ability to engage in "'inter-communication,' independent of conventional networks." Educators must also account for how television would be used in the future, the committee noted. By the end of the 1970s, it was predicted, 40 to 60 percent of U.S. households would have cable and interactive capability, yet no government entity was giving much thought to the *quality* of future content. National policy discussions focused instead on management issues, or on apportioning ownership of the expanded number of channels.

How each institution conceptualized its "essential mission," the committee concluded, would influence how it might engage in broadcasting. If an organization regarded itself as "priest and prophet to the new democracy," then one type of television presence would be appropriate; if a "mission of investigation and of communicating the results of investigation" was embraced, then other paths might be better.[15] If the organization believed that broadcasting should serve the general public, "then the needs of that public ought to be reconsidered" (and defined), with "novel and superior ways" devised to meet those needs.[16]

In Washington, the Smithsonian engaged in similar debate, with one distinct difference—"diffusion" of knowledge to the general public had formed part of its mission from the outset, although the institution had never been very daring in the means chosen to pursue that goal. Other than exhibitions, print remained the preferred medium. Radio had been tried in the 1920s and again in the 1940s, at the instigation of individual staff members or outside agencies. Films had been produced to support exhibitions or research, not for commercial distribution. The Smithsonian had once maintained an extensive film library; by 1972, even that operation was moribund. A television studio, originally intended for internal broadcasts, had been built within the Smithsonian's new Museum of History and Technology but never used. Mass-market television offered opportunities requiring ostensibly less institutional investment. Dozens of producers, directors, and studios were submitting proposals and requests for filming on site. The NBC and CBS experiences had demonstrated that the Smithsonian was ill-prepared to think in the "novel and superior ways" that might bring successful outcomes. A pan-institutional review, led by William W. Warner and Telecommunications Study Group chair Julian T. Euell, attempted to identify more appropriate parameters for choice.

Before choosing projects, Euell observed, the Smithsonian must define its goals and assess its organizational capacity. What did the institution want to achieve through the use of television: extend public outreach beyond museum walls, generate revenue, or exploit innovative delivery systems for home education, such as video playback devices?[17] To "enter into public television while simultaneously doing commercial television" would "overtax the time, ability, and undoubtedly the patience" of curatorial staff.[18] On two aspects, scientists, curators, and managers agreed: the Smithsonian must "insist upon accuracy and a high-quality production" and "proceed with dignity."[19]

Questions about money permeated these discussions. Other Smithsonian ventures into mass media—such as the 1940s radio series *The World Is Yours* and the 1970s public service announcements produced by Radio Smithsonian—had not even recovered their costs. Jacques-Yves Cousteau's organization and the National Geographic Society (both private, nongovernmental entities) had been able to control all aspects of their television ventures and break even by acting as their own executive producers: they invested in the filming and editing, and (betting on sufficient revenue from aftermarket sales) paid the difference even if sale of a program for broadcast did not cover the initial cost. Because the Smithsonian received federal funding and most scientific research projects and collections were publicly funded, such entrepreneurial activity was deemed inappropriate. The institution could not afford to invest the estimated $175,000 to $275,000 for production of one high-quality hourly special (later estimates for unsponsored specials ran into the millions) or to subsidize production, management, and marketing.[20] The Smithsonian would have to contract with potential investors, such as an independent producer or a commercial network, and establish firm guidelines for quality control, staff involvement, and program review. Rights issues would need to be reevaluated— CBS, for example, was now claiming that it possessed "all rights in all media in perpetuity" to the completed *Smithsonian Adventure* documentaries after the contract had been terminated. And no contract, the review advised, should include overly restrictive "exclusivity" clauses that might prevent scientists and staff from involvement in other television ventures (such as giving interviews, allowing objects and research operations to be filmed, and offering advice and assistance) or that might inhibit valuable publicity through news coverage.[21]

"Exclusivity" began to emerge as a critical factor in how scientists and their institutions reacted to television proposals. If a network filmed a special about a research project (or beetle collection or even a single iconic object), should that contractual arrangement be allowed to prevent any other network or production company from filming a documentary on the same research collection or object? At the National Academy of Sciences, "exclusivity" clauses were rejected as inappropriate because, it was argued, academy members were free to make their own arrangements with television representatives and the academy itself did not conduct scientific research (that is, did not generate the "content" being discussed). Institutions like the Smithsonian were being pushed to define where the essence of popularization resided. Within a topic? In the specific research results, researcher, or laboratory? Or in how a topic might be framed by a creative person like a documentary filmmaker? How could an institution profit from and control commercial discussion of its scientific work without agreeing to "exclusivity"? Should public institutions enter into such agreements at all? When, after an earlier Smithsonian deal had fallen through, MGM produced a documentary featuring the work of three prominent Smithsonian archaeologists (*In Search for the Lost World*), with no financial compensation to (or official approval by and mention of) the institution, Ripley and his staff realized that new policies must account for potential conflicts of interest. The institution would have well-defined public relations goals, but individual scientists should not be prohibited from describing their work to the public.

The question became no longer *whether* to produce programs, but *how*. Could television be used to advance the institution's goals without demeaning its reputation, puncturing its dignity, or violating principles of accuracy, both factual and interpretive?[22] The likely benefits of reaching an international audience in the millions seemed to justify the risks.[23]

For many others within the institution, the real choice lay between commercial and nonprofit media, with one group arguing for negotiating only with educational television. The decision was not easy. The 1968 NBC special *Man, Beast, and the Land* had attracted almost thirty million viewers, an audience rarely achievable by most public broadcasting programs at the time.

The pressure became intense. The Smithsonian possessed "a virtual intellectual storehouse" of content ready to be tapped for an expanding media marketplace.[24] Telecommunications experts were forecasting a fivefold growth in cable subscribers within the next five years, widening the technology's geographic and demographic reach and increasing the demand for content.[25]

Whether the arrangement was with a public or a commercial entity, however, the institution must retain "control within reasonable limits of the content and manner of presentation" and ensure that internal costs (time, people, space, and actual expenditures) would be fully reimbursed. Programs would have to be entertaining as well as educational. There was nothing essentially wrong, one curator admitted, with a television program starring "Jack Benny in the Musical Instruments Hall—as long as it's handled with taste." The comedian could hold viewers' attention while Smithsonian experts engaged in "diffusion of knowledge" about the history of violins or the physics of sound.[26]

Continuing past institutional practices—that is, reacting "piecemeal" to each new proposal rather than developing explicit institutional policies—would perpetuate inconsistency and disappointing outcomes. News coverage and individual filming would continue, and the Smithsonian would lose the opportunity to create a significant television presence through which to shape its public image and institutional identity, assigning that role to others by default. The National Geographic Society, for example, had recently produced a special (*The Violent Earth*) about the Smithsonian's own Center for Short-Lived Phenomena. It was in that context, then, that two quite different television ventures proceeded during 1972—and eventually collided.

HISTORIANS CONFRONT TELEVISION

Prompted by White House pressure to plan activities celebrating the American Bicentennial in 1976, the National Science Foundation (NSF) had provided initial funding for the Museum of History and Technology (MHT) to develop "blockbuster" television documentaries about American science ("from Benjamin Franklin to Enrico Fermi"). The ambitious joint venture proposed to engage "scientists with an interest in history, historians with an interest in science, and anthropologists who understand the connection," and to demolish "myths that serve to bring science and technology into disrepute by imputing to it the power to solve social as well as technical problems," yet do so without the "self-serving attitudes and objectives sometimes apparent in the efforts of scientists themselves in the field of public understanding of science."[27] Draft descriptions framed science as instrumental knowledge for a technocratic world.[28] NSF gave the Smithsonian $100,000 for "conceptual planning," but neither the government agency nor the museum had sufficient funds to produce the ambitious commercial network series.[29] To be ready for broadcast in 1976, several million dollars had to be raised through congressional appropriation and/or foundation support.

Museum director Daniel Boorstin and senior curators appointed a "Bicentennial advisory committee" of external experts (all historians) and hired historian of technology Eugene Ferguson to serve as principal consultant. Aside from one or two people who attended an advisory meeting, no participant in the project had experience in writing for or producing television.

WOLPER (THE NEGOTIATION)

The other project involved a legendary television producer. When S. Dillon Ripley first met David L. Wolper in 1972, the two men engaged in enthusiastic discussions about a potential series.[30] By then, Wolper was one of the most successful producers working outside the networks, and just the type of creative person that Ripley appreciated. In the first three months of 1968 alone, nine hours of Wolper-produced documentaries had appeared on commercial television, including the first of a dozen Jacques-Yves Cousteau specials for ABC, a National Geographic Society "Amazon" special on CBS, and an acclaimed three-part adaptation of William L. Shirer's *The Rise and Fall of the Third Reich*. Later that year, Wolper partnered with historian Theodore H. White on a television version of *The Making of a President*.[31] In his forties, with energy, substantial financial resources, and a track record of hundreds of completed films, Wolper perceived television as a medium for reaching vast audiences—thirty million or more viewers at a time—with interesting, informational, entertaining, and occasionally provocative content.

Wolper Productions and the Smithsonian began negotiating a contract for "development, production and programming of a prime-time commercially-oriented documentary series" centered on the institution's "activities and concerns."[32] Wolper wanted a four-year arrangement, with exclusive rights to produce programs associated with the Smithsonian, and ten-year distribution rights (he initially asked for perpetual rights). Although Wolper agreed to expert review by the Smithsonian, the contract also attempted to reserve crucial creative decisions for the producer.

The next phase of negotiations involved securing sponsorship. During the 1970s, American businesses were restructuring their advertising approaches and increasingly underwriting media projects to link corporate identities with prestigious nonprofit institutions.[33] Corporations had initiated science and educational television projects like *Our Mr. Sun* and *The March of Medicine* in the past, but now documentary producers like Wolper began to pitch ideas directly to specific sponsors so that financing would be in place before production planning (or even the choice of themes) began. Blockbuster series such as those starring Kenneth Clark (*Civilization*) and Alistair Cooke (*America*) had attracted corporate underwriting exceeding $5 million. Six months before the contract with the Smithsonian was in final negotiations, Wolper sought clearance to discuss the potential series with North American Rockwell and other likely underwriters. Eventually, he convinced DuPont Company to be the sole sponsor of three half-hour specials and expressed his intention to sell the films to the network program departments, which would value entertainment over informational or educational content.[34]

Exclusivity remained a thorn in the flesh. When Wolper first expressed concern that "competitive aspects of other television arrangements" made by the Smithsonian might affect the project, Ripley had directed curators and scientific staff to keep senior officials notified of any potentially competing arrangements, whether with commercial or educational outlets. This pressure undoubtedly dampened enthusiasm for the CBS News attempt to restart the Smithsonian documentary project.[35] Wolper requested "protection for subjects treated in programs he would produce," essentially attempting to fence off certain topics altogether. Ripley's deputies, mindful of the Bicentennial project moving forward at MHT, discussed the "obvious need to liberalize the Wolper view of his requested 'exclusivity,' to give us more freedom of action in moving ahead with non-Wolper television projects."[36] They had little success in achieving this goal.

Wolper's demand represented an altogether new situation, quite unlike the rights issues associated with other media, and one that posed a dilemma for scientific institutions. An outside author writing a popular book describing a Smithsonian butterfly collection or marine biology research project, for example, would not have demanded that no other person be allowed to write about the same collection or research. No radio or motion picture producer had ever been granted (or probably ever requested) such exclusive access. For television, however, the amount of money invested in production became justification for requesting exclusivity ("It is doubtful that a sponsor will commit $4 or $5 million per year for the presentation of the Smithsonian programs if he cannot be assured that another competing series will not be permitted by [the] Smithsonian").[37] The institution, planning to step boldly into a world defined as "'commercialization' to meet public needs that can best be served, most economically and efficiently, by the private sector," was being forced to choose between the marketplace and traditional scholarly attitudes toward intellectual material.[38] Commercial network television seemed the best option for reaching the public. Gilbert M. Grosvenor, head of the National Geographic Society, was urging his friend Ripley "not to spend the time and money on educational television, which . . . has an extremely limited public, but rather to program for prime time commercial television."[39] Ripley seemed aware that collaboration with a commercial organization might require the Smithsonian "to make unacceptable concessions to popular taste," as had happened with NBC and CBS.[40] Nevertheless, the potential of educating millions of viewers about the institution's work proved hard to resist.

In January 1973, several Smithsonian regents expressed "doubts" about the Wolper contract, echoing concerns they had raised about any involvement with commercial television.[41] Some questioned whether productions would overburden staff, but most worried about the "degree of exclusivity" and the implications of a Smithsonian connection with commercial advertising.[42] Among the regents were several business executives who emphasized that no contract would ever be foolproof. The institution could not realistically expect to retain complete control over what was produced, yet the exclusivity Wolper understandably sought should not be allowed to inhibit other legitimate television coverage of exhibits, collections, research, and staff. Open access reinforced the institution's public mission. When the regents established a subcommittee to review the proposed contract, Wolper was informed he must wait for final approval.

That spring, external advisers (including several New York advertising firms) reinforced Ripley's conviction that Wolper was the best choice, even if he was notoriously independent and occasionally controversial.[43] Wolper had recently been criticized by the National Rifle Association and major environmental groups like the Audubon Society and National Wildlife Federation for allegedly inaccurate and misleading dramatizations in his film *Say Goodbye*.[44] He was also becoming embroiled in a major political fight with the FCC, whose proposed new rules for prime-time access would effectively reduce the amount of network time available for documentary specials. Fighting on behalf of all independent producers, Wolper had filed a challenge to exempt independently produced documentaries and to require networks to "favor status for scientific, cultural, historical, anthropological or educational specials."[45]

WOLPER (THE OUTCOME)

The contract for three to five one-hour prime-time network specials, signed in late May 1973, allowed the Smithsonian to become engaged in production of other programs for public television (or to license projects for commercial outlets) as long as those films were funded by "non-commercial sources" and on subjects that would not "compete" with the Wolper series.[46] Ripley reminded all professional staff that the Wolper organization had "exclusive rights to most Smithsonian commercial television activity" and that no one should arrange a project or appearance without clearance from the top.[47] The following month, in a news interview, Wolper seemed to acknowledge that the institution would play *some* intellectual role in production: "it has always been my theory that if you want to make a factual program about a certain subject, you go to the absolute authority on that subject to be your partner. We did it eight years ago with National Geographic. . . . The same practice was followed for our series with the Smithsonian."[48]

Prior to approving the contract, the regents had been assured that programs would be "non-topical, non-political, and hopefully educational" and that the Smithsonian would have "absolute approval of the subjects, scripts, the rough and final cuts of the films" with review for "factual accuracy, taste, and reflection upon or detriment to" the institution's reputation.[49] Nevertheless, content negotiations quickly became stormy. Smithsonian staff suggested science-related topics like invasive and endangered species, continental drift, meteorites, forensic anthropology, human ecology, Stonehenge, and insect behavior, as well as arts and humanities topics like painter George Catlin, but by November, they were being asked to react to nine different topics, which included such titles as "The Curse of the Hope Diamond," "Mysteries, Monsters, and Madness," "Diamonds Are a Man's Best Friend," and "Up, Up and Away" (about the history of flight). Reactions were especially negative to the Hope diamond idea ("not prime SI material," "strictly Sunday supplement stuff") and to monsters ("What crap!" "I wonder if we have anything new to say"). The next month, Wolper's staff narrowed the list to three (monsters, flight, and the Hope diamond).

DuPont had agreed to commit up to $2 million for the first year (including production and network charges), and the company became heavily engaged, through its advertising firm BBD&O, in selecting topics. Meanwhile, internal Smithsonian opposition to Wolper's style of popularization grew. Porter M. Kier, director of the National Museum of Natural History,

FIGURE 13. Television producer David L. Wolper standing next to the Hope Diamond in its display vault at the Smithsonian's National Museum of Natural History, April 1974. Wolper had just begun production of his CBS special *The Legendary Curse of the Hope Diamond*. Courtesy of Smithsonian Institution Archives.

was outraged at the proposed themes of the monsters and Hope diamond shows, at first refusing to allow his staff to cooperate at all with Wolper's scriptwriters and then (facetiously) demanding thousands of dollars for research costs in advance of any discussion, and other senior officials also expressed concern that sensational topics would erode the institution's dignity.[50] Euell felt compelled to remind his colleagues that, while "there are some risks in attempting to sell the Smithsonian's research and collection interests to a mass audience," the regents and senior officials had "considered this question at length and agreed to our taking these risks."[51]

Even Ripley, however, began to express reservations privately about the Wolper deal (although not about Smithsonian participation in television generally), telling Euell that if the Wolper group were not "prepared to listen seriously to ideas" then "they might just as well give up the idea of doing programs for us"—"We are not a commercial organization and are not about to sell ourselves for a mess of potage."[52] Euell's assistants attempted to soothe Ripley. They assured him that communications were, indeed, flowing and that they had selected the topics "from among Wolper's proposals . . . because several of them were directly relevant to Smithsonian collections and research."[53] The institution was now caught up, Ripley countered, in a "dialogue with . . . ritual purveyors of sensationalism" and must remain impassioned in the defense of accuracy, authenticity, and the importance of evidence.[54] To guard its interests, the Smithsonian hired an experienced researcher to verify scripts (the National Geographic employed a similar fact-checker for its specials).[55]

Good advance publicity for the first Wolper special, *Monsters! Mysteries or Myths?*, narrated by Rod Serling, lured unexpectedly large audiences during prime time in late November 1974. *TV Guide* invited science writer Isaac Asimov to contribute an essay about legendary monsters and promoted the special on the cover.[56] With an excellent rating of 31.2 (a 44 percent share, or about fifty million viewers), the show was one of the top three Nielsen-rated programs that week (along with situation comedies *Sanford and Son* and *Maude*) and broke all existing records for documentaries shown in the New York and Los Angeles television markets.[57]

Success proved a mixed blessing—gratifying to those who had pushed for a television series but deflating to the internal critics. Curators who had strenuously objected to the emphasis on "monsters" were appalled to see an actor playing "Bigfoot," to watch "eyewitness" interviews of alleged encounters with various monsters, and to hear scientists described as "non-believers . . . sifting the evidence for clues" to monsters' existence.[58] Even critics who praised the program as exciting, engrossing, and entertaining wondered what the Abominable Snowman, Bigfoot, and Loch Ness monster had "to do with the venerable Smithsonian Institution." "I'm not quite sure," one speculated: "maybe they're the only three things on earth that aren't stored somewhere within its vast halls."[59]

Curators had feared just such sensationalism. During development, many had complained that script materials contained "pro-monster" and "anti-science" bias, with no clear connection to Smithsonian research; scientists were used principally as "bad guys" to deny that monsters exist.[60] Wolper had called such complaints misplaced. This was a shooting script, not a "word script," he had replied: "We are interviewing people all around the world and we are not writing a script on what they are saying [because] we are only learning exactly how they feel about different subjects." Criticism should be reserved for the rough cut (to be viewed by Smithsonian staff two weeks before airtime). The program, he wrote, has been "proposed, discussed, re-discussed, re-mentioned, and has been finally approved." The "quality of the script is something I wasn't aware that the Smithsonian Institution was even going to get involved in. . . . I cannot start letting Museum Personnel, and Museum Curators . . . [tell] us how we should write our scripts. That's completely ridiculous."[61] And then, in a passage emblematic of how postmodern popularizers redefined their roles as science's semiofficial "interpreters," Wolper had added:

> One question is why have you, as an Institution full of scientists, not been able to convince people that there are not those Monsters. I think that we will show you a better way to convince people. . . . My business is communication to large numbers of people. I think I know how to communicate a message to large numbers of people better than you. Obviously, the scientific community with all its scientists have still not been able to convince the public-at-large that these monsters do not exist. You've had your shot at it, and now let me have mine.[62]

CBS publicity for the program left little doubt about erosion of science's cultural authority: "This program will . . . leave it to the viewers to draw their own conclusions as to whether these creatures are in fact a myth—or a reality."[63] Only the *New York Times* dared to call the show "impossibly manipulative," complaining that promotion of prime-time specials such as the monster program reflected an irresponsible trend of titillating rather than informing viewers.[64]

After the ratings success, Wolper did not hesitate to crow. He reminded Ripley that the special had "accomplished your goal to bring the Smithsonian to a large cross-section of the American people." Now, try to keep your eyes on that goal, he urged: "Sometimes the people at your organization have a tendency to try to depart from it. We must be very careful and be sure to choose the right subject matter."[65] Ripley replied that accuracy, authenticity, and relevance to the Smithsonian's interests must still come first.[66]

Similar internal disagreement swirled around the second special, *Flight: The Sky's the Limit*, in part because of the Smithsonian's reluctance to include references to the *Enola Gay* (which is in its collections) and the plane's role in carrying the first atomic bomb dropped on Japan. Although assured that errors had been corrected, curators discovered that problematic sequences remained in the script six months later: "After seeing shots of hundreds of planes in the air, we dissolve back to a shed in [the Smithsonian storage facility] containing the fuselage of the Enola Gay. Then we dissolve to a sequence of the flight of the Enola Gay. One plane, not thousands, would end the war."[67] Tepid critical and audience response ("fairly uplifting," "relatively straightforward," and "a pretty good survey course") to the January special undoubtedly disappointed all parties.[68] Praised as good family viewing, and more cheerful than the competition (an NBC special titled *The Nuclear Threat to You*), the program drew a Nielsen rating of 12.1, less than half the audience of the monsters special.[69]

To win back viewers, Wolper and DuPont pressed for dramatization and sensationalism in the Hope diamond show. At first, the producer attempted to assuage scientists by suggesting that dramatic vignettes illustrating the so-called curse ("tragic and violent events which befell owners of the diamond upon their acquiring it") would be interspersed with discussions of the related chemistry, mineralogy, geology, and mining history, but initial drafts included more drama than documentary.[70] The narrative became dominated by melodramatic "historical" reenactments. Curators responded with resounding condemnation of the script as a "tale of the supernatural which emphasizes the sensationalism associated with the current exorcism and occult frenzy."[71]

To be fair, the diamond was (and is) one of the Smithsonian's most famous objects, with both scientific and cultural significance.[72] Donated in 1958 by its last private owner, jeweler Harry Winston, this is the only diamond known to fluoresce red under ultraviolet light and then to phosphoresce (that is, continue to glow after the light source is removed). The diamond's clouded origins made the topic attractive for television drama, for the gem had "appeared" in Europe during the seventeenth century, was purchased by King Louis XIV of France in 1668, and then centuries (and many owners) later had been sold by luxury jewelry firm Cartier to socialite Evalyn Walsh McLean, whose tragic personal life fueled the curse legend. With famous actors playing the roles of Louis XIV, Marie-Antoinette, and McLean, and with dramatic recreation of an "exorcism" of the diamond inside a Washington, D.C., church, the program strayed far from the scientists' ideal of an educational-scientific series.[73] Objections to "approach, inaccuracies, taste, over dramatization, lurid yellow journalism, unnecessary and objectionable sex and ethnic references, emphasis on 'the curse,' [and] lack of informational material on diamonds," and complaints that it was a "'soap opera' account of the imagined trials and tribulations of previous owners" and "beyond the limits of good taste" were all ignored by the producers. A few months before the March 1975 broadcast, Porter M. Kier had even suggested that the script centering on the "curse" be scrapped altogether and a different film be created "using the Hope Diamond as the 'star' to show our wonderful gem and mineral collection and how diamonds are formed, mined, and cut."[74] But no one could convince Wolper "that the curse isn't completely interwoven with the legend" or that fictionalized dramatic episodes should be abandoned in favor of accurate but dry scientific discussions.[75]

The ratings deflated all criticism. *The Legendary Curse of the Hope Diamond*, narrated by Rod Serling, attracted around thirty-five million viewers.

Internal objections to the three Wolper productions reflected the evolving debates over popularization. In each case, experts had reviewed scripts line by line, and the Wolper organization had responded whenever criticisms bore on accuracy or authenticity. What the scientists found consistently offensive, however, were not specific technical errors but the sensationalistic tone, dramatic approaches, and minimal educational content. By 1975, Smithsonian museum directors seemed united in the conviction that the benefits "from prime time exposure were not sufficient to counter-balance the injury which the professional image of the Institution would suffer in the long run from identification with such 'sensationalist' presentations."[76]

One notable gray area had become "relevance"—a problem extendable to much of television's popular science. Was it sufficient to call something a "science" series if the content focused more on cultural, political, social, or ethical aspects (which were assumed relevant to the audience's interests) than on technical explanations (which were more interesting to the scientists)? Wolper (and other producers) contended that they knew which themes and emphases would attract substantial audiences, so they increasingly embraced a model that eschewed tutorials (even for complex subjects) in favor of "social relevance." Ripley defended these choices, arguing that "[t]he higher the rating received by our shows, the more people we expose to our lures" (to information publicizing Smithsonian educational exhibitions and publications).[77]

All parties continued to negotiate the next group of specials, offering suggestions, debating "appropriateness" and "salability," and proposing finalists. Wolper rejected many of the historical themes, and attention began to settle on pandas, short-lived phenomena, archaeological mysteries, animal behavior, and the Bermuda Triangle.[78] The sensationalistic Hope diamond show made matters worse. Nazaret Cherkezian, head of the Smithsonian's telecommunications activities, told Wolper: "A principal source of the dissatisfaction and opposition which . . . persists despite the favorable public response to our programs is the widespread feeling that the only role so far accorded the Smithsonian in these programs has been to respond negatively to the themes on which the shows have been based."[79]

EXCLUSIVITY (REDUX)

Although the "exclusivity" clause had not been not intended to preclude all other Smithsonian television efforts, there is evidence that it did play such a role in regard to the MHT Bicentennial project. By fall 1973, curators had developed general outlines for two modest series, one topical and the other biographical. MHT officials had also begun negotiations with the BBC on coproduction of a prime-time miniseries about "how American ingenuity, improvisation, knowledge, and love of machinery" fueled the nation's industrial development and influenced cultural lifestyles.[80] The BBC group, which had just successfully produced *The Ascent of Man*, was working with Hugh Odishaw (former head of the IGY programs at the National Academy of Sciences) and looking for a third American partner. With funding to be provided by BBC and Time-Life, and with the NSF's blessing, the new project could incorporate ideas already researched for the Bicentennial project. The Smithsonian hierarchy, however, worried about conflict with the Wolper contract and invited the Hollywood firm to become involved, agreeing to a $20,000 feasibility study.[81] MHT officials informed the BBC that, because of the exclusivity clause in Wolper's contract, cooperation in the Bicentennial project would be impossible.[82]

Negotiations again stumbled over intellectual property issues. Wolper's company wanted to be able to use the ideas developed with the NSF money even if no television program were produced, hinting that *any* MHT or Smithsonian science program might be blocked altogether in favor of nonscience topics.[83] Enthusiasm for science waxed and waned at the networks throughout the winter. In December, a Wolper Productions staff member stated that national networks had little interest in "science and technology . . . for prime-time showing," yet a few months later that same person reported that, because the networks were becoming "turned off" on Bicentennial projects, they were now "interested" in science.[84] At a March 1974 meeting (when DuPont was still onboard as underwriter), company representatives emphasized that they wanted more science and technology topics in the next group of specials and preferred not to have "a rival Smithsonian series on the air."[85] The institution's contractual obligation to Wolper and the sponsor's insistence thus forced the Smithsonian to suspend work on the history series altogether, award the feasibility contract to Wolper, and hope that MHT's ideas might be incorporated within the next specials.[86]

Interactions with the Wolper group sputtered along, with more misunderstandings and rejected program ideas.[87] The treatments submitted by Wolper were, in fact, typical for commercial producers—enthusiastic but lacking in scientific or scholarly detail. For television professionals, the vital questions involved creativity and "feasibility"—could a science series get airtime, be sold to sponsors, and attract a sufficiently large audience to justify the investment? For curators and scientists, the essential question was what such programs would say. Eugene Ferguson later said that he had been "dismayed from the beginning by the apparent pattern in the television industry of pinning down the sale of a program before any serious thought had been given to the substance of the program. . . . When I was ready to talk with television people, I found that [they were] not interested in talking about substance, and in fact dismissed the general subject of science and technology as unsuitable for 'prime time' television."[88]

Then, for reasons relating to the business climate, DuPont withdrew its sponsorship. Recent court rulings had also shifted control of some of the prime-time schedule from the networks to local stations. Presidential election-year coverage, Olympics coverage, and a glut of Bicentennial documentaries chilled the market for all independent productions.[89] Rather than redirect efforts toward public broadcasting (which, with underwriting from Gulf Oil Corporation, the National Geographic had already begun to do), the Smithsonian canceled Wolper's contract in summer 1975.[90] Two years later, Wolper won international acclaim with his groundbreaking miniseries *Roots*.

PIPERS AND TUNES

Whenever an academically oriented institution reaches out to the public, the University of Chicago broadcasting committee had observed, "it is likely to have to behave, to some extent, 'uncharacteristically'" because it must rely on techniques and expertise "which are not traditionally academic but which are to be employed in a manner compatible with and useful to the academic enterprise." "Such a dependency," the committee warned, "has its dangers."[91]

As the Smithsonian's relationship with Wolper grew more contentious,

Julian T. Euell outlined three possible options for action. To continue along the present path—attempting to develop programs saleable to a commercial network for prime time, in order to achieve the greatest possible cultural "presence" and largest possible audience—would "inevitably disappoint our bureau directors who will continue to feel that their professional reputations are being jeopardized by identification with sensational interpretations."[92] In that case, expert advice on issues of "accuracy and authenticity of material" should be welcomed, but extracurricular criticism should not be allowed to derail multi-million-dollar productions. A second option would be to develop future shows for broadcast outside prime time, with no hope of attracting large audiences. With greater technical content, lower budgets, and no star-power narrator, such shows might draw fifteen to eighteen million viewers, in contrast to the fifty million who had watched the monsters special. Euell's third option became the route adopted for the majority of science-related specials from the mid-1970s on, including the next Smithsonian series: to work with public broadcasting. Such projects still involved distinguished creative professionals like Wolper and required major underwriting but were "predominately educational," with audiences of "no more than two to three million," and did not expand the Smithsonian's constituency. Euell cautioned that moving toward public television offered no guarantee of resolving arguments about appropriate content, however: "As long as someone else is paying the bills, the insistence upon certain rights in the preparation of the program will always be present."[93]

CHAPTER NINE

In Splendid Isolation: The Public's Television

[C]ollaboration has its possibilities and they are real. It also has its dangers. . . . Let there by all means be collaboration . . . but not too much.

HUW WHELDON, 1974[1]

Although science production units in the United States eventually developed their own distinctive styles, they drew considerable inspiration during the 1970s from the other side of the Atlantic Ocean, where the British Broadcasting Corporation (BBC) had become a training ground for documentarians.[2] BBC Television operated two national networks, and direct funding from subscription fees (annual payments by viewers, collected by the British Post Office) provided some insulation from commercial and political interference. By 1974, about 3 percent of all BBC programming consisted of science, not counting segments within news and public affairs shows.[3] Two weekly series, *Tomorrow's World* and *Horizon*, had been on the air since the 1960s, and all science programs together consistently attracted between 10 and 20 percent of BBC viewers.

In addition to selling programs abroad, the BBC also engaged in co-productions with U.S., Canadian, and Australian television stations and networks and with commercial groups like Time-Life Inc. Huw Wheldon, BBC managing director of television, in a 1974 lecture at the Smithsonian Institution, acknowledged that collaborative arrangements offered exciting possibilities, but he emphasized that they also contained dangers, not the least of which was loss of independence: "An organization which has invested money or facilities in a production could be forgiven . . . for wanting to be in a position to influence the story or the style."[4]

ANGELS AND EVOLUTION

While Smithsonian officials wrangled with Wolper during the winter of 1973–1974, the institution hosted advance screenings of one of American television's most anticipated miniseries. *The Ascent of Man*, a joint production of BBC and Time-Life, had premiered in the United Kingdom during the 1972–1973 season, and featured Jacob Bronowski as narrator and intellectual connective tissue. Trained in mathematics at Cambridge University, Bronowski had first worked as a government statistician and econometrician and, during the 1950s, had written several plays for radio production, including the award-winning *The Face of Violence*. After publishing *The Common Sense of Science* (1951), he spent a year at MIT as visiting professor of history and began writing newspaper essays tackling the moral issues raised by development of nuclear weapons. These pieces eventually formed the core of his influential book *Science and Human Values* (1958).[5]

Through the years, Bronowski had appeared on British television, including as regular panelist on *Brains' Trust*, and was described as "completely at home in front of the camera."[6] For *The Ascent of Man*, Bronowski worked closely with producers Adrian Malone and Dick Gilling on the scripts, and the program themes reflected Bronowski's vision of the relationship of science to the humanities, arts, culture, and moral philosophy.

The Smithsonian screenings of *The Ascent of Man* attracted enthusiastic audiences. Crowds gathered outside the auditorium two and a half hours before the doors opened in late November 1973, and six hundred people were turned away, so officials quickly added twelve more showings.[7] Public demand was so great that two screenings were added on New Year's Day, and the final event, on Sunday, February 24, 1974, with a special appearance by Bronowski, was mobbed.[8]

Bronowski's intellectual passion dominated the thirteen episodes. In "Lower than the Angels," a lyrical account of evolution and cultural change, Bronowski is first glimpsed as a small figure dwarfed by enormous hills. Programs combined poetry, literature, art, and music with technical information in order to explain advances in metallurgy, chemistry, genetics, and atomic physics; they discussed how scientific tools had altered human relationships with the natural world. Kenneth Clark praised *The Ascent of Man* as the exception to the rule that television ordinarily avoids dealing with ideas. A few academics had attacked the series as watered-down scholarship, but Clark emphasized that Bronowski (who died in August 1974, six months before the U.S. broadcast premiere) had, in fact, fulfilled a scholar's duty by communicating "in such a way as to be comprehensible to a large number of people."[9]

COLLABORATION AND CORPORATIONS

Critical and audience reaction to *The Ascent of Man* marked a welcome new phase for public television. Until then, noncommercial, educational channels in the United States had struggled to compete for viewers. During the 1950s, noncommercial stations (most owned by educational institutions) had been treated as either podiums for pedagogy or training environments for television professionals; government and foundation funding was insufficient to allow these educational stations to compete with commercial operations. In 1962, the Educational Television Facilities Act finally provided more funds for constructing stations, and in 1967, the Public Broadcasting Act led to establishment of a national distribution system and reoriented noncommercial programming toward entertainment and public affairs. The Corporation for Public Broadcasting (CPB), formally created in 1969, allocated federal funding to National Public Radio (NPR), the Public Broadcasting System (PBS), and individual stations, yet emphasized development of noninstructional programming, in part because larger national audiences helped to justify federal subsidy.[10] In 1970, there were 184 public stations in the United States, but vast areas (primarily in the West and Midwest) could not receive any educational channels.[11] Moreover, viewers had begun to settle into consumption routines, tuning in familiar programs rather than watching something out of the prime-time mainstream, such as a science show on a public station. All documentary television, Raymond Williams has observed, turned from education toward "experience" during the 1970s; even the most serious projects anxiously monitored every hint of change in their potential audiences.[12]

As public television productions became extravagant, expensive ventures, they required a new patronage model. The costs overwhelmed the resources of most individual corporate investors, foundations, and government agencies. This need to attract multiple wealthy funders—and please them all—opened the door to compromises in content and tone and reflected an era promoting the privatization of culture.[13] In the early 1970s, the entire PBS system also came under fire from the political right (spurred by Richard Nixon's complaints about Watergate coverage). As a result, CPB assumed tighter control over funding, and PBS was transformed into a "clearinghouse" for distribution. Creative initiatives had to come from independent producers, individual stations, and corporate or foundation underwriters, or via collaborative ventures. Without a central science production group, or incubator science unit comparable to that at the BBC, American producers were forced to shop each fledgling idea to potential funders, a burdensome process for one-shot documentaries or miniseries.

In the Boston area, WGBH-TV, established by a consortium of local philanthropists and academic and cultural leaders, was becoming one of the country's premier public broadcasters, with a VHF channel that had sufficient power to compete with local commercial stations. Run by a well-funded, nonprofit educational foundation, the station could afford to take creative risks.[14]

Michael Ambrosino, who had joined WGBH as associate director of programming in 1964, had long been interested in developing some type of science production group. After returning to the station in 1971 from a CPB fellowship year at the BBC, he created a science unit and began soliciting long-term funding to subsidize an expert production staff.[15]

Around the same time, NSF started funneling more money toward television through its Public Understanding of Science program, including a grant to AAAS to explore planning of a public television program about science. AAAS first used the seed money to leverage additional funds from private foundations and to conduct surveys of the state of science on television, but had not embarked on any production ventures. Instead, the association decided to endorse the WGBH proposal. Ambrosino's original design had six parts; in addition to a regular science series and separate archaeology, social science, and children's projects, he proposed a public affairs series and another devoted solely to science films acquired from abroad. *NOVA* eventually evolved into an all-inclusive project (natural and physical sciences, social sciences, archaeology, technology, and public affairs) designed not to build on a unifying narrative week after week but to serve as a platform for independently created coproductions and acquired films, under the same title and logo.

With a $40,000 AAAS planning grant and another $1 million in grants from CPB, NSF, Polaroid Corporation, and Carnegie Corporation, Ambrosino lured BBC-trained producers to Boston. He also arranged to purchase British programs to fill out the schedule while the unit developed its own production capability. As a result, *NOVA* premiered in March 1974 with an award-winning program created originally for the BBC *Horizon* series. "The Making of a Natural History Film" described the photography of the Oxford Scientific Film Unit, much as "Filming Nature's Mysteries" had revealed nature filmmaking secrets on *Disneyland* in 1956.[16] Original *NOVA* programs during that first season included "The Search for Life" (about the Mars lander) and "Strange Sleep" (discovery of anesthesia), but most episodes had been made elsewhere.

The scientific community reacted positively, with only occasional complaints about minor inaccuracies or particular segments. Appreciation for *NOVA* far outweighed criticism. The series had strong internal support from prestigious advisers and a growing number of fans, most of whom embraced the dignified tone, the politely provocative approach to social issues, and the presentation of what *NOVA* called "science adventures for curious grown-ups."[17] Within a few years, television commentators seemed enamored with *NOVA*'s quality ("one of the most consistently interesting and intelligent programs on the air") and ability to convey science without "being patronizing or boring."[18] From the outset, *NOVA* was unashamedly an insiders' program, populated with scientific elite, unlikely to approach any topic from outside the lines. Yet it also represented the new wave in popularization by declaring adamantly, following the lead of *The Ascent of Man*, that science is part of, not apart from, culture.[19]

Ambrosino recruited an intelligent, experienced staff. Graham Chedd, who trained in chemistry at Cambridge University, served as *NOVA*'s initial science editor; John Angier, who replaced Ambrosino as *NOVA*'s executive producer in 1977, had studied physics at the University of Essex, and he supervised over eighty *NOVA* productions before forming an independent production firm with Chedd. Subsequent producers included British filmmakers Mick Rhodes and Josephine Gladstone. Longtime staff member Paula Apsell became executive producer in 1984 and continued to lead the unit into the twenty-first century.

The British influence on *NOVA* was strong at the beginning. In the first season, six out of the thirteen programs were original BBC productions (on whales, the Crab Nebula, bird navigation), sometimes retitled or with soundtrack narration re-recorded by American announcers.[20] The second season featured seventeen new programs, eleven of which were produced or coproduced by British television companies. In *NOVA*'s fifth season, 45 percent of its new programs were BBC coproductions; in the eighth season, over half. By the tenth season, participation by American science producers, directors, and scriptwriters had grown, and *NOVA* was casting a wider net—less than one-quarter were BBC coproductions. Even so, a book published to celebrate the series' tenth anniversary opened with an homage to British documentary filmmaking and to the BBC's role in nurturing

NOVA's first producers.[21] In 1985, *NOVA* had an "acquisition" staff of four, who primarily traveled to festivals, previewed likely candidates, handled the transition of films into the *NOVA* format, and worked with independent producers to assure conformity of style and approach.[22] By 1989, WGBH was producing slightly over half of each year's programs on its own, with the remainder coproductions with other groups, including the BBC.

Even though international sharing of natural history films had been a common practice for decades, criticism of "foreign" influence could prompt defensive reactions from *NOVA*'s staff, who argued that collaborations reflected economic necessity rather than cultural agendas.[23] It cost *NOVA* ten times as much to produce a new program as to buy one from the BBC, they explained; without such purchases, WGBH could not afford to supply PBS network affiliates with a full season's worth of programs.[24] Nevertheless, it is fair to ask whether (and, if so, how) this substantial proportion of coproductions influenced *NOVA*'s overall message, especially during those early decades. The United States and Great Britain have intertwined histories, but there are significant differences in the two nations' popular cultures, cultural memory, social values, political agendas, and scholarly traditions, including in the radical critique of science. The British connection may have helped inadvertently to propel *NOVA* toward more politicized topics. There was also homegrown support for such an approach among American academics in the humanities and social sciences.

BALANCING ACTS

Georgine M. Pion and Mark Lipsey, summarizing two decades of opinion polls, have concluded that the 1970s response to science did not, as many scientists claimed at the time, "constitute an anti-science backlash"; public attitudes, in fact, remained "overwhelmingly positive."[25] That era did, however, represent a watershed in science's relations with the public, with increased attention to geographical apportionment of research funding, contentious new government regulations applied to research, and intensified scrutiny by journalists.[26] Faith in scientists now coexisted with growing public awareness that unanticipated, undesirable impacts from discoveries could be uncovered down the road. Postwar expansion of government-funded research systems had brought economic bounty and linked university, public, and commercial laboratories more closely to national goals. Contemporary social turmoil—the civil rights movement, women's voices demanding reproductive choice and workplace equality,

Vietnam war protests—also reached into the laboratory. Protest groups like Scientists and Engineers for Social and Political Action criticized military-funded research at universities; Science for the People focused on equity in the technical workforce; physicists debated the efficacy of the proposed U.S. antiballistic missile system; and the animal rights movement attacked biomedical testing.

Among social scientists in the United States and Europe, various streams of scholarly analysis of science's social, ethical, and political contexts merged into what became known as "science and technology studies" (STS). Emmanuel Mesthene's Program on Technology and Society, William Stephenson's group at Washington University, the Harvard Program on Public Conceptions of Science, and subsequent programs created the intellectual foundations for new professional societies and for journals like *Social Studies of Science* and *Science, Technology & Human Values*. As historians, sociologists, philosophers, and political scientists explored the interface of science, technology, and society, questioned traditional notions of scientific "authority" and "objectivity," and dissected controversies over research fraud, human and animal experimentation, and genetic manipulation, their work influenced the writings of science journalists. Television scriptwriters found similar inspiration among academic discussions that reframed science in terms beyond the technical.

The STS debates also reshaped popularizers' perceptions of their own roles. Was it "irresponsible" if a single technical detail or minor qualification had been omitted? Should attention to controversy be automatically interpreted as sensationalism? Journalist Robert C. Cowen explained that the mass media's expanded reach, television's tendency to condense (and thereby exaggerate) expert disagreements, and the scientists' own naiveté about the mass media were compounding the challenge of communicating in a responsible manner. How, Cowen asked, can communicators "put across the moral and social issues of science and technology in correct perspective when many experts seem unaware of the public effect of what they say?"[27]

Science documentaries on public television increasingly attempted to take on the responsibility for discussing such topics. Designed as an anthology series, without continuing characters and sets, *NOVA* never promised cumulative or comprehensive discussion of science. The series was unified only by its title and format; each episode had to be "complete in itself."[28] Relevant technical concepts were redefined, as if not explained in previous broadcasts. Continuity came instead from *NOVA*'s consistent attention to science's social, political, and moral aspects. Whenever an episode explained how a cyclotron worked, the script reached beyond the "black box," attending to the external world outside the machine and how the research might affect society. Program ideas frequently began with social relevance and then proceeded to focus on the laboratory bench, rather than the reverse.

Even though *NOVA*'s initial federal underwriting had been predicated on a goal of promoting basic research, content also eventually concentrated far more on medicine, engineering, and politics than on research in physics, biology, and chemistry. On the occasion of the series' tenth anniversary, executive producer John Mansfield wrote that "*NOVA*'s charge is to report both accurately and responsibly on the world of science and technology. It combines the best of investigative journalism with a sense of show biz, hoping, like any other TV program, to lure viewers back week after week."[29] Mansfield acknowledged that achieving these goals required a delicate "act on the high wire": "wobble too much in one direction, be overdidactic, and you plunge your viewers into the bog of boredom; teeter too far toward easy superficiality and glitzy gimmicks, and you invite the wrath of the academic community."[30] The series did not want to alienate scientific supporters by shining too bright a light on science's flaws and foibles, but, first and foremost, it could not risk boring viewers. Maintaining a loyal audience (essential for pleasing current and prospective underwriters) required, as in commercial television, continually fine-tuning content to suit viewer preferences. Play to your strengths and survive.

VISUAL FEASTS

In the first sixteen seasons (through 1989), *NOVA* presented over three hundred new programs, about one-quarter of them on medicine, human biology, or public health; one-tenth on animal biology; and another one-tenth on some aspect of technology, including aeronautics, engineering, energy, and computers (from "Fusion: The Energy of Promise" in the first season to "Back to Chernobyl" in the sixteenth).[31] In only one season were there no programs focusing on animal biology. Other major areas of interest included anthropology, archaeology, paleontology, environment and ecology themes, and astronomy and space. Attention to physics through the years rarely exceeded one-tenth of a season's offerings.

Biography might have been expected to be a favored format, but scientists' personal histories and careers figured far less prominently than did exploration of how science was being applied to solve contemporary problems. During the first sixteen years, about one-tenth of new *NOVA* programs focused on individual scientists, most of them biologists or medical researchers, such as Denton Cooley and Lewis Thomas. When scientists were interviewed in programs, they appeared primarily as experts on a topic, rather than as individuals with personal lives and families outside the laboratory. And in those first decades, *NOVA* ran far more biographical profiles of men than of women, reflecting gender biases throughout the scientific community as well as in popularization of science.

From the beginning, *NOVA* paid significant attention to topics relating to airplanes, military technology, and weapons, with the proportion increasing during the 1980s, including "War from the Air" (second season), "Hitler's Secret Weapon" (fourth season), "Confessions of a Weaponeer" (fourteenth season), and programs about the Reagan administration's "Star Wars" defense initiative. The series also mimicked commercial network fads. One critic wondered if "competition" explained why *NOVA* had produced a 1984 "science test" episode, noting that *NOVA* was "no longer the only show in town." "When Nova started, it was truly unique," executive producer Paula Apsell explained. "All that's changed. The commercial networks are paying more attention to science. 'Nova' even has plenty of competition on public television."[32] The three most popular PBS shows in the previous season had been *NOVA*, *Nature*, and *Smithsonian World*: "Although we're popular," Apsell added, "we can't sit back and rest on our laurels."

The greatest proportion of episodes through the years looked at medicine, genetics, and human biology: the first season included "Strange Sleep" and "Are You Doing This for Me, Doctor?" (on human experimentation); in the second season, "The Secrets of Sleep" was followed by programs on meditation, the double helix, blindness, and cancer; in the eighth season, "The Malady of Health Care" compared American and British health care systems; in the fourteenth and fifteenth seasons, programs examined the AIDS epidemic, leprosy cures, rheumatology, and whether polio had finally been eradicated. Human reproduction was a predictable viewer favorite. One of *NOVA*'s most watched episodes (thanks to multiple reruns) was "Miracle of Life," a film by Swedish filmmaker Lennart Nilsson. Decades later, *NOVA* commissioned an updated version from Nilsson ("Life's Greatest Miracle"), incorporating advances in genetics and the latest cinematography techniques.

The popularity of National Geographic's television specials reinforced

NOVA's scheduling of programs about anthropology, archaeology, and paleontology: about the Cuiva, Anasazi, Tuareg, and Maya, and about many, many dinosaurs. Dramatic re-creations described a famous fraud ("The Case of the Midwife Toad"); explained the accidents at Brown's Ferry, Chernobyl, and Three Mile Island nuclear plants; and recounted the discovery of penicillin ("The Rise of a Wonder Drug"). To its credit, *NOVA* did not shy away from controversial topics, and even ardent critics called most presentations balanced. A seventh-season profile of Edward Teller ("A Is for Atom, B Is for Bomb") was praised for its "fairly rounded portrait of a complex and controversial life"—"neither a character assassination nor a puff."[33]

Mild sensation did follow the 1975 broadcast of "The Plutonium Connection," about the challenge of controlling plutonium used for domestic power plants. The filmmakers asked a chemistry student with no prior training in nuclear physics to design a "practicable" atom bomb within five weeks and then filmed his research in unclassified sources and libraries: "Boom! . . . you got a bomb—really, it's that simple." A Swedish physicist (asked to review the plans) called the device workable and "shocking." Response to the program varied. The *New York Times* found the cookbook approach disturbing: "At what point does a detailed warning become a primer in the very subject it is supposedly warning against?"[34] Others called the filmmakers courageous, and an admiring U.S. Senator entered the *NOVA* transcript into the *Congressional Record*.[35]

By 1981, *NOVA*'s generally supportive advisory board had begun to express concern about how intellectual integrity could be insured when so many films were acquired from elsewhere. When a BBC-produced program, "Did Darwin Get It Wrong?," aired in 1981 (coincidentally, in the midst of an ongoing legal case about creationism), advisory board member and evolutionary biologist Stephen Jay Gould was livid, calling the program "a total and unmitigated disaster," riddled with errors (such as referring to Darwin as "Sir Charles") and giving undue weight to the creationists' cause and arguments.[36] As Mansfield explained to Gould and other advisers, reliance on acquisitions posed a dilemma. Collaboration enabled them to present more programs at lower overall cost, but required them to trust the integrity of the original filmmaker's research.

BASIC ECONOMICS

Comparison of the topics in *NOVA*'s first and sixteenth seasons demonstrates both consistency and change. In *NOVA*'s first season (1974), human and animal biology dominated six of thirteen new productions. These were gentle representations of science, praising basic research in fields like astronomy, biology, and physics while acknowledging science's practical side.

Medical advances opened *NOVA*'s sixteenth season. The four-part "Pioneers of Surgery" described heart and organ transplants, and how innovations were reducing invasive surgery. Seven of twenty-one new productions that season explored some aspect of medicine or medical education. Programs pushed viewers to wonder "Can the Next President Win the Space Race?," "Can the Vatican Save the Sistine Chapel?," "Do Scientists Cheat?," "Who Shot President Kennedy?," and whether it was "Hot Enough for You?" (on climatology). *NOVA* confronted the "Killer Gene," traveled to Easter Island, and described "The All-American Bear." Responding to changing viewer preferences, *NOVA* had also begun to include more anthropology, archaeology, and paleontology, and to reference Hollywood films like *Indiana Jones* and *Jurassic Park*. By then, the BBC's *Horizon* had also shifted toward more ratings-friendly topics like aviation, space, earthquakes, volcanoes, and charismatic megafauna.[37]

Covering a broad range of subjects took both time and money. *NOVA*'s production process far exceeded the several days that Lynn Poole's team had taken to create their live show. In 1973, John Mansfield described the "ideal" schedule: "research the idea (say, two months), film it (say, three weeks), edit the film (allow about ten weeks), and air it on PBS."[38] In later years, that process lengthened further. By 1983, the *NOVA* group was making ten in-house episodes a year; five full-time production teams were working simultaneously, year-round. Each film that *NOVA* purchased or coproduced also required editing to conform to American television formats and lengths.

The decision to proceed with a particular program began with fundraising years in advance of the intended premiere. When federal spending for cultural programs declined in the late 1970s, PBS groups turned to corporate sources for support.[39] For *NOVA*, slightly over half the unit's funding initially came from public sources, the rest from corporations or foundations. Increasingly elaborate and expensive projects required additional donors. By the sixth season, *NOVA* was receiving major funding from engineering company TRW Inc., and NSF's on-screen credit for under-writing was shared with the corporation. TRW's involvement continued until 1992, when it was replaced by pharmaceutical and medical-supply company Johnson & Johnson. Other funding partners included the Arthur Vining Davis Foundation (seasons ten and eleven), Allied Corporation (seasons twelve through fifteen), and Prime Computer Inc. (sixteenth season); during the 1990s, support came from the Lockheed Corporation, Raytheon Company, Merck & Company, and Northwestern Mutual Life Insurance Company. The independent films acquired by *NOVA* were also often under-written by several different foundations and organizations; for example, "The Fragile Mountain," broadcast in the tenth season, was financed by the Ford Foundation, World Bank, Atlantic Richfield Company, and several United Nations agencies. Although programs in the early seasons acknowl-edged funders and underwriters, *NOVA* (like other public television series) eventually thanked donors by describing each organization's mission, ser-vices, and products and showing corporate slogans and logos.

Did this external funding shape presentation approaches? By the late 1990s, some observers were asking whether the programs that prospered on public television were those that espoused "conventional" moral and political views and stayed "comfortably behind the times in order to avoid alienating anyone."[40] Producers also voiced concern about "censorship by apprehension—a tendency to not propose shows that would make under-writers unhappy."[41] Although donors argued that their only interest was in supporting a great public resource, contributions often coincided with cor-porate public relations goals, and public television's financial crisis in the early 1980s made these sources essential.[42] As telecommunications analyst Oscar Gandy explains, the underwriting of public television constituted an "information subsidy," part of overall public relations campaigns to serve each donor's interests, whether that be a federal science agency or an oil company.[43]

Public television's prohibition on selling advertising time may have mitigated immediate pressure from commercial interests, but the stations still had to compete with for-profit networks. Over a quarter century after *NOVA*'s creation, the executive producer joked about the "ratings race" in a speech to science writers, explaining that her staff assigned internal code names to topics under consideration, such as "old boys and their toys" (programs on "jet fighters, lost aircraft, and random military hardware") and "TRSH" (for "transparent ratings seeking hype," like programs on shipwrecks, pirate gold, or debunking the Bermuda Triangle), all intended to please their core audience of relatively well educated, relatively affluent males.[44]

CONNECTIVITY

NOVA's success paved the way for dozens of innovative projects. In 1976, for example, the six-part miniseries *Cousteau/Oasis in Space*, produced by the Cousteau Society in cooperation with Texas A&M University, explored global environmental problems like industrial pollution, population pressure, and famine, and featured such celebrities as Norman Borlaug, Barry Commoner, Paul R. Ehrlich, Buckminster Fuller, Margaret Mead, and Carl Sagan.

A series created by Michael Ambrosino typified the complexity of such efforts. In 1976, Ambrosino received a $58,000 National Endowment for the Humanities planning grant and then, after two years of development, secured more substantial underwriting for an anthropology and archaeology series, with additional money from Polaroid Corporation for the development of educational materials. The *Odyssey* series included both original and acquired productions. For the first season, Ambrosino's group produced six of the twelve films; the remainder were purchased from BBC, Grenada Television, and National Film Board of Canada. The original productions focused on topics related to the American context. "Seeking the First Americans," about the first humans inhabiting North America, for example, featured two archaeologists with radically different interpretations of the Paleo-Indian site at Clovis, New Mexico, and offered a characteristic "controversy" scenario to sustain viewer interest. Another *Odyssey* program explored the career of anthropologist Franz Boas, who had emigrated to the United States in 1887 and achieved prominence in part because of his work on Native American cultures. "Other People's Garbage" followed historical archaeologists who dig in urban middens and construction sites, and other episodes looked at the lives of the !Kung, Maasai, Cree, and Incas. The second season included productions about Donald Johanson and discovery of the fossil remains of "Lucy," a biography of Margaret Mead, nomads in Afghanistan, women in Marrakech, Roman constructions at Bath, England, and anthropologist Laura Nader's study of consumer complaints as cultural artifacts. In addition to the films, the *Odyssey* project produced a full-color magazine (distributed to 115,000 educators), 40,000 posters, 30,000 library brochures, 30,000 study guides, and videotapes for teachers.

By 1979, PBS had established an identity as a "network of science."[45] Another blockbuster import, the BBC/Time-Life miniseries *Connections: An Alternative View of Change*, starred television personality James Burke, cohost of the BBC series *Tomorrow's World*. Like the Bronowski miniseries, *Connections* combined multiple presentation formats: new and archival film, historical objects, computer graphics techniques, demonstrations, a celebrity introduction, and panels of experts dissecting each film. Production cost over $10 million, and involved filming at 150 different locations in nineteen countries.[46] The ten episodes traced chains of circumstances and ingenuity—from ancient technologies to nuclear weapons and space travel—identifying such triggering ideas as plowing, writing, taxation, and astronomy, while unpacking contemporary problems like overpopulation and dwindling natural resources. The narrative referenced themes of uncontrollable technology and science-created threats. Burke's undisguised disdain for unrestrained, technology-driven policy decisions permeated the discussion of energy and environment.

Thanks to additional underwriting from the Bell Telephone System, American broadcast of *Connections* was accompanied by elaborate educational materials. For the eighth episode, "Eat, Drink and Be Merry," the teacher's guide suggested that students debate whether economic growth must "essentially be restricted because of declining resources." Throughout the series, Burke dispensed dire warnings about the world's future, suggesting that the 1965 New York City blackout, medieval plagues, slavery, V-2 rockets, and nuclear power were typical "generic triggers" provoking technological change. One critic compared Burke to "a one-man interactive media machine, relying on the originality of his ideas, and all the naturalness and enthusiasm of his narration, to keep viewers interested."[47] Other critics condemned Burke's "horizontal slice of history" and "irreverent views" as a "technological Monty Python" in which the narrator "thumps, bangs and whistles his eclectic way through invention after 'triggering' invention."[48] Audiences, however, found the programs engaging, and Burke followed the project with equally successful sequels like *Connections²* on the Learning Channel.

REPEAT STORIES

To paraphrase analyst Todd Gitlin, *NOVA* and similar presentations relied on a repeated formula and dependable consistency in order to "express and cement" the idea of science with unquestionable and immutable authority.[49] The productions also exploited an older popularization technique: the use of storytelling. John Tyndall's nineteenth-century lectures on light and the scientists' radio interviews in the 1930s had all relied on a speaker's ability to enthrall, to paint images with words. Television provided a bigger platform augmentable with films and faces. As Huw Wheldon emphasized, television's stories allowed viewers to examine themselves and the world around them and thereby learn "what the world is like to live in."[50]

During the 1970s, American television's science popularizers perfected those narrative techniques to build the case for science. Paula Apsell credited Michael Ambrosino with instigating this "new approach" in that "he believed that science is a story, and if told with visual flare and strong characters, people would watch."[51] *NOVA* learned to emphasize the "blood, sweat and tears" that precede the moment of discovery, to tell the "soap opera" sometimes hidden within the black box, to make every episode a drama with "conflict, climax, and resolution." As Apsell admitted, sustaining these narratives posed a constant challenge: "People watch television voluntarily. We'd like to tie them to their chairs Tuesday at 8:00, but we can't. We have to entice them with good stories that they can understand without an advanced degree. Striking the right balance between education and entertainment is the essence of what we do."[52] Other documentary filmmakers have seconded Apsell's frank summary of the elements of decision making. Even when producers embrace the "moral obligation" to explore socially relevant aspects of science, Jon Palfreman writes, they instinctively choose narrative arcs, "bankable topics," and other approaches designed to gain "emotional leverage" with their audiences.[53] In the end, the fanciest visualization techniques could not sustain millions of viewers without assistance from skills perfected by generations of storytellers.

CHAPTER TEN

Defining What's New(s) about Science

[A]ny decision taken in regard to the content of a *mass* medium is . . . political . . . in that the decisions will tend to influence the nature of the information and attitudes contained in society.

ANTHONY SMITH, 1973[1]

Television executive Reuven Frank once attempted to explain network news decision making to a group of scientists: "You want us to cover more science on the evening news, and I'm not sure we should. We are not teachers. The agenda of news is news. . . . Almost all scientific 'news' is the discovery of something that is there all along."[2] Frank's characterization of scientists' work as insufficiently novel to warrant mandatory attention must have rankled his audience, but the comments encapsulated how the television business pigeonholed science. An unrelenting flow of university, government, and industry press releases, each trumpeting research "breakthroughs," had made scientific advances seem routine. Television decision makers saw no reason to cover "science for science's sake." Laboratories and researchers had to demonstrate that they deserved time on the air.

The choice not to cover a topic on the network evening news was consequential. Simply mentioning an accomplishment, controversy, issue, or person could kindle other media attention. Likewise, ignoring something implied insignificance or irrelevance. Network news operations treated science news as neither privileged nor essential. Science was not like the stock market. Scientists received far less attention than criminals, movie stars, and political celebrities. Nevertheless, television news, especially from the 1960s through the 1980s, played an important role in reframing science's public image. By presenting scientific inquiries in the context of rough-and-tumble politics rather than isolated in academe's ivory tower, by addressing both research risks and benefits, and by interpreting science as part of, rather than apart from, culture, television news made important statements about how science fitted into modern life. Messages entwined within the news—reinforced and referenced by popular entertainment—also assisted in adjusting science's public image to conform to postmodernist assumptions about its vulnerability, culpability, and social responsiveness.

PARSING THE DEFINITION

What exactly is "the news"? Historian Anthony Smith has pointed out that "chains of information"—transmitting and filtering different types of knowledge—have existed in all societies, all ages.[3] In 1712, the *Spectator* celebrated the roles played by personal perception, gossip, and experience: "All matters of fact, which a man did not know before, are news to him."[4] Early in the twentieth century, newspaperman John B. Bogart declared famously that only novelty matters: "When a dog bites a man, that is not news; but when a man bites a dog, *that* is news."[5] Science writer Deems Taylor concluded in 1917 that newspapers emphasized the controversial aspects of events like scientific conferences because "[f]ew people are passionately interested in chemistry, but everybody likes a row."[6] As historian J. Herbert Altschull explained, the news is a "dynamic phenomenon" shaped by context and time—"Darwin's information about natural selection" would have been news in its day, he observed, even though the evolutionary processes themselves were not new.[7] By the end of the twentieth century, however, editorial selection was being dismissed as elitist, and audiences were encouraged to set their own personal preferences for incoming information. Internet sites and blogs (coupled with economic pressures on traditional print newspapers) further decentralized the filtering processes, returning more power to the individual news consumer and eroding the influence of science's anointed interpreters.

In each era, one aspect has remained the same: what is defined as "news" today becomes redefined as "old news" tomorrow. Once a story has been reported and received, neither journalists, their organizations, nor their audiences perceive it as news. And per the *Spectator*'s definition, almost any television report about current scientific research would be "news" to viewers who are not experts on the topic discussed.

Attempts to develop standards for selecting science news have provoked considerable debate through the years.[8] What, for example, constitutes "science"? Which scientists should be interviewed? Only those with doctorates and employed by certain institutions? Who should decide what is or is not the latest news of science—journalists or scientists? What must be included and what can be omitted? In discussing the popularization of science, Warren Weaver once explained that he would only eat eggplant if the vegetable were cooked so that "it doesn't taste like eggplant" and was "worried that the same thing may happen to science; that when you make science so that the public likes the taste of it, it may turn out that it isn't science."[9]

Scientists and journalists have also disagreed over *when* a science story should be considered news. The intervals between idea, funding, data collection, analysis, and publication can stretch for years. By tradition, publicity follows rather than anticipates journal publication, a custom re-inforced since the 1930s by informal agreements between reporters and journal editors.[10] Authors (who perceive publication in a journal as essential to career advancement and professional status) agree to forgo publicity until peer review provides "an authoritative stamp of reliability"; in exchange for advance access to journal content, reporters agree to temporary embargoes until an issue is published.[11] During the first decades of television, its science correspondents generally cooperated in the same embargo policies by delaying broadcast of filmed interviews with researchers until the print publication dates.

Editing and business practices, the marketplace, and external events have also shaped how television has treated science news. Plane crashes and political upheavals receive priority, pushing other stories, no matter how significant, to the back of the queue; exclusive news from high-status sources is favored over topics of long-term significance.[12] Each network then factors the results of audience research into its choices. For audiences in the tens of millions, these selection biases, over the decades, resulted in science coverage far more reductionist than nuanced.

DECONSTRUCTING THE NIGHTLY NEWS

Content analyses of television news can be tedious, complicated, and expensive. In the ideal, such studies should examine time allocations, visual images, and sound tracks as well as scripts. For reasons relating to economics and practicality, most studies of television's science news have used small samples, tracked coverage of a single story, or counted the numbers of "science" and "technology" stories without differentiating among fields.[13] Two broad surveys from the 1990s, however, do provide some insight to how the media attended to science during the last decades of the twentieth century. The Project for Excellence in Journalism found that science, technology, and personal health stories increased substantially in U.S. newspapers, magazines, and television between 1977 and 1997, and that science and technology stories averaged around 3.5 percent of all news broadcasts on ABC, CBS, and NBC, with the coverage almost doubling by the end of the 1990s.[14] A Kaiser Foundation study found that health was the fifth most popular topic on local television news programs.[15]

The online summaries of *ABC Evening News, CBS Evening News,* and *NBC Nightly News* broadcasts, compiled by the Vanderbilt University News Archive, provide similarly useful information about historical trends in network coverage of science, scientists, and related political and ethical controversies. Analysis of those summaries reveals that network attention to science was, in fact, disappointingly meager during a time when science itself was flourishing.[16] From 1968 to 1989, ABC, CBS, and NBC consistently covered topics in science and science policy yet rarely carried more than one science-related story per night (733 science-related news items within a sample of 868 half-hour broadcasts).[17] Countries invaded other countries; people saved or murdered other people; accidents and natural disasters happened; politicians grabbed and abused power; celebrities divorced; businesses succeeded and failed; stock markets and hemlines went up (and down); and scientists made discoveries. During an era of dynamic growth in university, government, and industry research worldwide, the news of the day *included* references to science but did not track the world of science all that closely. Reuven Frank's dismissive comments apparently reflected conventional attitudes at all three networks.

During moments when major global conflicts, political crises, or catastrophes dominated the headlines, science news sometimes disappeared altogether from the nightly television agenda. Between 1972 and 1975, as the Vietnam conflict raged and the Watergate political crisis simmered, the number of science stories declined significantly. In 1974, for example, there were only five science-related stories in the twenty-nine broadcasts sampled for that year, compared to triple and quadruple that amount in the samples for years preceding and following.

During the mid-1970s, the networks began to hire journalists to report exclusively on science, medicine, and space, and within less than a decade, all major networks employed several health/medicine and science correspondents. Robert J. Bazell, for example, who has a Ph.D. in immunology, began his decades-long career at NBC in 1976. CNN's launch of a daily science report and a half-hour weekend series (*Science and Technology Week*) in 1985 may have also prompted the rival broadcast networks to increase their coverage after that time.

The science news that appeared, like the rest of broadcast content, adhered to each network's assumptions about viewer preferences.[18] Science news tends to establish, Martin W. Bauer and his colleagues point out, a "dramatic narrative" that retells but does not necessarily "mirror" the science of the day.[19] Most national television stories were fact-based (major research conclusions or discoveries; Nobel Prizes or deaths of famous scientists) or issue-based (controversies over ethics and safety, research applications, and laboratory practices; the politics and economics of funding and regulation). Over time, emphasis shifted from the fact-based stories toward reports dominated by social and political aspects. Although preoccupation with celebrities and personal tragedies predated the development of broadcasting, the amount of "featurized and people-oriented" science stories and attention to sensational issues like scientific fraud increased from the 1970s on, following television trends for other human-interest topics ("lifestyle," accident, crime, and scandal).[20]

Three major subjects (space, environment, and medicine) dominated science-related coverage, with varying intensity. Attention to space research and space exploration, an important part of all television coverage during the 1970s, declined slightly in the 1980s. Environmental pollution and policy stories appeared less often but remained subjects of continuing news interest. Attention to the biomedical sciences and public health increased steadily from 1968 through 1989; almost one-third of the news stories sampled discussed some aspect of human biology or medicine, with the proportion rising throughout the 1970s and 1980s and with continual attention to cancer research and the latest medical spectaculars like heart transplants and "test tube babies."[21] Public health topics received ample discussion, especially at the onset of the influenza season. Far less attention was paid to physics and astronomy unrelated to the space program (including nuclear, chemical, and biological weapons) and to science education, botany, anthropology, or engineering research. Most "science" news was thus related to space, medicine, or environmental quality (mirroring trends in entertainment programming) or was connected to government actions (for example, vaccine safety or pollution controls) or to social, political, and economic events, such as war.

Science *could* be found enfolded within network news, though, whenever reports on nonscientific topics included technical references, explanations, and expert interviews. In 1984, during a record-breaking summer heat wave, CBS rejected the clichéd approach of sending photographers to film polar bears sitting on ice at a zoo. Instead, a reporter explained how urban trees act as "natural air conditioners" and mentioned that researchers were attempting to breed specimens resistant to Dutch elm disease because so many valuable shade trees were being lost.[22] *Incorporation* of science gradually increased within television news, following the practices of entertainment programming, where science was being casually mentioned in jokes, in bits of dialogue, or as the occupations of fictional characters.

Television routinely exploited symbolic backdrops and location shots to convey authenticity and actuality, and so major oil spills were usually covered by showing film of oil-soaked wildlife and beaches rather than by interviewing scientists.[23] Stories about environmental quality—in the air (Los Angeles smog alerts), in water (mercury contamination of fish), and in food (agricultural pesticides)—similarly featured stock footage (belching smokestacks, dying swordfish, and stunted crops) while voice-over narration summarized the political debate and occasionally referenced related research.

The cultural significance or insignificance of science could also be emphasized via a practice common in print as well as visual media. Television news managers choose the lead segments for every broadcast as carefully as newspaper editors plan the front page.[24] In 1968 and 1969, science stories led about one in every five news programs sampled. That number dropped to about one in every fifty by the early 1970s, and rose to one in every fourteen during the 1980s. Most lead stories, from the 1960s on, related to the U.S. or USSR space programs (including accidents involving launches), followed by the Chernobyl nuclear power plant accident, *Exxon Valdez* oil spill, and medical innovations like artificial hearts. Science was most likely to be discussed in a lead story if connected to a sensational accident, epidemic, or crime.

Television news thus supplied a complex but incomplete portrait of contemporary science, ignoring many aspects and overemphasizing others. By the late 1970s, the news focused more on ethical controversies and the limitations of expertise. Although some news reports glorified scientific contributions to social and economic progress, it became far more common for stories to discuss the risks associated with that progress, to address the dangers lurking within what scientists knew or sought to know.

CONSENSUS AND TIMING

From 1968 through 1971, all three networks would often choose to run the same science story almost half the time, often with remarkable similarity in how the stories were framed and in the individuals (or types of individuals) interviewed. Many consensus stories were, predictably, about space missions, Nobel Prizes, and science-related accidents. And yet accidents in Pennsylvania (Three Mile Island nuclear plant, 1979), India (Bhopal chemical facility, 1984), and Ukraine (Chernobyl nuclear plant, 1986) did not receive equal notice from network television, perhaps because of their timing and perceived relevance to American audiences.[25] The 1979 Pennsylvania incident coincided with environmental protests and Middle East oil embargoes, so the news media were primed to pay heed.[26] Network news had ignored two previous nuclear power plant incidents in the United States because they were not initially framed as potential catastrophes, while coverage of Three Mile Island consumed almost 40 percent of evening news broadcasts during that event's first week.[27] Network news programs also allotted more time to the Chernobyl accident than to the Bhopal chemical-plant explosion, even though the number of people killed, injured, and displaced was far greater in the India event and even though that accident also had long-term economic and regulatory impacts on a major international industry.[28]

Timing matters. News coverage of a minor issue could, accurately or not, imply importance and could prompt reports on other news outlets, especially if no other event distracted the pack. The greater the coverage, the more that viewers (and other journalists and politicians) probably assumed a topic was significant. As Doris Graber has explained, "*How* an issue is framed and presented determines not only whether it will or will not be covered by the media, but also whether it will be noticed and learned and whether it potentially influences people's opinions and ultimately, in a multi-step process, generates governmental action."[29] Simultaneous coverage, though, did not necessarily mean equivalent emphasis. On January 25, 1978, all three networks carried stories about federal water-purification regulations—CBS devoted almost five minutes and NBC, one minute, thirty seconds.[30] In July 1978, an eight-minute NBC report about the birth of a "test tube" baby mentioned "baby factories" and asked priests to comment on how laboratory fertilization techniques differed from abortion.[31] ABC's report on the same event ran half as long (four minutes) and included reaction from religious leaders.[32] CBS ran two reports: an initial story (three minutes, ten seconds) focused on the publicity circus and a second segment (five minutes, forty seconds) centered on religious reactions.[33]

In the years before the first lunar landing, NASA would time many of the astronaut transmissions to coincide with evening news programs. Like news coverage of scientific expeditions in the 1930s, the story lines favored excitement over education. As physicist Philip Morrison observed, "TV played it badly. Just adventure and mystification: No one ever explained what was happening."[34] Viewers learned endless details about astronauts' diets, Carl Sagan noted, and little about the research that enabled the space missions.[35] Television's defenders, like newscaster Walter Cronkite, brushed off such criticism, and until the 1986 *Challenger* accident, network coverage tended to glorify the visual (vehicles poised on launch pads), emphasize the drama, and minimize potential risks.[36]

IN THE BALANCE

Communication about risks identified by or related to scientific research posed special challenges for television. The Fairness Doctrine, a regulatory policy first articulated by the FCC in 1949, attempted to ensure reasonable balance in how broadcasters covered public controversies. Between 1969 (when the policy was upheld by the U.S. Supreme Court) and 1987 (when the FCC abolished it), the most convenient quantitative measure of presumed fairness—the amount of time offered to representatives for each side in a dispute—introduced unintended consequences for discussion of science-related controversies. Securing an authoritative spokesperson for each position, regardless of the strength of support (or validity of supporting evidence), was offered as proof of fairness, with "authoritative" defined through job title (government official or university professor) or membership in a related interest group.[37] Interviewing sources from different organizations or, in the case of science-related issues, juxtaposing an academic scientist with an industry spokesperson, or a political official with an environmental activist, became de facto practice in television.[38] This approach coincided with an explosion of litigation—alleged breaches of environmental regulations, land-use disputes, medical and pharmaceutical liability cases—in which scientists and engineers served as expert witnesses. If research appeared to implicate a cosmetic as potentially carcinogenic, then the manufacturer, attempting to litigate in the court of public opinion, would hold a press conference, and well-credentialed experts would cite studies that indicated the opposite.

As a result, viewers saw expert pitted against expert, scientist against scientist, and heard a narrative continuum of unevaluated ideas, proposals, and statements. A thirty-second sound bite from a nuclear physicist would be followed by a thirty-second sound bite from a nuclear energy executive; a statement from an evolutionary biologist would be followed by an interview with a school board member advocating adding creationism to the curriculum. Reports would be framed as if there were equal numbers of advocates and opponents, each mustering equally valid evidence. This same formula was also being adopted by science documentaries and informational series seeking to appear objective, fair, or balanced when discussing controversy.

Scientists increasingly functioned "as signs . . . signifying the presence of expertise" and stood before the cameras "not to impart information but to lend authority to the text."[39] "Which experts do we believe?" one local television news executive asked. "It's getting hard to figure out because we have experts coming out of our ears."[40] Television, of course, needed on-camera authorities to provide "faces" for otherwise dry explanations, and that circumstance was complicated by another common news reporting practice: to cite people rather than data as the authority for a story.[41]

Within journalism, objectivity has never been considered to be the same as neutrality; the term traditionally refers to reporting processes adopted to guard against or reduce bias.[42] Not only is "perfect objectivity" an illusion in science journalism, but the false "quest for objectivity," science editor Rudy Baum has argued, may even undermine the goal of informing the audience. Simply reporting "facts without interpretation of their context and meaning can lead to meaningless headlines like 'eminent scientists disagree.'"[43] Science journalism, Baum adds, should never ignore "the credentials and credibility of scientists being interviewed"; a news story that states "critics exist" should explain "the extent to which those critics' views are accepted" by their peers.[44] Historian John Burnham observes that, by indiscriminately applying the "equal time" policy to science coverage, the mass media seemed to imply that astrologers have equivalent authority to astronomers in matters relating to the universe.[45]

Even when environmental and public health disputes involved questions about the validity and reliability of scientific evidence, television's news programs frequently omitted relevant technical explanations. Coverage was, instead, constructed around choice and trade-off, thereby recoding scientists from experts into surrogates for a political position. In 1971, for example, public health officials wanted to ban several caustic household cleaning products related to accidental poisonings of children. One news story included interviews with scientists describing a workable alternative chemical (phosphate) and other experts warning about potential adverse environmental impacts from its use.[46] Likewise, efforts to ban the artificial sweetener saccharin from soft drinks (because of research linking it to certain cancers) prompted complaints from diabetics and warnings from dentists that natural sugar posed dangers to children's teeth. Aiming for a tone of fairness, news programs presented equal numbers of experts for every possible side to an argument. When Wyoming ranchers wanted to kill coyotes threatening their livestock, biologists pointed out that poisoned bait might inadvertently harm nesting eagles.[47] Arctic wildlife biologists explained that development of Alaskan oil resources could threaten polar bear habitats.[48] In each news story (ban saccharin, control coyotes, avoid poison, exploit oil reserves, protect bears), people labeled as experts were interviewed, the political arguments were outlined, but relevant research (or the standards for assessing the various arguments) would be ignored. Television news rarely spent valuable time explaining the science.

When promising medical approaches or treatments fell outside the boundaries of traditional medicine, media organizations struggled to keep viewers informed while ensuring balanced, responsible coverage. During the 1970s, hundreds of cancer patients traveled to Mexico to obtain Laetrile (which was banned in the United States). Some journalists, aware of skepticism within the medical community about the drug's effectiveness, argued that the sheer number of American patients crossing the border had made the topic news.[49] In 1977, *NBC Nightly News* solved the problem by pairing an interview with a university-based cancer specialist who opposed government licensing of the drug and an interview with a Mexican physician whose clinic sold Laetrile, thereby implying that each person's statements had equal scientific merit and objectivity.[50]

During the early stages of the AIDS epidemic, the networks sought interviews with prominent biologists like Robert Gallo and Anthony Fauci, who explained attempts to identify the cause and, later, the search for a cure. When political debate over public health strategies intensified, and homophobic attacks and rhetoric inflamed discussion of the human tragedy, the news reporting gradually changed. Many journalists admitted that they had promoted false hopes of a vaccine, and television news began to devote less attention to the internal scientific debates (or bulletins about research progress) and more to human-interest aspects, describing the disease's physical and emotional toll on patients and families.

By the 1980s, television news routinely framed most science in terms of *both* benefits and risks, *both* certainty and uncertainty. Science is not a sure thing, ABC reporter Jules Bergman declared at the 1978 AAAS annual meeting, "and it's about time, gentlemen, that we stop telling the people that it is."[51]

COINCIDENTAL DRAMA

Television news references to novels or movies continually blurred "the conventional boundary between fact and fiction."[52] In 1978, for example, television coverage of in vitro fertilization referenced the fictional "artificial births" in Aldous Huxley's *Brave New World,* and one network report even included film clips from a forthcoming made-for-television movie based on the novel. Such content exemplified the "continuum" and permeability of entertainment, news, and public affairs content, where enjoyment enhanced effect.[53] The news reinforced moral lessons being dramatized in television's fiction, and vice versa.

The tendency to overdramatize medicine—a legacy of the "Kildare" era—was reinforced by cinematographic styles emphasizing "realism." After an NBC camera crew followed one physician for months—filming him in emergency rooms, operating rooms, and hospital rounds—the *Lifeline* producers then edited the footage and imposed an artificial story line, a format mimicked in many subsequent news, documentaries, and fictional dramas. Journalistic descriptions of real experimental medicine would play out against the backdrop of fictional series, often telecast on the same channel an hour or two later. As television drama transitioned from the sentimental "Marcus Welby" era (friendly family physicians) to unflinching realism (death, dying, and incompetence in urban hospitals), the fictional plots referenced the same themes found in the news: economic pressures, triage, patient rights, and access to experimental treatments.[54]

Sustained sensationalism surrounded news coverage of biotechnology, both inspiring and drawing inspiration from popular fiction. In the summer of 1974, an international group of microbiologists proposed a voluntary moratorium on research using recombined genetic material until they could assess the potential hazards. The serendipity of popular culture played a role in how the media initially framed this event. The film version of Michael Crichton's best-selling novel *The Andromeda Strain* (1969) was first shown on television in 1972 and had offered colorful visions of runaway microorganisms and mutant microbes. The first press reports about something called "recombinant DNA," coupled with biologists' cautious public statements and air of mystery, brought such fictional scenarios to mind. When scientists met the following spring to reevaluate the moratorium, journalists rather naturally wanted to be present. The meeting's organizers at first wanted to admit only two reporters (eventually compromising on a pool of sixteen) and, reflecting scientists' distrust of television, only permitted still photography of the event. Resumption of the research, with self-monitored regulation, and subsequent establishment of collaborations between universities and newly formed commercial biotechnology corporations, failed to dampen the controversy. Community protests against proposed laboratories provided lively opportunities for television coverage throughout the 1970s.[55]

Some molecular biologists later excoriated the media for "emphasizing the more spectacular alternative scenarios and the clash of the more charismatic personalities."[56] Given fictionalized images of "Andromeda strain technology out of control" and the stark differences between research opponents and proponents, sensationalistic coverage may have been unavoidable.[57] And biologist David Baltimore (himself a frequent target of protest) argued that the technical debate not only underlined scientists' sensitivity "to the problems that can arise out of scientific research," but also engaged public interest and helped to outline political choices in a more concentrated, "organized," and "responsible" way than might have happened otherwise.[58] The controversy raised awareness of the importance of proactive public education. To participate effectively in future political choices and to place alleged benefits and risks in proper perspective, citizens needed contextual explanations, needed more than recitals of facts.[59]

Whenever sensational events coincided with related fictional drama, network news unfortunately did little to separate facts from fictions, in part because most television professionals lacked the necessary expertise. *The China Syndrome*, a Hollywood movie in which operators of a nuclear power plant attempt to downplay a potentially disastrous accident, had opened in theaters nationwide on the weekend before a real-life accident occurred in Pennsylvania.[60] Initial television reports on March 28, 1979, screamed that the ongoing crisis at the Three Mile Island nuclear plant could jeopardize the safety of hundreds of thousands of people, and many viewers probably found that consequence plausible.[61]

Although the reactor did experience a meltdown, the accident was contained, nothing "exploded," and no significant radioactivity was released to the atmosphere.[62] Television initially offered only weak reassurances, however. On Friday, March 30, Walter Cronkite introduced his broadcast with the ominous statement that the world had just "faced the considerable uncertainties and dangers of the worst nuclear power plant accident of the atomic age. And the horror tonight is that it could get much worse."[63] Misinformation, rumors, and lack of coordination among public officials exaggerated the structural weaknesses of rapid-fire electronic journalism.[64] The President's Commission on the Accident at Three Mile Island later concluded that most journalists acted responsibly, but television's need for speed and its addiction to drama had fueled a sense of crisis.[65] One television reporter admitted, "Who I reached on the telephone determined how I was going to report the story in an hour."[66] Television abhors silence. There was a tendency to fill the air with speculation, to get a story on the air quickly, rather than locate and interview the best, most knowledgeable experts.

FIGURE 14. Media coverage of a press conference about the Three Mile Island nuclear accident, March 1979. Scientists caught up in newsworthy projects or events could face a bewildering array of television cameras and journalists. Nuclear Regulatory Commission file photo. Courtesy of Nuclear Regulatory Commission.

The Three Mile Island incident emphasized that, in a world of complex, interconnected technological systems, where breakdowns or disasters could affect millions, the task of delivering information to the public should be a shared responsibility among media, scientific experts, and government officials. When the crisis erupted, the switchboard of a small public-interest group, the Scientists' Institute for Public Information (SIPI), known as a reliable source of scientific information, was swamped with telephone calls from reporters who did not routinely cover science.[67] SIPI began connecting journalists seeking information with experts willing to be interviewed or to provide explanations on background. As a result of this experience, SIPI established a permanent Media Resource Service that, until 1995, maintained a roster of tens of thousands of experts and provided valuable assistance to professionals in all parts of the media. Over 90 percent of its inquiries eventually came from journalists who were not science specialists and worked in places like local television stations.

The event also underscored the relationship between television and the state of public knowledge about complex scientific topics. Few viewers could be assumed to understand how a nuclear reactor works. Even experienced reporters floundered over where to begin. "Where is Mr. Wizard now that America needs him?" columnist Clarence Page asked, explaining that, with a few basic materials, the children's show host could "unravel the secrets of the universe and still have time for commercials."[68] After all, Page pointed out, Mr. Wizard had once explained a nuclear reaction by resting fifty ping-pong balls on fifty preset mousetraps.

Communicating about twentieth-century science also required subtlety and nuance, which dramas could do with greater effect, if not necessarily more accuracy, than the news. By the end of the 1970s, television scriptwriters were weaving plots around microbiology and biotechnology, echoing dramatic narratives constructed in the news. In a CBS made-for-television movie *The Henderson Monster* (1980), the central character, Nobel Prize–winning biologist Thomas Debs Henderson, attempts risky experiments with recombinant DNA to create a microbe that might mutate and "escape" from the laboratory but that could also result in a lucrative cancer treatment.[69] Publicity for the movie invited comparison with the recent Cambridge, Massachusetts, "town and gown" confrontations over local regulation of university-based research. Rather than construct a "horror" scenario, however, the film ridiculed the "mercenary halls of academe." Critics observed that the screenwriter had had "a field day hooting at a scientific community more eager to win a Nobel Prize than to take reasonable precautions."[70] And the script underscored scientists' moral equivocating: the principal character claimed to be working for the public good but had private financial motives.[71] In interviews connected with advance publicity for the film, the actor who played Henderson said, "I think we broke through the stereotypes. He isn't a guy in a white coat, formal and saintly. He's in an arena of fierce scientific competition."[72] Yet many television critics referenced those same stereotypes in reviewing the film, referring to "science amok" and "the monster created by these modern scientific Frankensteins."[73]

When discussing nuclear energy and nuclear weapons, a combination of timeliness and drama could also reinforce fear. ABC's made-for-television *The Day After* aired in 1983, in the midst of intense political debate over the plausibility of a climatological disaster dubbed "nuclear winter" (that a nuclear exchange between superpowers would create sufficient dust, smoke, and soot to block most sunlight from reaching Earth).[74] The movie's plot imagined that a real city (Lawrence, Kansas) had been destroyed in a nuclear strike, and offered horrific images of postcatastrophe life.

The producers insisted that they had not intended *The Day After* as a political statement, but several antinuclear lobbying groups (Ground Zero, Coalition for a Nuclear Freeze, and Campaign against Nuclear War) obtained contraband copies of the unfinished film and launched a public relations campaign to which political conservatives, led by the Reagan administration (in the midst of trying to deploy more nuclear missiles in Europe), responded with anger and hostility.[75] Then, in late October, Carl Sagan described the idea of nuclear winter in the Sunday newspaper supplement *Parade* and, the following week, addressed a high-profile Washington, D.C., conference, "The World after Nuclear War," and debated the issue with physicist Edward Teller and biologist Paul R. Ehrlich on ABC's news show *Nightline*.[76]

Nuclear critics turned political animosity to their advantage. The resulting news coverage of the controversy, Allan M. Winkler documents, reinforced the network's own publicity plans, making the film "an event even before it was aired" and playing into the plans of the antinuclear lobby.[77] The broadcast made the covers of *Newsweek* and *TV Guide* and received substantial coverage on morning and evening news shows. ABC held advance screenings for prominent civic and religious leaders and government officials; some right-wing groups threatened to boycott advertisers.[78]

The Day After turned out to be the highest-rated made-for-television movie broadcast up through the 1980s, drawing about a hundred million viewers for all or part of its two hours.[79] Recalling the reaction to Orson Welles's "War of the Worlds" radio broadcast and concerned that the film's graphic depictions of nuclear aftermath might trigger similar panic, ABC arranged for psychiatric personnel to be available to calm viewers. Few people called the network's hotlines, however, and, as measured in over thirty postbroadcast surveys, the film failed to generate the anticipated adverse reaction or any significant shifts in viewer attitudes.[80] Analysts later speculated that advance publicity may have inoculated viewers against intense negative reactions and reinforced their sense of having acquired "knowledge" about nuclear war. Learning about nuclear weapons, even if the context was labeled as fiction, may have stimulated a sense of empowerment rather than alarm. The conflation of fact and fiction continued in the news. When *The Day After* was shown on European and Japanese television, *NBC Nightly News* ran follow-up stories about West German antinuclear protests, the reaction of Hiroshima survivors, and the British defense secretary's postbroadcast appearance to argue for maintaining his country's nuclear arsenal.[81]

ENGINEERING THE NEWS ABOUT SCIENCE

During the late 1970s and early 1980s, data from George Gerbner's Cultural Indicators project and similar media research prompted many analysts to link declining enrollments in science courses to the state of science on television.[82] The inference seemed reasonable. Television offered one of the few cultural contexts in which people might *accidentally* encounter science; the amount of factual science programming appeared to be shrinking with each season; and the fictional portrayals, which Gerbner had showed constituted the largest proportion of attention to science, increasingly diverged from scientists' ideal of who they were and what they did. Although additional technical explanations might have helped viewers better assimilate and interpret what they saw, television executives resisted demands to engage in what they perceived as "education" and often interpreted such suggestions as pressure from special-interest groups seeking more publicity. A 1987 exchange between Carl Sagan and NBC news anchor John Chancellor epitomized the deepening divide between scientists and broadcasters. Chancellor framed the issue in terms of social change: society's values and goals were shifting, Americans were "less concerned with the public good than the private gain," and these trends influenced attitudes to science as well.[83] Television responded to society's wants and needs; it reflected and reported on but should not lead society. Change must come from within the culture, not be arranged by the television industry, Chancellor insisted. To argue that "if only television were better, people would be better and the country would be better . . . inevitably conflicts with one of our cherished freedoms"—that of a free press.[84]

To Sagan, such statements represented an abdication of responsibility. The two speakers went back and forth: Sagan asserting that "presenting more science on commercial television" was *not* social engineering, and the journalist countering that it was.[85] Chancellor pointed out that the scientific community had two choices: to "manipulate the media" by demanding that it send a particular message" or "be smart enough and mature enough to expect the lack of science education to be addressed by politicians, the clergy," and other influential Americans. If those people take a stand, he argued, then television will report it—"but if they don't say it, the media should not be expected to create the interest and promote the agenda."[86]

CHAPTER ELEVEN

Entrepreneurial Popularization

Television destroys our comfortable preconceptions by showing us just
enough to prove them wrong, but not enough to replace them with the
certainty of first-hand experience.

JAMES BURKE, 1978[1]

By 1980, American adults were spending, on average, about three hours
a day watching television.[2] The most popular science-related programs—
fictional dramas like *Battlestar Galactica*, *The Bionic Woman*, *The Incredible
Hulk*, *The Six Million Dollar Man*, and *Quincy, M.E.*—bore little resemblance
to scientists' popularization paradigm. Carl Sagan, among others, accused
the television industry of thoughtless bias and a lack of social conscience.
Cartoon scientists ("moral cripples driven by a lust of power") and pseudoscientific nature specials, he complained in *TV Guide*, ignored "the joys of
science, the delights in discovering how the universe is put together, the
exhilaration in knowing a deep thing well."[3]

Other indicators, though, seemed to suggest a less dire situation. In
1978 alone, the *New York Times* had created its weekly "Science Times" section, new science magazines were being founded (including *Discover* and
Penthouse Publications' *Omni*), and the circulations of established publications like *Scientific American* and *The Sciences* began to rise substantially.[4]
Many of these newer ventures embraced a constructivist approach. Americans still liked to read about science, explained the publisher of *Mother
Jones*, but they increasingly preferred analyses of the "interface" between
science, politics, and society rather than simply scientists' "happy talk."[5]

The evidence of increased public interest did not provoke development
of more science programs on either commercial or public television, however, for reasons relating to how the industry made its programming decisions. As analyst Ien Ang explains, "[T]here is no way to know in advance
whether the audience will tune in and stay tuned," so broadcasters have

long opted to copy winning formulas, to predict future performance by extrapolating from quantitative analyses of past behavior.[6] Without a string of ratings successes, science shows could seem risky options. During the 1980s, with the expansion of cable outlets, advertisers had begun to buy airtime according to estimates of *who* might be watching (age, gender, income, spending patterns) not just *how many*, and the audiences for science were not well-defined.[7] Moreover, the underwriters and managers of public television had began to scrutinize *their* ratings more closely and to plot how to compete better against the networks.

CHARISMA AND THE COSMOS

Carl Sagan had failed to mention in his *TV Guide* article that, with business partner Gentry Lee, he was developing his own television celebration of "the joys of science." When *Cosmos* premiered in 1980, the thirteen-part miniseries opened a wide lens on the universe, oscillating from cosmology to earthly environmentalism. Titles like "Blues for a Red Planet" and "The Edge of Forever" beckoned to youthful audiences, while discussions of climate change, environmental pollution, and nuclear proliferation reflected the host's political agenda. Sagan sought to reach rather than "teach." Well-crafted television programs might rekindle enthusiasm for space research, "make science seem like a cause, a calling, an ideal," and might "lure the public away from . . . unhealthy fascination with pseudoscience and irrational belief systems."[8]

Cosmos's spectacular computer-generated imagery influenced countless subsequent projects, thanks to the formidable creative skills of producer Adrian Malone, who had also guided *The Ascent of Man*. Born in Liverpool in 1937, Malone attended public school in England but never finished college. After serving in the Royal Air Force, he began a broadcasting career in radio in 1958, eventually working for the BBC. From 1965 to 1977, Malone was involved in some of British television's most exciting projects, including *The Life Game*, *A Plague on Your Children* (about chemical and biological warfare), the *Horizon* series, and *The Age of Uncertainty*, hosted by economist John Kenneth Galbraith. Malone moved to the United States to assist in building an educational television program at the University of Pennsylvania, and when that project collapsed, he founded an independent production group with another former BBC director.[9] Determined to outshine the recent hit movie *Star Wars*, Malone devised imaginative special effects, such as making Sagan appear to be traveling in a spaceship, a device that became one of the project's signature images.[10] *Cosmos*'s cinematic vision of the future connected viscerally with viewers. Sagan already had modest fame outside academe. *Cosmos* now propelled him to international stardom.

Sagan's celebrity machine was not fueled by idealism alone. His initial television appearances had been planned as a publicity campaign for his book *The Cosmic Connection*.[11] In 1973, following appearances on *Today* and

FIGURE 15. Carl Sagan, April 17, 1978. Sagan had just received a Pulitzer Prize for his nonfiction book *The Dragons of Eden*. Two years later, his miniseries *Cosmos* premiered on public television to rave reviews. Courtesy of Smithsonian Institution Archives.

a Dick Cavett special, Sagan was invited to *The Tonight Show*. The earnest professor proved a perfect foil for Johnny Carson's deadpan style, and Sagan's witty repartee led to appearances two to three times a year for the next thirteen years.[12] The television gamble proved successful. After the first Carson appearance, Sagan received extensive media attention; *The Cosmic Connection* went through twenty printings and sold over half a million copies; sales of Sagan's subsequent books did as well or better.[13] Sagan's fame thus represented a mutual construction by a charismatic scientist, book publishers, and the television industry. His controversial political views (and occasionally abrasive personality) may have attracted criticism from other scientists, but the television appearances brought financial wealth and independence.[14] A few weeks after the first *Cosmos* episode aired, Sagan made the cover of *Time*.[15]

Coproduced by KCET-TV (Los Angeles), Carl Sagan Productions, BBC, and West Germany's Polytel International, *Cosmos* was completely underwritten by private foundations and by corporations like Atlantic Richfield Company (which gave $4 million for production and another $2 million for advertising and promotion).[16] Such lavish funding allowed programs to be offered free to PBS stations. Without such subsidy, local stations would have had to pay a full or partial fee (purchasing the right to air programs), or CPB would have had to underwrite broadcast. To the chagrin of executives concerned about length and production budget (eventually over $8 million), Sagan and his staff insisted on expensive special effects and extensive location segments (they filmed in Europe and Asia as well as the United States). Fortunately, *Cosmos* was a critical and financial success, wildly popular with audiences and recipient of fifteen Emmys and a Peabody Award. It broke all ratings records for PBS, and the reruns attracted even more viewers.[17] Twenty years after its premiere, and years after Sagan's death, the programs remained so popular that PBS broadcast an hour-long synthesis, *The Best of Cosmos*, and the Discovery Channel acquired the rights to rerun the entire series.

ADDITIONAL INFORMATION, USEFUL INCOME

Understanding the impact of *Connections*, *Cosmos*, and similar miniseries requires looking behind their ratings. Each was conceived as a project intended to influence beyond the initial telecast. Building on practices used for 1930s educational radio programs, each series promoted coordinated print materials for classroom use (such as teaching guides and reading lists), related books, and video copies for classroom and home replay. Time-Life Inc., for example, sold *The Ascent of Man* to schools in 16 mm and videotape formats, along with a study guide, grossing almost a million dollars by 1975.[18]

Development and marketing of these outreach products required substantial investments, so public broadcasting increasingly turned to corporations for underwriting. The U.S. premiere of *Connections* was coordinated with publication of two books and fifteen specially commissioned expert essays placed in over four hundred newspapers.[19] The National Endowment for the Humanities and the Bell Telephone System also subsidized educational kits containing teacher's guides, posters, source books, and reprints of the newspaper essays. Rather than donating film copies, as earlier Bell science series had done, the *Connections* project encouraged teachers and schools to videotape programs themselves off the air with no copyright restrictions.

Companion books, films, and videos could generate valuable revenue streams. Sagan's *Cosmos* book sold over 400,000 copies within four months of the broadcast premiere, eventually earning millions of dollars in royalties.[20] Book versions of *The Ascent of Man* had sold over 750,000 copies and grossed $11.4 million by 1977.[21] *NOVA* engaged in similar marketing of videotapes for home and classroom viewing.

BOOM AND BUST

Commercial prime-time success remained the elusive prize. During the early 1980s, two science series seemed poised to achieve that goal.

Universe—later retitled *Walter Cronkite's Universe* after its well-known host—premiered as a four-week, Saturday-night miniseries in summer 1980. After rave reviews, CBS commissioned twenty-six additional half-hour episodes. Each *Universe* program contained three segments, with network correspondents like Diane Sawyer and Charles Osgood reporting on current topics in a style reminiscent of CBS's public affairs program *60 Minutes*. Complicated topics like bioluminescence or Antarctic seal vocalization were compressed into five- or ten-minute segments. Executive producer Jonathan Ward argued that even these snippets of science contributed to public understanding: "Anything . . . that gives a sense of how science is done is legitimate. . . . Maybe we can't document the entire process. But we can say it's going on."[22]

Interesting topics, a famous host, and fast-paced scripts brought excellent

Nielsen ratings for the premiere (34 percent of the potential audience for the time slot). In "Crystallization," *Universe* introduced a regular feature exploring the natural world through scientists' eyes and, much as earlier television shows had done, aimed a camera through a microscope. Subsequent programs discussed wildlife (pandas, dusky sparrows, whales), energy and environment (the greenhouse effect, garbage recycling, rain forest destruction), and film and media technologies (special effects used in *Star Wars*). Programs discussed the search for extraterrestrial life (interviews with Carl Sagan and Philip Morrison) and featured the Charles and Ray Eames short film *Powers of Ten*. "Gene Factory" focused on commercial work at the new biotechnology firm Genentech; "Lucy" featured paleoanthropologists Donald Johanson and Richard Leakey; other programs looked at obesity, animal testing, and "smart cars." *Universe* could never duplicate that first season's ratings. CBS canceled the series in 1982.

Another syndicated series launched around the same time, *Omni: The New Frontier*, also seemed like a good bet because both ABC and NBC purchased the programs for their owned and operated stations and the project was linked to *Omni*, a new science magazine created by *Penthouse*'s publishers. Even the adult magazine connection and a celebrity narrator (actor Peter Ustinov) could not salvage a flawed concept, however. Critics condemned *Omni: The New Frontier* as a "corruption of science" that "defrauded its audience" and "offered neither solid information nor any sense of what the scientific world is all about."[23] Viewers apparently agreed.

The high-visibility cancellations of *Universe* and *Omni*, and the coincidental collapse of an AAAS-sponsored popular magazine, led to speculation that the media's "science boom" had burst even though other popularization ventures were thriving.[24] In the same year that *Universe* was canceled, PBS carried several highly acclaimed miniseries and two exceptional documentaries about physicist J. Robert Oppenheimer. Two other science and nature series premiered that year. Science museum attendance was up, as were sales of popular science books. As journalist Barbara Culliton noted, "[T]here's no evidence that the public's appetite for science is sated, at least not yet. Rather, there is a demand that it be interesting."[25]

A project initiated in 1982 by two *NOVA* alumni, Graham Chedd and John Angier, demonstrated the volatility of the television marketplace. Chedd-Angier Productions created *Discover: The World of Science* in association with the science magazine *Discover*, then owned by Time Inc. The multipart "magazine" format included segments on such diverse topics as art forgery, robotic arms for quadriplegics, laser surgery, and regrowth of vegetation on Mount St. Helens. The series was marketed directly to the magazine's subscriber base and (with a tie-in mailing of sixty thousand study guides) to educators and students, had full sponsorship from video-game manufacturer Atari, and was hosted by Peter Graves, star of *Mission: Impossible*. During the first season, *Discover: The World of Science* aired in the top sixteen media markets in prime time, received enthusiastic critical reviews, and attracted twice as many viewers as *Universe* and four times more than *NOVA*.

Despite good ratings and reviews, the series failed to attract sufficient advertisers. As Angier observed at the time, "[B]y its nature, corporate sponsorship is a fickle business," leaving television producers "at the mercy of shifting fortunes" in the industry: "The marketing and public relations executives who decide our fate are not considering whether or not science programming is in the public interest; instead, they are calculating, quite properly from their point of view, whether a show fits into their publicity or sales campaigns of the moment. This calculation applies just as much to the corporate underwriter of public television as to the corporate sponsor of commercial television."[26] By 1985, *Discover: The World of Science* had moved to PBS. New episodes, supported by underwriters rather than commercial advertisers, continued to be filmed through 1989 and—reflecting their production quality and audience appeal—programs ran in commercial syndication for many more years.

With the help of corporate underwriting, some science programs appeared on prime-time commercial television during the early 1980s, most with cultural or dramatic themes. *I, Leonardo*, a one-hour special produced by CBS and wholly funded by IBM, was narrated by Richard Burton and starred Frank Langella as Leonardo da Vinci, scientist, artist, engineer, architect, and military planner.[27] The program breaks carried no advertisements for IBM products. Instead, Walter Cronkite described the recent "Olympics of the Mind" competition (two thousand grade-school teams had built Leonardo-inspired spring-driven cars using a $35 budget and local community resources). IBM also worked with AAAS to develop teacher guides and student handouts, distribute filmstrips and audiotapes, and encourage teachers to videotape the program for their classes.

The emergence, decline, and resurgence of newspaper science sections provided a parallel example of the strong connection between successful popularization and economics. From 1982 to 1985, at least seventeen U.S. newspapers (with a combined circulation of about eight million readers) had established "science pages."[28] Newspapers tended to cite market research data as the rationale for creation of such sections, and to deny that they were thereby "ghettoizing" science.[29] From 1984 to 1986, the number of health sections doubled, and by late 1986, the number of weekly science sections had tripled (sixty-six U.S. newspapers had a regular science section, with a combined circulation over eleven million, and eighty papers had started a weekly science page).[30] In 1990, over ninety-five daily newspapers carried weekly science sections; the number of sections devoted "entirely or mostly" to health and medicine more than doubled; and inauguration of science sections may have boosted attention to science elsewhere in the newspapers as well as on television.[31]

CHOICES, QUESTS, VOYAGES, AND TRUTH

What appeared in print or on the screen could never please every scientist, especially if a researcher's ideal narrative contained no hint of sorrow (or joy) in her field, only an unembellished summary of her research conclusions. At the local level, programs like Jack Horkheimer's *Star Gazer* survived by offering accessible nature or astronomy lessons, but most viewers wanted more complexity and more sophisticated entertainment.[32] Filmmakers increasingly delved deeper into the life of science, analyzing ethical challenges and shedding light on the stresses and strains rippling throughout the research system. Television programs looked beneath and behind the technical, emphasizing mystery and adventure without denying that science posed new challenges (and potential risks) for human society and the global environment.

Hard Choices, hosted by bioethicist Willard Gaylin in 1981, typified the nascent skepticism. Underwritten by the National Endowment for the Humanities, the KCTS-TV (Seattle) project explored moral and social challenges raised by developments in biology and medicine, such as genetic screening and rationing of experimental treatments. The program's central theme became a public television mantra: now that science has created this knowledge, what should society do with it?

Quest for the Killers, a 1985 miniseries produced by June Goodfield, examined topics in epidemiology and public health.[33] Trained initially as a zoologist, Goodfield had refocused her work toward the history and philosophy of science and its popularization. Even after the disappointing attempt to create a series with Smithsonian help, she remained convinced that television could facilitate communication about research in ways not possible in print. Science, she emphasized, "doesn't exist in a vacuum—

whether or not discoveries are used is largely determined by nonscientific considerations" like money, politics, religion, and tribal jealousies.[34] Television could exhibit scientists in "the venal, flawed world that science hopes to serve."

Funding for *Quest for the Killers* came from four international pharmaceutical firms (CIBA-Geigy Ltd., Merck & Company, Pfizer Inc., and Squibb Corporation). Each hour addressed a different stage in an epidemiological investigation. "The Kuru Mystery" described how pediatrician Carleton Gajdusek identified the cause of a horrifying central nervous system disease attacking native communities in Papua New Guinea. "Vaccine on Trial" followed human trials conducted in New York City during the mid-1970s, which led to a vaccine for hepatitis B. "The Three Valleys of St. Lucia" showed how public health researchers used different approaches to control naturally occurring diseases like schistosomiasis. "The Last Outcasts" described the challenge of treating illnesses like leprosy that are associated with centuries of cultural taboos and superstitions. The final program, "The Last Wild Virus," emphasized the intricate relationship between science and politics in the international attempt to eradicate smallpox through surveillance and containment.

The National Academy of Sciences also embarked on two multi-million-dollar television ventures during the 1980s, both funded by the Annenberg CPB Project, IBM, Digital Equipment Corporation, and other corporate and private foundations. The seven-part *Planet Earth* combined spectacular film footage and elaborate computer reconstructions to discuss geology and earth science topics like plate tectonics, earthquakes, volcanoes, and solar wind.[35] *The Infinite Voyage*, which attempted to visualize the excitement of basic research, premiered on public television and then aired on selected commercial stations in twelve markets. In the first episode, "Unseen Worlds," physicist Leon Lederman guided viewers through a special effects rendering of the Fermilab particle accelerator, demonstrating breakthroughs in scientific imaging techniques.

Extensive underwriting did not guarantee that projects would proceed smoothly. *The Brain*, produced by a consortium of stations in the United States, France, Japan, and Canada, focused on neuroscience research. Originally scheduled to air in 1983, the project had actually begun in 1978 with about $200,000 in NSF planning grants.[36] A $2 million production grant from Annenberg and CPB in 1981 was insufficient to complete filming; cutbacks in federal funding for PBS delayed the project further, which prompted the BBC to withdraw as a coproduction partner. By the time the eight-part miniseries premiered in 1984, it had cost $5.5 million.

Most of these projects used actors as the narrators or on-camera hosts, in part because few scientists could carry off that role as well as Bronowski or Sagan. For the 1987 miniseries *The Ring of Truth*, MIT physicist Philip Morrison provided both memorable narration and intellectual inspiration. Principally funded by Polaroid Corporation, and with supplemental funding from NSF and private foundations, *The Ring of Truth* offered Morrison's personal interpretation of science, combining a celebration of creativity with emphasis on social responsibility. Produced by Michael Ambrosino, the programs moved from "Looking" (which examined how human vision influences understanding) to "Doubt" (which explored the universe), from the infinitesimally small to the unimaginably large. *The Ring of Truth* represented the heights to which television could soar when aimed at a well-educated, attentive audience. Morrison cautioned, though, that viewers had "a role to play" in the television experience as well as in science: "They must attend to what's going on. Then they can decide whether what we present has the ring of truth."[37]

ENVIRONMENTAL AWARENESS

Critics rushed to explain each project's success or failure. Had a program reflected increased public interest in science, or another mothlike television fad? Michael Arlen characterized the newest shows as "humanistic science," speculating that they were popular because they emphasized the "merely colorful" and "charming" rather than research's tedious side.[38] Jonathan Weiner contended that television remained "uncomfortable with science" because commercial networks still hesitated "to place subject matter with such a high specific gravity on the tenuous and flimsy web of the national attention."[39] The successful commercial science shows of the early 1980s, he wrote, built on audience fascination with "reality programming"—science is, after all, about the ultimate reality—while assuaging network executives' guilt about the plethora of comedies and lightweight programs in prime time.[40] Neil Postman was even less kind, asserting that any authentic information and ideas within these programs had been obscured by their entertainment approach and by a "torrent of short-duration images" aimed at dwindling attention spans.[41]

Ongoing trends within the media both redefined and limited the audiences for such programs. Belying long-standing predictions that viewers would not "take to any pay-as-you-look scheme," the cable industry had begun to flourish.[42] When Lynn Poole stepped in front of the television camera in 1948, Baltimore viewers could choose from only a few channels; three decades later, there were dozens, including cable channels focused exclusively on science, technology, and medicine. In 1981, a heretofore free educational cable network established by NASA and the U.S. Department of Health, Education, and Welfare became privatized as the Learning Channel and began offering educational and instructional programs. In the mid-1980s, Nickelodeon, an educational and informational channel created in 1979 by MTV Networks Inc., expanded into the evening hours. Creation of the Cable Health Network in 1982, to provide independent content on health and medical science, offered additional specialized content.

Starting in the late 1970s, the National Geographic Society had placed its specials on public television. Gulf Oil Corporation had been the sole underwriter, on both commercial and public television, spending over $30 million on production and promotion from 1975 through 1983. When the society announced in 1985 that it was moving *National Geographic Explorer* to Nickelodeon, that decision signaled a change in the organization's communication strategy and eventually resulted in creation of its own separate cable channel. As National Geographic president Gilbert M. Grosvenor explained: "After 20 years on broadcast TV, eight of the top ten highest-rated programs on PBS were National Geographic Specials. We're ready now to break new records on cable."[43]

The commercial Discovery Channel was also launched in 1985, providing a suite of specialized spin-off channels and establishing a new venue for science, health, and nature documentaries. One of Discovery's first science and environment programs was *The Nature of Things*, originally created for the Canadian Broadcasting System in 1960. When geneticist David Suzuki became the host in 1979, he expanded the series beyond celebration of natural history, increased the attention to environmental topics, and endorsed a more political approach. Even though many of his university colleagues dismissed television as "a vulgar activity," Suzuki believed the medium could improve the public's understanding of science and technology.[44] By the mid-1980s, he was producing additional programs, such as an eight-part miniseries *A Planet for the Taking*. To Suzuki, however, the challenge lay in the *context* as much as the *content*. Where was the *real* science on television? How, he asked, could viewers be helped to distinguish the illusion from the authentic, the "background 'noise' from meaningful 'signal'?"[45]

Another international television personality who became familiar to American audiences during the 1980s was British naturalist David Attenborough, whose self-assured reticence and urbane voice attracted both praise and satire. James Burke once commented that Attenborough "talks to you almost with a certain shyness," exhibiting a diffidence that never assumes viewers will be interested but seems intent on convincing them to take heed.[46] Social scientist Donna Haraway characterized Attenborough as a showman, for he "whispers the audience into the film, constantly turning his head from viewer to animal scene and back, actively drawing back the curtain like a theater master [of ceremonies]."[47] The style established an intimate, personal relationship between viewers and the nature narrator; it anthropomorphized animals by removing them from reality into a fantasy space that existed only on television. Spectacular cinematographic techniques and, eventually, computer-generated image manipulation shifted the naturalist's narratives from the lecture-hall podium to a Hollywood simulacrum.[48]

Although not a scientist, Attenborough had devoted most of his career to assisting public communication. After reading zoology at Cambridge University, he had begun a career in publishing; then he shifted into television, as a BBC producer and host of the popular *Zoo Quest*.[49] While American commercial television was experimenting in the 1950s with shows like *Meet Me at the Zoo* and Disney's nature films, the BBC had established a separate natural history unit and was developing education-oriented programming like *Looking at Animals* (1951).[50] Attenborough's long-lasting *Zoo Quest* series premiered in 1954, initially as a studio production and then, with portable recording equipment, filmed on-site around the world, pioneering the model of the intrepid adventurer in search of nature's exotica.[51]

Attenborough's various natural history miniseries—such as *The Living Planet*, broadcast in the United States in 1985, and *The First Eden*, shown two years later—reflected his conviction that television could stimulate public concern for the environment, expand knowledge of nature beyond what is shown in zoos, and make viewers "more sophisticated and sensitive" to ecological challenges.[52] One of his most successful projects was *Life on Earth*, an eight-part series about planetary evolution, produced by the BBC in association with Warner Brothers Studios, funded by Mobil Oil Corporation, and first shown on PBS in 1982. Other successful projects followed, from *David Attenborough's Natural World* and *The Life of Birds* to *State of the Planet*, the latter examining biodiversity, habitat destruction, overharvesting, and the effects of unrestrained pollution.

Nature series now routinely extended their reach via secondary and tertiary marketing, and for the entrepreneurial popularizers, cable broadcast rights, video sales, and rentals brought profits that could be plowed back into new projects. The video version of *Trials of Life* earned over $21 million in five years for its distributor and producer.[53] Reruns of nature programs soon peppered the schedules of cable networks as well as independent broadcast channels.

One of the most long-lasting and well-regarded PBS natural history series, *Nature*, premiered in 1982 thanks to the efforts of conservation biologist Thomas E. Lovejoy. Dismayed at the lack of a regular, science-based nature series on television, Lovejoy approached a New York public station about developing such a show.[54] Without substantial funding in its early years, the series relied primarily on films purchased from the BBC but consistently eschewed a sentimental approach in favor of scientific honesty. *Nature*, Lovejoy later recalled, sought from the beginning to offer viewers "a weekly glimpse of something interesting and beautiful in the world" along with "a reminder that not all was well."[55] Like Suzuki and other politically astute popularizers, Lovejoy knew that, to mobilize public opinion around conservation issues, Americans needed to be inspired rather than frightened or depressed: "If we had gone with gloomy, we probably wouldn't have survived more than one season."

Environmental advocacy groups also incorporated television in their campaigns to influence public attitudes and political action. The National Audubon Society, for example, had been publishing the popular magazine *Audubon* for over forty years when the society's president persuaded Ted Turner, owner of CNN, to donate $250,000 to initiate production of a television program.[56] Turner, already concerned about the interconnectedness of global problems, had founded the Better World Society to use television for public education on population, nuclear proliferation, environment, and poverty issues.[57] *The World of Audubon* specials premiered on the Turner Broadcasting System in 1984 and did not shy away from controversial topics. In 1989, "Rage over Trees," narrated by actor Paul Newman, described how Pacific Northwest logging companies were cutting down old growth forests. In 1991, "The New Range Wars," narrated by actor Peter Coyote, looked at the environmental impacts of western ranching practices. In response to this last program, a group of ranchers organized a boycott of the series, and when the General Electric Company withdrew as underwriter, Ted Turner stepped in to cover production costs.[58]

In subsequent years, hundreds of more modest projects followed similar approaches, combining spectacular film, moderate amounts of technical information, and celebrity hosts or narrators. *Living Wild*, a 1984 PBS series produced by John Mansfield, promised "a rare selection of some of the most spectacular wildlife films ever made" and "a tribute both to the earth's vanishing wildlife and to the wildlife photographer's art."[59] Episodes emphasized the fragility of ecosystems, the damage wrought by human development, and the role that science played in documenting change; filmmakers tracked Adélie penguins in Antarctica, sharks off Baja California, insects in the Chihuahua desert, and golden jackals on the Serengeti Plain of Tanzania. *Living Wild*'s narrators included actors Jeremy Irons, Maureen Stapleton, and Orson Welles, marine biologist Alastair Birtles, and David Attenborough (who described the relationship of a New Guinea tribe to the dugong sea mammal). Publicity stressed both the attractiveness of each beleaguered species and the magnitude of any potential loss (that is to say, the rate at which a species is vanishing, or the acreage of tropical rain forests destroyed by logging). Natural history specials like these reinforced one consistent message during the 1980s: the connection between scientific research and environmental conservation.

HOSTING SCIENCE

Despite positive responses from both viewers and critics, public television did not provide a secure haven for science's documentary makers. The system was under fire both from political conservatives who detected liberal bias in programming choices and claimed that PBS cared little about middle-class Americans, and from liberal groups chagrined at public television's increased acceptance of corporate contributions.[60] This debate became further entangled in partisan wrangling over federally subsidized arts programs, and in reaction, both PBS and its governing board, CPB, began to exercise tighter control over program content. When Congress attempted to reduce federal appropriations for CPB in the 1990, public broadcasters responded by gathering even more private money and allowing underwriters to show corporate logos and offer "descriptions" of their products and services (which are not technically considered advertisements because airtime is not purchased). PBS also began to accept underwriting for programs on topics related to a corporation's business—for example, computer firms subsidized series about information technology, and pharmaceutical firms funded shows about medicine.[61]

By the 1980s, most science documentaries included similar features:

somber message, precise explanations of relevant scientific principles or research, careful vetting of scripts, interviews with many different technical experts, and dramatic recreations of historical events (for example, Darwin's voyage on the *Beagle*). Reconstructive techniques—whether applied to science's own history or to speculation about Paleolithic creatures—became more common, with narration blending seamlessly into fictional re-creation. These newer forms of "picturizing" science and nature, several analysts point out, established false authenticity.[62] What viewers saw was simply not real.

It was, nevertheless, entertaining. Documentaries offered visual feasts: tropical jungles populated by rainbows of birds, colorful fish streaming through the ocean, intimate moments of maternal animal care, and intrepid scientists tramping across rugged terrain. Unlike the pioneers of the early 1950s, directors and technical personnel knew how to televise crisp images from microscopes and telescopes, and could use computers to manipulate those images (edit, alter, stop action, or speed up what was shown). The contribution of the second wave of popularizers was to exploit those techniques in order to enliven social and political discussions, in order to visualize moral dilemmas and ethical choices.

The scientists who hosted the programs of the late 1970s and 1980s were also no longer "talking heads." They did not stand stiffly behind a podium and point to a blackboard; they wandered around while they spoke to the camera, picking up objects and gesturing. They appeared relaxed, human, and occasionally amusing; they spoke authoritatively yet without academic jargon. The other experts interviewed for the programs, however, often seemed drained of spontaneity, in part because each one- or two-minute sound bite was edited from hours of formal interviews. *These* experts and their expertise now mattered little to a program's success or failure. Instead, it became common wisdom in documentary television that only the presence of a "star" would attract viewers to science. The hosts were introduced to the audiences not as accomplished experts but as attractive "personalities, entertainers, and storytellers" who rode bicycles or hiked up volcanoes.[63] Neil Postman, who had earlier dismissed the lecture format ("professor on screen . . . behind a desk") as boring, decried this latest trend as turning science into Las Vegas. Science on the television screen, he complained, was not simply "obstructed by entertainment" but fast becoming "indistinguishable from it."[64]

CHAPTER TWELVE

Warning: Children in the Audience

We'll finish him off with the emergency de-magnetizer! We'll show him
we also have drawing power. Jumping electrodes!

Dialogue from *Frankenstein Jr.*, 1976

During television's early years, American networks underwrote children's
programs on the assumption that the content might encourage families
to purchase television sets and thereby expand the viewer base. In 1949,
42 percent of children's programs were sustaining.[1] Television was even
initially marketed as a *social necessity* for children. In 1950, print advertisements from the American Television Dealers and Manufacturers Association declared that, without access to a set at home, children might feel "left
out" if they had not seen the shows watched by schoolmates.[2] Although
some parents reacted negatively to the suggestion that "without TV, no
child could hold his own," others welcomed television as a promising window onto experiences and cultures beyond their neighborhoods, as a beneficial technology that responsible, aspirational parents should provide in
the home. In a contemporaneous analysis of reaction to the manufacturers'
advertising campaign, Margaret Midas argued that those well-educated,
affluent parents who rejected television were actually conforming to peer
pressure within their social circles, where the medium was dismissed as a
low-culture opiate for the masses and unreflective of elite values.[3]

Such debates represented skirmishes in culture wars that, while influencing academic study and critique of children's programming for decades, never impeded the march of commercialization and offered a familiar footnote to the history of science on American television. In 1950,
broadcaster and playwright Robert Lewis Shayon warned parents and
policymakers that they ignored television's power at their peril.[4] To char-
acterize this technology as just another communications tool, he wrote,

was to miss the "pied piper" effect on children, enticing them away from more productive activities. He and others attempted to persuade parents to restrict viewing and to pressure networks into increasing educational content, but such pleas were drowned out by the giggles and laughter. It was too easy to turn on the set and let the kids be entertained.

The need to hold the attention of impatient little viewers encouraged creativity—fantasy and magic tricks as well as comedy. One of the first science series for children, *Science Circus*, ran on ABC during summer 1949. Every Monday evening, in front of a live studio audience in Chicago, Bob Brown performed stunts in the character of an absent-minded science professor. On another evening program, *Mr. I. Magination*, comedian Paul Tripp recreated historical events and imaginative situations and took viewers, via innovative camera techniques and rear projections, to fantasy locations like Ambitionville. When CBS canceled *Mr. I. Magination* in 1951 in favor of more westerns and science fiction dramas, critic Jack Gould mourned the loss of an original, "meticulous" production and condemned the expanding "cultural totalitarianism" that served advertisers first and carried "worship of ratings" to the extreme.[5]

With shows like *Animal Clinic* and *Acrobat Ranch*, ABC became the first network to create Saturday-morning programming specifically for children. By 1951, all four networks together were showing twenty-seven hours of children's programs each week, most of it sustaining and, at first, most scheduled on weekdays between 6:00 and 8:00 p.m.[6]

NBC's *Kaleidoscope* typified what the industry assumed would keep families entertained. *New York Times* sports columnist John Kieran had achieved fame as a wisecracking "resident genius" on radio's *Information Please*. *Kaleidoscope*, a fifteen-minute show that ran from 1948 until 1952, consisted primarily of Kieran's narration of instructional films on science, nature, and technology topics and had the distinction of being one of the first television series of any type to be placed in syndication.[7] Most early experiments in educational programming tended to be shorter-lived and far more serious in tone. NBC's *Watch the World* consisted of films (including about science) introduced by host John Cameron Swayze, and it lasted only a few months in 1950. *It's Fun to Know*, a CBS weekday afternoon program in 1951, featured a teacher speaking to a group of children sitting in the studio. Jack Gould disparaged that program's lack of imagination, excessive formality, and unsophisticated production values. The instructors, he wrote, "all suffer from the jitters" and lack "stage presence."[8] Gould suggested that if the teachers aimed to have more "fun" on the show, then the audience might enjoy it too.

Not surprisingly, ratings helped justify the shift toward entertainment and away from anything that seemed too instructional. Children were offered animal shows—*Sunday at the Bronx Zoo, Animal Clinic, Zoo Parade*, and *Meet Me at the Zoo*—or space heroes—*Captain Video, Tom Corbett*, and *Commander Cody*. In 1953, they could even watch adventures that referenced Cold War reality: on NBC's *Operation Neptune*, the captain and crew of an American submarine fought "deep-sea evildoers."[9] And beginning in 1952, on a locally produced Los Angeles series that was nationally syndicated in 1955, actor Roy Steffens combined educational, military, and space themes in *Captain Z-Ro*, whose inventions fused the attributes of rocket ships and time machines.

WIZARDRY

No fictional hero, though, could ever rival the popularity and longevity of the Mr. Wizard character—a friendly, neighborly scientist who demonstrated simple experiments on television for almost half a century.[10] Midwesterner Don Herbert was trying out for dramatic roles in New York City when World War II began.[11] After fifty-six missions as an army air force pilot, he returned to radio work in 1945, performing on *Captain Midnight* and *Tom Mix*, selling scripts to other shows, and, in 1949, coproducing a health series in Chicago. He later explained that, while devising a new children's science show, he learned that the Cereal Institute, a trade organization, wanted to sponsor an educational program.[12] It was the advertiser, he said, who suggested calling the character a "wizard"—Herbert added the "Mr." to soften any "negative connotations."[13] NBC bought the idea, and *Watch Mr. Wizard* premiered March 3, 1951, on a fourteen-station network, live from a Chicago studio. Within less than a year, *Watch Mr. Wizard* was carried by forty-three stations nationwide (with audiences around 800,000) and by the mid-1950s on over one hundred stations (with audiences in the millions).

Watch Mr. Wizard differed from *Johns Hopkins Science Review* and *The Nature of Things* because Herbert designed his show for children ages eight to thirteen, rather than for their parents. If adults tuned in, they saw a nonthreatening, easygoing, intelligent man with a smiling face, carefully guiding youthful assistants through simple experiments. In June 1951, the *Chicago Tribune* wrote that Herbert "looks more like a bomber pilot . . . than a science teacher" and "doesn't resemble the conventional television conception of a man of science, with beard, and white jacket, surrounded by laboratory gear, test tubes, retorts, and Bunsen burners."[14]

Each week, in shirtsleeves and tie (and occasionally wearing a lab coat), Herbert would use ordinary household items—milk bottles, coffee cans, knitting needles—to explain scientific principles like gravity, magnetism, and oxidation. Every experiment was carefully scripted, and, although Herbert's first assistant was a young neighbor, the series eventually employed child actors because they could memorize their lines and exclaim "Gee, Mr. Wizard!" with the required spontaneous sincerity.[15]

NBC scheduled Herbert's live show for the early evening and distributed it via kinescope to dozens of other stations. In 1955, the series was rescheduled for Saturday afternoons and weekend mornings, and then production was moved to New York. By the 1960s, shows were being taped in advance so teachers could coordinate episodes with classroom lessons. Herbert also created a brilliant promotion and marketing scheme, whereby viewers were encouraged to join local Mr. Wizard Science Clubs, each having at least five members and named after a common element (for example, the Minneapolis Aluminum Club). Expanding from 3,200 clubs in 1953 to over 5,000 in 1956, the organization helped interest tens of thousands of children in science.[16]

Even though he had only minimal scientific training in college, Herbert managed to convey a sense of authority and expertise, bragging to reporters about his extensive research files and his ability to work with local university professors. During the 1960s, for example, New York University physicist Morris H. Shamos was employed as the show's science adviser to "ensure the authenticity of the demonstrations."[17]

Herbert had starred in over five hundred broadcasts when NBC rather abruptly canceled the series in 1965.[18] Herbert's next venture involved a miniseries for adults. Each of the eight half-hour episodes of *Experiment* (1966) centered on an individual scientist (such as a geologist or primatologist) and examined the "pitfalls" as well as accomplishments of a "man with a struggle."[19] The middle-aged Herbert played a far more "professorial" host in jacket and tie, with elaborate demonstrations not meant to be tried in the kitchen. An advisory council headed by Warren Weaver helped Herbert select the subjects. Grants from NSF and philanthropic organizations subsidized broadcast via National Educational Television. In 1966, Herbert and the advisory committee even began planning another series (apparently never filmed but its theme representing the spirit of the times) to "portray the cultural and humanistic aspects of science."[20]

In 1971, under assault from parents' groups and other civic organizations for flooding Saturday-morning schedules with cartoon shows and commercial advertisements, the networks attempted to regain the high ground. ABC declared that it would "upgrade the informational and educational elements" of existing shows, while NBC added new programs like an adventure show called *Barrier Reef* and a revival of *Watch Mr. Wizard*.[21] In an assessment of the newest version, critic Cleveland Amory marveled at Herbert's consistency: he was still "doing the same kind of show" and would probably "be doing it in 2051."[22] The revival was brief. Within one season, the networks drifted away from education and once again banished the wizard.

Throughout his career, Herbert maintained close relationships with commercial sponsors, and he used those connections to support his projects and sustain some type of presence on the air for almost forty years, even when his network series were canceled. From 1954 to 1962, Herbert (in the character of Mr. Wizard and sometimes wearing a white lab coat) delivered "Progress Reports" during the commercial breaks of *General Electric Theater*, explaining such things as "how electric motors work" or describing General Electric's jet engine manufacturing operation.[23] In 1975, General Electric employed him to lobby on Capitol Hill for the controversial program to develop a liquid-metal fast breeder reactor, with advocacy commercials explaining the technology, and Herbert continued appearing in advertisements for technology and utility firms through the 1970s.[24]

By the late 1970s, he was collaborating with NSF and General Motors Research Laboratories to create "How About . . . ," a series of eighty-second syndicated science and technology news "inserts." Initial grants of $500,000 from NSF and $312,000 from General Motors allowed production and free distribution of fifty-two segments that eventually appeared on over one hundred local stations and won several awards for public service.[25] Herbert made frequent appearances on late-night television, sparring good-naturedly with Johnny Carson, and then eventually regained the commercial platform that he preferred. *Mr. Wizard's World* ran from 1983 to 1990 on the Nickelodeon cable channel, and in reruns for a decade afterward.[26] Other than moving from black and white to color, and adopting the popular "magazine" format, that show retained the original formula. Herbert remained a "kindly uncle" demonstrating science to youthful admirers. He simply adapted his old egg-into-a-bottle trick into a water-filled-balloon-and-soda-bottle one. As the Wizard explained, "Most kids today have never seen a milk bottle."[27]

Prepared by Prime Time School Television for NICKELODEON, The First Channel For Kids.
NICKELODEON is available only on cable television.
Call your cable operator or check your TV listings for the NICKELODEON channel in your area.

FIGURE 16. Cover of educational booklet distributed for Don Herbert's *Mr. Wizard's World*, 1983. The guide included quizzes, program summaries, and suggested classroom activities. Courtesy of Smithsonian Institution Archives.

DEFENSIVE PROGRAMMING

Serious educational programs for children became rare on commercial television during the 1960s and 1970s. Scientific associations made little or no effort to underwrite or encourage production of such shows, and government funding agencies had insufficient resources to subsidize expensive national series. Commercial broadcasters tended to initiate educational projects primarily as defenses against social or political criticism and then to cancel them, as with *Watch Mr. Wizard*, when the campaigns weakened.

During the 1960s, social science research had begun to tie human behavior to the violence observed on television, and many political analysts blamed television drama (rather than social fragmentation, cultural permissiveness, or parental neglect) for rising levels of urban violence and adolescent crime. Liberal groups bemoaned the sex and violence in programs shown during early evening; political conservatives accused the major networks of "ideological plugola" in promoting the liberal bias, "provincialism," and "parochialism" of the northeastern United States.[28] Assaulted from all sides, the television industry responded by developing educational programs to be shown on Saturday mornings or weekday afternoons along with the regular cartoons.[29] ABC's *Discovery* blended science and cultural experiences, occasionally visiting observatories and other science facilities. That network's *Science All-Stars*, which lasted only one season, featured children demonstrating their science projects to eminent scientists.

Each new show represented a costly response to critics, especially if the program failed to attract audiences and advertisers. Network data showed that viewers, young and old alike, preferred entertainment. Cartoon shows like *The Jetsons*, *Astro Boy*, or *The Hector Heathcote Show*, which had fictional characters who were scientists, had higher ratings than instructional ones. Without government or public pressure to increase science content for children as a matter of responsible public service, the commercial television industry had little reason to change.

Major restructuring of public broadcasting during the 1960s eventually created an alternative venue for children's programming. PBS was not, strictly speaking, a network like CBS with independent production capability. PBS depended on its nonprofit station partners to support projects, marketing, and viewer relationships. Nevertheless, the establishment of a parallel public broadcasting system opened opportunities for innovative educational programs. Soon significant numbers of children were being weaned away from network cartoons—usually with alternative sets of cuddly fictional characters and sophisticated animation.

As social and political criticism of commercial television continued, the networks responded by creating educational "spots."[30] One CBS project in the 1960s, *In the News*, had consisted of short films interspersed like commercials within children's series. So ABC created a variation on this format, *Schoolhouse Rock*, which ran on Saturday mornings from 1973 until 1985 (and was revived during the 1990s). Featuring a rock-music sound track, the three-minute animated lessons in math, language, civics, and science were developed by an advertising executive who had noticed that his own child struggled to learn the multiplication tables yet had no difficulty in remembering song lyrics.[31] Encouraged by a young ABC vice president (Michael Eisner, who later ran the Disney corporate empire), the first series, *Multiplication Rock*, eventually spawned three others under the Schoolhouse umbrella: *Grammar Rock*, *America Rock*, and *Science Rock*. Their premise was that "knowledge is power," especially when "knowledge" could be set to music ("every triangle has three sides—no more, no less / You don't have to guess / When it's three, you can see"). *Science Rock* celebrated the physical sciences, health, and the environment; its characters sang about "Newton and the Apple" and "Interplanet Janet" and explained that gravity "keeps us in our place" ("I'm a victim of gravity / Everything keeps falling down on me"). The popularity of *Schoolhouse Rock* paved the way for similar combinations of educational content, pop music, satire, and animation.

In addition to attempting to deflect federal pressure, these projects responded to mounting social science evidence documenting television's negative impacts on children.[32] As the FCC emphasized, improving the programming aimed at children would be in the public interest because children's "ideas and concepts are largely not yet crystallized and are therefore open to suggestion" and children "do not yet have the experience and judgment . . . to distinguish the real from the fanciful."[33] A number of private advocacy groups, impatient with the pace of federal action and the broadcasters' unenthusiastic responses, intensified the political pressure. Action for Children's Television, for example, emerged as a national force for change during the 1970s, petitioning Congress and the White House, persuading the FCC to require more serious programming, calling for networks to refuse commercial advertising for programs aimed at children, and condemning the practice of linking performers to specific products.[34] To counter such criticism, the National Association of Broadcasters

acknowledged that television had "special responsibility" in the socializa-
tion of children ("Programs designed primarily for children should take
into account the range of interests and needs of children, from instructional and cultural material to a wide variety of entertainment material"),
but association members did little to initiate real change.[35]

A TROUBLING STATUS QUO

One major difficulty for reformers, broadcasters, regulators, and educators
was that it made no sense to pour millions of dollars into an educational
project if few children would watch the resulting programs. As Sonia Livingstone explains, both the family and television coevolved during the
medium's first half century, and one important change over that time was
the shift "toward privatized viewing."[36] The rising proportion of multiset
households allowed children to consume television as individuals, with
less parental monitoring and with television no longer automatically a
"shared experience" within a household.[37] Because even young children
were watching substantial amounts of commercial television and the vast
majority of science-related content was in documentaries, dramas, and
comedies created for adults, one had to look beyond the usual Saturday-
morning schedule to understand the science *available* on television to children, not just the science in programs *directed* at children.[38]

Commercial and public television in the Boston area during the 1970s
typified what was happening throughout the country.[39] The region's academic, economic, scientific, and cultural resources might have been expected to nurture high-quality science programming, especially since
public station WGBH had created *NOVA*. Nevertheless, nonfictional science
content (discussions of results, discoveries, and applications; accounts of
the history and social institutions of science; and biographies of or interviews with scientists) constituted less than 1 percent of Boston-area television in 1977.[40] Aside from *NOVA* and *Infinity Factory*, most of that science
content was in segments of children's programs like *Captain Kangaroo*,
Kidsworld, *Mister Rogers' Neighborhood*, *The Wonderful World of Disney*, and
Zoom!, and in natural history series with adult themes like *Wild Kingdom*.
The majority of attention to science or scientists occurred within fictional
programs (cartoons, comedies, and dramas).

On Boston television, during Saturday mornings in 1977, over one-third
of all children's programs included some science-related content, primarily as fictional characters, dialogue within cartoons or live-action dramas,

and vague references to technology or scientific knowledge. The Sunday-evening schedule contained *The Six Million Dollar Man*, three nature programs (all broadcast on an independent station with a relatively weak signal), and *Star Trek* reruns. On weekday afternoons (from 4:30 to 7:30 p.m., the hours in which Nielsen surveys indicated that the child viewing audience reached its daily peak), science was conspicuously absent from Boston television, with the exception of the PBS nature show *Hodge Podge Lodge* and reruns of science fiction series like *Lost in Space*.[41] On occasion, *Captain Kangaroo* might devote an entire show to explaining a science topic, or *The Electric Company* would include a scientist as a character in a skit, but in 1977 even Mr. Wizard had temporarily left the screen.

The science that *was* incorporated within educational series exemplified the medium's bias toward simplification, terseness, and condensation, with critical steps eliminated in experiments and basic principles ignored in explanations. Commercial nature series—*Animal World*, *The World of Survival*, *Lorne Greene's Last of the Wild*, and *Animals, Animals, Animals*—emphasized facts about natural history and tended to ignore the research that had created that knowledge. In fictional programs, "detect-along" features designed to discourage channel switching allowed children to follow clues and "solve" a mystery, but had little educational value and often would link science with magic. On *Scooby-Doo*, *The Krofft Supershow*, *Space Ghost*, and *Super Friends*, arsenals of fantastic "immobilizers" were deployed to combat the "forces of evil throughout the solar system."[42] Fictional weaponry could facilitate a character's escape, yet the same character might be felled in the next episode by magic or homemade tricks: a car outfitted with dashboard computers might be stopped by simply scattering thumbtacks in its path. In an episode of the cartoon *Frankenstein Jr.*, the "boy-scientist" Buzz Conroy instructed his robot, Frankenstein Jr., to combat a bald evil (adult) scientist (Dr. Shock) and his monster Igor.[43] Shock claimed that his massive "electrical monster" would drain all the energy out of the world: "The more electricity I feed him the bigger he gets." So Buzz Conroy declared, "We'll finish him off with the emergency de-magnetizer!"

Heroes and villains alike exploited scientific knowledge as a weapon, yet even the heroes rarely succeeded without violence. In 1977, *Boston Globe* critic William Henry deplored the high concentration of violent incidents on television and noted that, on weekday afternoons, stations would rerun "shows that used to be on in primetime, often dating from an era more permissive about violence."[44] Nature specials, too, exploited violence to entertain viewers. In *Death Trap*, Vincent Price's horror-movie narration gleefully described carnivorous plants as "the many little death traps that may surround us . . . 'things which see without eyes and devour without mouths.'"

On Boston-area public stations, most science specials and miniseries were aimed at older viewers. With underwriting from pharmaceutical companies and private foundations, PBS rebroadcast such BBC films as *The Weather Machine, The Violent Universe, The Restless Earth, The Key to the Universe,* and Peter Goodchild's *Microbes and Men.* Specials produced by the Cousteau Society, National Geographic Society, and Survival Anglia Ltd. did offer family-friendly diversions during prime time, and on rare occasions, networks would produce specials for children, such as a CBS program about endangered species (*Saving Wild Animals—What's It All About?*) or an NBC special "illustrating Einstein's theory of relativity" (*The Day after Tomorrow*). Boston television's best science, however, was intended for adults.

ENTERTAINMENT TRICKS

By the 1980s, social science research had demonstrated convincingly that televised images could influence children's career choices, which raised further concern about the lack of competitive educational programming centered on science.[45] When Children's Television Workshop (developers of *Sesame Street*) asked thousands of children to name their favorite programs, the top five were either situation comedies or dramas. Only five science-related shows made the top twenty, all but one of which (*NOVA,* which ranked twentieth) were oriented around natural history and animals.[46] *Sesame Street*'s sassy, streetwise Muppets had already proven that an irreverent, musical approach could interest children yet satisfy demands for more educational content. This situation helped to justify development of *3-2-1 Contact,* which ran from 1980 to 1992 and used popular music and celebrity guests to "sell" science to eight- to twelve-year-olds.

Initially funded for $10 million by PBS, NSF, the U.S. Office of Education, and United Technologies, *3-2-1 Contact* starred a diverse group of teenage hosts who explained simple scientific concepts and interviewed experts while participating in fun-filled "adventures." To avoid public criticism that might endanger government funding or scare corporate and foundation underwriters, *3-2-1 Contact* engaged in self-censorship. Evolution was not mentioned lest stations in conservative areas of the United States refuse to carry the series.[47] Scripts also tiptoed gingerly around the subject of sex. A segment about the life cycle of corn and pigs was edited,

Jonathan Weiner writes, to remove the "amorous hogs of Iowa" (although the directors did retain film clips of pollen drifting languidly into corn tassels).[48]

Unfortunately, the success of such projects reinforced an attitude of "let PBS do it," especially for science. One television commentator minced no words in excoriating his colleagues within the networks: "It's nice to see that Mr. Wizard's children are alive and well and learning about science, thanks to the cable channel Nickelodeon, and even more to the Public Broadcasting System, which is rapidly turning science into a full-blown video art form," but "when it comes to science for kids," the commercial networks fail—"unless you consider the Galactic Guardians as Nobel candidates in physics."[49]

Like the Johns Hopkins and California Academy of Science programs in the 1950s and the revived Mr. Wizard franchise, the next stage of children's science series returned to the proven formula of revolving a show around a charismatic host. To keep children entertained while they were being educated, the new type of host exploited humor. Science journalist Ira Flatow created *Newton's Apple* in 1982 for Minnesota Public Television, and although Flatow left after six years, the formula remained intact, and the series continued on the air until 1999. Flatow led his viewers on themed adventures and would work alongside "assistants" (much like Mr. Wizard's helpers) and special guests. By its third year, *Newton's Apple* was carried by 276 public stations and had a regular audience of four million. The second host, scientist and museum administrator David Heil, added wildlife specialist Peggy Gibson and environmental educator Peggy Knapp to the cast, along with other visiting scientists, and expanded the commitment to providing female role models for young viewers.[50]

MR. WIZARD'S LEGACY

The success of *3-2-1 Contact* and *Newton's Apple* still did not prompt commercial networks to carry more education-based science programming for children, in part because cable-industry growth (and resulting audience segmentation) had complicated the marketplace. The amount of all broadcast network educational programming for children declined from eleven hours per week in 1980 to under two hours per week in 1990; passage of the Children's Television Act of 1990 did not reverse that trend.[51]

The next notable children's programs involved two natural showmen— and more comedy. Boeing Company mechanical engineer Bill Nye began his broadcasting career with a radio program about science in 1986, followed by television appearances as the "Science Guy" on a Seattle comedy show.[52] In 1993, he developed a *Bill Nye the Science Guy* pilot for public station KCTS-TV (Seattle) and obtained underwriting from NSF and the U.S. Department of Energy. Disney's Buena Vista Television then picked up the production, and the show premiered in commercial syndication. As often happens, timing proved crucial. Commercial networks and stations, once again under pressure from federal regulators, needed educational content, so *Bill Nye* became part of a package of syndicated series that local stations could schedule to fulfill Children's Television Act requirements. In 1994, NSF provided $4.5 million in support for program production and coordinated outreach efforts, and the Boeing Company provided additional funding. As a result, *Bill Nye* became the first program to run concurrently on public and commercial stations. The series was praised by Parents' Choice, *TV Guide*, and the National Education Association, and won several Emmys. Nye has said that while every show had "learning objectives," his goal was always to present science in an entertaining context, combining video special effects, outdoor adventures, and interviews with "way cool" guest scientists. No new programs were produced after 2000, but Nye continued on cable television with another series, and the original series remained in syndication. Nye's primary competitor, Paul Zaloom, played white-coated scientist Dr. Beakman, whose sidekick Lester Ratman wore a rat costume.[53] Based on a popular comic strip, *Beakman's World* premiered in 1992 on the Learning Channel and in commercial syndication. The next year, it was added to the CBS morning schedule (largely in response to regulatory pressure), but continued to run in syndication even after network cancellation in 1997.

At first, such programs seemed to indicate that the networks, prompted by the FCC, might finally welcome more educational content.[54] That hope proved short-lived, however, as demonstrated by the fate of two imaginative animated series: *Cro* (launched by ABC and Children's Television Workshop in 1993) and *The Magic School Bus* (which premiered on PBS in 1994).[55] Both projects required substantial subsidies from NSF and other groups and involved considerable creative talent. On *Cro*, the adventures of a Cro-Magnon boy were used as a platform to explain basic science, while on *The Magic School Bus*, the wild-eyed schoolteacher Ms. Frizzle (voiced by comedian Lily Tomlin) took her class on fantasy field trips. Neither series lasted more than a few seasons in production despite good reviews and strong support from educators and the scientific community. To survive for long on American television, children's science shows needed either a Wizard or an "emergency de-magnetizer."

CHAPTER THIRTEEN

Rarae Aves: Television's Female Scientists

[W]hereas 60 years ago most of what people knew was from their direct experiences, as the [mass] media developed, more and more of what we experienced was at second hand, highly prejudiced, and largely myth.

RICHARD LEACOCK, 1973[1]

Television slapped a human face on science yet, for decades, did little to spotlight or demythologize female scientists. Through how women were represented in dramas and documentaries, through repeated emphasis on male celebrity scientists, and through the choice of which scientists would be interviewed and discussed in the news, television transmitted potent statements about the role and status of women in science. Too often, that content simply perpetuated stereotypes of female scientists as superwomen whose achievements came at the expense of normal lives—images that discouraged young women from choosing scientific careers or mischaracterized those who did.

Historians of science like Margaret Rossiter, Marilyn Ogilvie, Joy D. Harvey, and Sally Gregory Kohlstedt have documented how female scientists were professionally marginalized and culturally stereotyped throughout the first half of the twentieth century. Nothing signaled perpetuation of that situation so clearly as the gendered titles of children's television shows, from *Mr. Wizard* to *Bill Nye*.[2] The default assumption remained stuck on an image of science as a predominantly (or, worse, *appropriately*) male occupation, where accomplished women were exceptions in a universe of male luminaries.[3] During the 1950s and 1960s, when the number of women with advanced degrees in science rose, the mass media continued to frame female scientists as, first, mothers and wives and then (and only then) successful researchers. None of early television's science popularizers and hosts were women. Few women appeared as program guests or were the subjects of documentaries, a neglect that continued into the *NOVA* era. During its first half century, American television reinforced an outdated cultural stereotype of science as an exclusively male domain and of female scientists as "rare birds," as either romantic, adventurous celebrities like Margaret Mead and Jane Goodall or impossibly adroit fictional superheroines who battled aliens or viruses while never smearing their mascara.

Such cultural invisibility matters. Viewers constantly compare representations on television to what they assume (or have been told) to be fact, interpreting cues in the light of what they already know about a topic.[4] All media images "present a first draft of reality . . . which audiences then edit, revise, and reformulate to fit" their own conceptual frameworks.[5] Preconceptions and default assumptions (such as whether most scientists are males) serve like frames around freshly poured concrete, shaping reactions to the content of popular culture.

Whenever television's spotlight turned toward male scientists, women tended to remain in the shadows. To some extent, while influenced by preexisting discrimination, that situation also reflected two extenuating circumstances—mass media fascination with certain older fictional stereotypes and the way (and reason) people achieved fame in the television age.

STEREOTYPES IN FICTION

What *does* a scientist look like? A "rather elderly gentleman, quite a gray beard with glasses, who spends all day looking in a microscope, living in a kind of cloister?"[6] A pleasant, glib Nobel laureate being interviewed on *60 Minutes* or trading jokes with Johnny Carson? Or an attractive, miniskirted female microbiologist pursuing a "runaway virus" in a made-for-television movie? The small number of female scientists shown on television, fictional as well as real, helped reinforce a narrow set of stereotypes. Especially in the early decades, whenever dramas or comedies featured a scientist (which was itself rare), the character was likely to be male.

Many of these fictional constructions were villainous or "mad" and based on characters, such as Victor Frankenstein or Dr. Jekyll, created in nineteenth-century novels. In addition to replaying Hollywood movies (from *The Bride of Frankenstein* to *Frankenstein Meets the Space Monster*), television produced countless variations on Mary Shelley's novel, from a 1957 "Frankenstein" on *Matinee Theater* to star-studded productions on both ABC and NBC in 1973. Robert Louis Stevenson's *Dr. Jekyll and Mr. Hyde* offered another ideal subject for dramatization on a small screen.[7] The

CBS anthology series *Suspense* featured Stevenson's plot in 1949, then re-staged it in 1951 with Basil Rathbone in the starring role. Writer Gore Vidal created a "Jekyll and Hyde" screenplay for CBS in 1955; ABC mounted a production in 1968 starring Jack Palance; and NBC's 1973 musical version starred Kirk Douglas. Even comedies celebrated that fictional evil scientist. In 1962, *The Many Loves of Dobie Gillis* had a "Dr. Jekyll and Mr. Gillis" episode, and years later, on *Gilligan's Island*, the hapless Gilligan dreamt that he had become a murderous Hyde. Screenwriters who adapted H. G. Wells's novel *The Invisible Man* tended to create a more sympathetic scientist, making the character an accident victim rather than a megalomaniac. *The Incredible Hulk* series then combined themes from all three books: a scientist accidentally exposed to radiation was transformed into a monstrous creature with superhuman strength whenever he was angered (that is, made "mad"), thereby referencing Frankenstein, monster, dueling personalities (Jekyll and Hyde), and transformative visibility.

No similar powerful roles, either villainous or heroic, were written for women during those early decades, perhaps because a female "Frankenstein" might have seemed even more implausible than an ordinary female scientist. Instead, television programs would perpetuate Hollywood's "standard characterization" of female scientists as either research assistants "permanently subordinate" to their male supervisors or else gentle, long-suffering saints burdened with "domestic and family responsibility."[8]

In 1954, under one-third of all fictional characters in network television drama, in all occupations, were women.[9] Through the 1980s, although women were more frequently cast as experts in law, criminal justice, and medicine, only a small proportion of science-related characters were played by female actors.[10] Two notable exceptions were the brainy and beautiful Dr. Helena Russell on *Space: 1999* and Andrea Thomas on *Isis*, the latter a superheroine masquerading as timid, unglamorous science teacher. Gender (as well as racial) bias in casting scientist characters persisted long past the time that television's other heroes and villains became a more diverse population.

Such "underrepresentation" is of more than incidental importance. When an occupation like science is underrepresented, George Gerbner has explained, the characters tend to be narrowly drawn, to reinforce prevailing stereotypes about race or gender, and to encourage the assumption that only a few people, or only certain people, can become scientists. In the 1970s, under 1 percent of all major and minor characters on prime-time television worked in scientific, engineering, medical, or other technical professions, and little change occurred during the 1980s and 1990s.[11] Almost three decades after Gerbner's first study of television drama, Anthony Dudo and his colleagues found that scientists still represented only 1 percent of all characters in prime-time network entertainment programming, and while described less often as evil or villainous, most characters were still played by white male actors.[12]

One change did eventually afford more opportunities for representing women. By the 1980s, television's fictional scientists had abandoned their movie predecessors' cobwebbed castles and had been assembled into teams that included more females and nonwhite males. Just like the scientists being described in the news, these characters managed multi-million-dollar budgets and debated moral choices (animal experimentation? human subjects?). Heroes fought political pressure, bureaucratic red tape, and corrupt industrialists; villains brokered corrupt deals with foreign governments, and seemed motivated more by money than by vanity. Fiction and fact merged in unending cycles of cultural reinforcement, and narratives acquired a certain authenticity because of their familiarity.[13] Three characters invented by writer Peter Benchley reflect the evolution of the stereotype. In *Jaws* (a 1976 movie shown frequently on television), the hero, Dr. Matt Hooper, was a marine biologist interested in sharks—young, eccentric, irreverent, unattached, independently wealthy. More than twenty years later, Benchley invented another socially responsible shark researcher for a made-for-television movie, *The Creature* (1998). This time, the hero, Dr. Simon Chase, was middle-aged, jaded, and woefully underfunded, and his attractive estranged wife (Dr. Amanda Mayson) was also a marine biologist.

In television drama, accomplished females began confronting personal obligations and expressing sexuality—yet female *scientists* rarely dominated the action. In her analysis of twentieth-century film portrayals, Eva Flicker observed that although "differentiation between the roles of women and men scientists" had begun to dissolve, with consequent "uniting of an intellectual and erotic person," most female characters remained "dependent on male characters and in this respect stand in the second row, behind their male colleagues."[14] One exception was what Flicker called the "lonely heroine" character (the antithesis of the Marie Curie stereotype), such as the accomplished, authoritative Eleanor Arroway (*Contact*), Jo Harding (*Twister*), and Rae Crane (*Medicine Man*).[15] Such movies paved the way for the creation of smart, assertive female researchers later seen on *CSI*, *NCIS*, and *Bones*. Contemporary dramas may still emphasize "femininity," but the newer female characters now appear less likely to compromise professional standards for the sake of romance.[16] Once all fictional scientists assembled in teams and admitted to fallibility, the female characters, too, could become more nuanced and complex.

"Character on television is form not content," Peter Conrad observes.[17] Costuming characters in the same universally recognizable outfit—a knee-length white lab coat—contextualized scientists in early drama and comedy. In 1950s science fiction films (replayed on television), scientists would be dressed in street clothes until they entered a laboratory, where they would assume "the role and guise of a researcher" simply by donning a white coat.[18] In 1970s children's cartoons, both heroic and villainous scientists and their assistants wore white coats. Once a female scientist put on that coat, it hid her breasts and curves; by wearing the uniform, women could join the ranks. Similarly, the subsequent standard costume for fictional scientists—a white biohazard suit covering the entire body— provided comprehensive disguise, ambiguous in gender as well as intention, obscuring the faces of villains and heroines alike.

CELEBRITY

In the television era, fame followed fashion. The medium adroitly exploited a signature status of the twentieth century, what Gilbert Seldes once called that "highly advertised commodity called 'personality,' which is not character and certainly is not talent."[19] Producers chose scientists to interview or profile not because profession or peers necessarily deemed them "most deserving of examination" but because they cooperated and conformed to television's prevailing stereotypes.[20] In these constructed "celebrity personas," Joshua Gamson writes, the "distinctions between fact and fiction" dissolved and, Richard Schickel adds, when simplified and scrubbed clean for the cameras, turned into "representations."[21] Scientists such as Edward Teller, Linus Pauling, and Carl Sagan willingly accommodated to the contextual pressures (often to advance political causes) and thereby offered new public "parables" of scientific identity.

Achieving fame in the television era differed significantly from how scientists had attracted public attention in the past. As late as the 1940s, a scientist could project a dignified tone of "world enough and time" while writing articles for *Saturday Evening Post* or being interviewed on the radio, could explain his work precisely, with appropriate caveats, and without divulging personal information. A reporter might mention appearance (height or hair color) in general terms, but fame tracked professional accomplishment rather than personal attractiveness. As the television industry perfected the business of creating celebrity for celebrity's sake, new relationships emerged between viewers and *their* celebrities, characterized by a sense of ownership, a demand to know, a presumption of a right to intrude, and a clamor for "human interest" stories.[22] Scientists achieved fame not for what they had discovered but because millions of people had seen them on television. As Rae Goodell described in her groundbreaking study of the first generation of television's "visible scientists," these celebrities could translate complex ideas for general audiences and were at ease in front of cameras.[23] Visibility reinforced celebrity. Each television appearance led to more invitations, greater fame.

Margaret Mead was one of the few female scientists in the United States to achieve fame during the print era and then to transition into television appearances. The anthropologist's work on adolescent sexuality among primitive tribes and her book *Coming of Age in Samoa* attracted considerable press attention during the 1930s, affording an attractive subject for journalists. Goodell concluded that Mead's fame rested on her accessible writings, excellent lecture skills, willingness to cooperate with the press, and the "relevancy" of her research rather than her gender.[24]

For primatologist Jane Goodall, who first came to public attention during the 1960s, gender probably did play a role. Donna Haraway has argued that Goodall's youth and attractiveness provided journalists with an alternative stereotype that conformed to the messages of the women's movement.[25] The first National Geographic television special portrayed Goodall as accessible rather than stuffy, enduring primitive conditions for the sake of research—and thus combining the roles of scientist, adventurer, *and* emancipated female. Goodall herself modestly attributed her popularity to the fact "that I happen to be a woman and I did something offbeat."[26]

The celebritization of scientists also drew strength from the late-twentieth-century practice of having the purpose of a public communication be to market an individual rather than to deliver a specific message, a circumstance favorable to charismatic people with thousands of fans. "It was as though they had been talked to by God," one friend remarked after observing the audience for a Goodall lecture.[27] Similar scenes occurred whenever Carl Sagan participated in public events. If Sagan left while other speakers were still on the platform, the astronomer's fans would often leave with him. To biologist Maxine Singer, such celebration of fame does not serve science well. By focusing on "personal flaws," she writes, the media has turned "the whole world into a soap opera," diminishing genuine accomplishment and transforming scientists into merely another set of characters on the screen.[28]

INVISIBILITY

Flora Rheta Schreiber, in contemplating television's future as "an instrument of public opinion," pointed out in 1950 that the ability to show the faces of those who "make the headlines" would not "remove the shadow of erroneous opinion" built on past experience, education, and bias; instead, it would most likely fortify an "illusion of knowledge."[29] What used to be merely names on paper had become "faces, vivid and realistic, recognizable." And, unfortunately, visualization reinforced the public's "natural disposition toward hero worship."[30] As Alberto Elena has concluded from his analysis of contemporary films, "Despite the growing participation of women in science and the development of the feminist movement," popular misconceptions refused "to yield to more balanced or realistic visions."[31] Whether framed as lab technicians or role models, as heroines or villains, the female scientists represented on television were chosen because they conformed to the *medium*'s needs and parameters.

Television thus made powerful statements about social relevance and importance by conferring "status" on some individuals and groups and ignoring others.[32] When women were not chosen for news interviews, or when dramas about scientists featured few female faces, those omissions conveyed messages about who mattered in the world of science: "Those who are made visible through television become worthy of attention and concern; those whom television ignores remain invisible."[33] The disconnect between accomplishment and celebrity meant that television's attention naturally flowed toward the male scientists who fit the prevalent contemporary cultural stereotypes and who could perhaps risk the criticism from colleagues that sometimes followed public engagement.

Throughout the first half of the twentieth century, women scientists were indeed less visible in popular science. In mass-market magazines, only a small proportion of all science articles, from the 1910s through the 1940s, were written by or about female scientists; few network radio programs in the 1920s through the 1940s explored the work of female scientists, and no woman served as the host of a major science radio series.[34] Even into the 1960s and 1970s, while the proportions of women in science classes, graduate programs, and professorial ranks were steadily rising, the scarcity of women scientists among television "talking heads" could reinforce the impression that women played only minor roles in the life of science.

On the first network television science programs, women made up only a tiny percentage of the guests. Of 459 individuals invited to appear on the four separate television series produced by Johns Hopkins University from 1948 to 1960, including the well-regarded *Johns Hopkins Science Review*, only thirty-two were women, and of these guests only eleven were scientists or women engaged in some science-related activity or profession.[35] While hundreds of males were the sole featured guests, only two women received such attention.[36] On the Johns Hopkins series, a woman was more likely to appear as a model in a bathing suit or fancy dress (reenacting some historical episode) than as an expert, a circumstance that mimicked the rest of television. From Betty Furness opening refrigerator doors to the décolletage of game-show panelists, women were being exploited for their attractiveness and sexuality and rarely celebrated for their intellectual accomplishments, in any field.

Several decades later, *NOVA* was little better. From its first season in 1974 through its twenty-eighth, *NOVA* presented thirty-two new biography programs about the lives and careers of individual scientists, yet only two of those focused on women—in 1976, "The Woman Rebel" described physician and birth-control advocate Margaret Sanger and in 1983, "To Live until You Die" explored the work of Elisabeth Kübler-Ross.[37] Contrast these two biographies to those of males during the same period. There were three separate *NOVA* programs about Albert Einstein and four on Richard Feynman (one of them a joint profile with Richard Lewontin and the other titled "The Best Mind since Einstein"). There were programs about male physicists ("A Is for Atom, B Is for Bomb: A Portrait of Dr. Edward Teller"), male chemists ("Confessions of a Weaponeer," about George Kistiakowsky), male biologists ("Stephen Jay Gould: This View of Life"), and male psychologists ("A World of Difference: B. F. Skinner and the Good Life"), as well as profiles of historical figures like Thomas Edison, Henry Ford, and Sigmund Freud. There was also, it should be noted, only one biography of a nonwhite scientist ("The Long Walk of Fred Young," about a Native American physicist). Perhaps many biographies of women or people of color were proposed (projects can certainly be canceled for reasons unrelated to a topic's significance), but *NOVA* and many other series nevertheless perpetuated an inauthentic image of who mattered in the world of science—and why.

Underrepresentation of female scientists continued long past the time that more women joined the ranks of television decision makers (producers and executives) and on-camera personalities. The average number of scientists interviewed each year within network television news programs increased from the 1970s through the late 1980s, but those scientists began to be chosen "generically" (as one of many available experts on a topic).[38] For the news, the choice of sound bites and interview subjects reflected mainstream values, including conventional assumptions about experts. A 1989 study of newspaper front pages found that while women journalists wrote 27 percent of the stories and women were in 24 percent of the photos, women were quoted as *sources* only 11 percent of the time.[39] On network television news from 1982 to 1984, one in seven on-camera sources were women, and in 1986, 15 percent of television news stories used women as on-camera sources, only a few of those scientists or similar technical experts.[40] On television news during 1989, women represented 9 to 15 percent of the correspondents (depending on the network) and delivered 10 to 22 percent of the stories.[41] A National Organization for Women study in the late 1990s found that men still provided about 87 percent of "expert" sound bites on television.[42]

Network television science coverage followed similar patterns, reflecting gender biases throughout journalism. Mwenya Chimba and Jenny Kitzinger have found that, in print news about science, profiles often defined female scientists' roles narrowly, focused on appearance and sexuality, and framed them as "exceptional," while male scientists were "represented as the norm."[43] From 1982 through 1986, the percentage of television network science stories introduced by a female anchor or with a female reporter shown on camera averaged around 17 percent. From 1968 through 1997, the average number of scientists shown on camera in science news stories almost doubled, but only 5 percent of the scientists interviewed were women.[44] In some years, no science story in a sample of broadcasts contained any interview with a woman scientist, at a time when women constituted at least one-fifth of all scientists in the United States. As Chimba and Kitzinger emphasize, even if the "scarcity of profiles of women scientists in the press . . . *reflects* the reality of gender inequality in the field," such omission also serves to "*perpetuate* it."[45]

After 1990, women scientists began to represent about one-tenth of all on-camera network news sources for science. The press, of course, chooses interview subjects on the basis of expertise (or relevance to a story's focus), so it is always possible that no women were experts on the topics of these stories. Bias undoubtedly played a role, however. Good reporters routinely ask sources for recommendations of whom else to interview and, when covering controversial and contentious topics, they interview not merely the appointed "spokespersons" for science, or heads of academies and laboratories, but also representative dissenters, critics, and minority voices.[46]

For science, the political context influencing coverage did slowly change. During the 1990s, with renewed political attention to women's health policy and increased National Institutes of Health (NIH) funding of related research, network news began to cover breast cancer, menopause, and estrogen therapy and to invite female scientists, such as NIH director Bernadine Healey, to discuss these topics on camera. Space coverage, too, offered new opportunities for female reporters. Marya McLaughlin, who covered the Apollo program for CBS, is credited with being one of the first female space correspondents on television, along with Lesley Stahl, Martha Teichner, Lynn Sherr, and Jill Dougherty, and in 1990, Jacqueline Adams became the first woman and first African-American to anchor network space coverage.[47]

Educational programming might have been expected to be more sensitive to gender equity but, as Jocelyn Steinke and Marilee Long found, even the 1980s and 1990s children's series featured twice as many male scientists as female scientists.[48] Of eighty-six women appearing in one group of children's educational series, sixty-eight were shown in secondary roles as students or assistants. *Mr. Wizard's World*, for example, contained no adult females or female scientists, although young girls did appear as Mr. Wizard's assistants on almost every program analyzed (as had been true on Don Herbert's first series). On *Beakman's World*, about one in eight scientists was female, while on *Newton's Apple*, which made a conscious effort to provide role models for young girls, two of the regular characters (a reporter and a naturalist) were female, as were 57 percent of scientists shown in the series. Nevertheless, Steinke and Long concluded, the program titles projected a contradictory message, declaring "that science is part of a man's world" and implying "either literally or figuratively, male possession or ownership of the world of science."[49]

One need not belabor the obvious. Such overt bias continued in documentaries and science specials through the 1990s, when more change began to occur, thanks in part to increased numbers of female television producers like June Goodfield, Josephine Gladstone, and Paula Apsell. Eight of the twelve films in the PBS miniseries *Living Wild* were produced by women, and at least three focused on the work of female scientists.

Nevertheless, the theme of cultural "invisibility" remained alive on American television. A 1995 PBS miniseries, for example, conscientiously documented the educational, attitudinal, and social barriers dissuading girls and young women from pursuing scientific careers, including persistent gender-role stereotypes within the culture, but its title (*Discovering Women*) conveyed a different message, implying that female scientists worked in obscurity and perhaps needed to be "discovered" in order to be considered real.[50]

CHAPTER FOURTEEN

The Smithsonian's World: Exclusivity and Power

That set—a Cyclopean eye staring out, gray and vacuous, or colored and dancing as if seeing stars after a blow to the head—seems to me to have failed its purpose. . . . the producers of television programs often pile sensation upon sensation in order to ensure that the viewer will remain tuned to their channel.

S. DILLON RIPLEY, 1982[1]

Science writer Wil Lepkowski once described the 1980s as a decade during which popularizers moved beyond translation and became "critics of the culture." In that new role scrutinizing science's political and ethical implications, he wrote, "different skills and sensitivities come into play, not always easy to acquire. I tell [cub reporters] that if they want a scientific metaphor, then they should go after a policy story much the same way a biochemist disengages the cell—for the purpose of understanding the wholeness of the dynamic."[2] Such a systemic approach—focusing on the synergy, connections, and moral reverberations within the world of science—redefined the popularizer's client and role. Popular science should first serve the audience, should help readers or viewers understand how science affects their lives. This new approach licensed a deconstructive, almost subversive retelling of science's stories, influenced by postmodernist academic studies.[3]

The evolution of *Smithsonian World*, broadcast on PBS from 1984 to 1991, exemplified how these changing attitudes reshaped science on television. Analysis of the project offers a glimpse of the intense content negotiations usually hidden from view. By the series' fifth season, in response to marketplace pressures, *Smithsonian World* had veered away from an initial focus on knowledge creation and had begun to explore socially relevant, controversial topics, with these later programs often constructed as aesthetic "reflections" of what viewers sought to understand rather than authoritative summaries of what experts knew. As such, the history of the series demonstrates well what television eventually made of science, and provides some insight to how Lynn Poole's inspiring "illusions of actuality" became fragmented and formulaic entertainments based on illusion.

UNPROFITABLE REALITY

While *Cosmos* was demonstrating that high-quality productions could attract both viewers and critical acclaim, the Smithsonian struggled to achieve a suitable television "presence." Despite the Wolper contretemps, S. Dillon Ripley had resisted all suggestions that the Smithsonian abstain from engagement with mass-market television, citing a strong internal rationale and continued encouragement from the board of regents and powerful members of Congress.[4] The Smithsonian engaged in extraordinary research. Its museums possessed millions of treasured objects reflecting human civilization, art, and science, with comprehensive records of the world's biological, geological, and cultural history. Television could take viewers behind the scenes to observe research in progress and could display galleries, exhibitions, and collections to those unable to visit in person.[5]

Propelling internal decision making during the 1970s was Nazaret Cherkezian, director of the Office of Telecommunications, whose past television experience included work on *Sunrise Semester*. As the Wolper project disintegrated, Cherkezian began discussions with Corporation for Public Broadcasting (CPB) officials about a new magazine-format show.[6] Spurred by a conversation with CPB president Henry Loomis (during which Loomis raved about "the critical and audience successes" of a recent National Geographic special), Secretary Ripley approved an aggressive search for funding and a public broadcasting partner.[7] Early in 1976, the first comprehensive proposal for a series called *Smithsonian World* circulated outside the institution. That outline emulated miniseries like *Civilization* and *The Ascent of Man* and focused on the institution's science assets—from beetle collections to the Marine Mammal Salvage Program.[8] The show would celebrate the "adventure" of knowledge creation and research, "personalize the presentation of knowledge . . . using Smithsonian professionals," and explore such topics as invasive species, environmental pollution, and telescopes, with all scripts written in-house.

Acquiring planning money constituted the first big hurdle. Unlike the previous CBS and Wolper projects, the new venture required the Smithsonian to raise the immense sums required for high-quality productions, identifying major donors and arranging to fund the conceptual phase, including salary for a professional television producer and a scriptwriter. Cherkezian estimated that the project would require an initial investment of $150,000 and that a complete series of twenty hour-long films would cost around $152,000/year (including staff salaries, travel, and Smithsonian overhead). Because television salaries, costs, and station overhead rates were rising, estimates soon ballooned to around $3 million for just one year of production. With such amounts on the table, Ripley began contacting national corporations like IBM, Mobil Oil, and Procter & Gamble, and the Smithsonian initiated a partnership with Washington, D.C., public station WETA-TV to co-own and coproduce *Smithsonian World*. Each organization would be a full participant in production, with right of veto on content, but the Smithsonian would bear "no financial responsibility for the series," and all private funding would be funneled through (and controlled by) WETA.[9]

In late August 1977, the Procter & Gamble Company said no, explaining that the company preferred to fund "quality programming which has the potential of drawing substantial audiences on commercial television."[10] Through friendship with executive Thomas J. Watson, Ripley persuaded IBM to donate $125,000 in seed money, and the project planning moved forward while the funding search continued. During 1979, the partners approached over seventy large corporations and foundations, with no success.[11] They submitted a proposal to NSF (which had underwritten *NOVA*'s first phase), but the agency offered only $20,000 (promising an additional $180,000 if $3 million could be raised elsewhere). Many corporations explained that their public broadcasting contributions were on hold; others offered to contribute only one-third of the requested amount. CPB, whose cooperation was essential for securing corporate contributions, also hesitated, in part because of budget reductions affecting the entire public broadcasting system.[12] "Our pressing need," one Smithsonian official wrote, "is for a specific breakthrough, a commitment from some organization . . . determined to get into the game."[13] If not, "realism might dictate the abandonment of the quest."

Internal Smithsonian dissent added to delays. Five years after the first *Smithsonian World* proposal had been circulated, Cherkezian complained to Ripley that "if it were not for the 'naysayers' and the timid at heart, the Institution would now be in the forefront in providing quality programming for television—*Smithsonian World* would be in its third year."[14] Whenever a willing underwriter or partner had been located, the Smithsonian refused to commit matching funds. Cherkezian reminded Ripley that "Outside producers who provide their own funding are uniformly unwilling to relinquish any sizable degree of editorial control over the productions. In almost all cases, this puts the Smithsonian in an untenable position, for we must be able to ensure historic and scientific accuracy and substance in programs bearing our name. . . . If the Institution is to make any meaningful contribution through the media, it must begin with a solid financial commitment of its own."[15]

Cherkezian's sense of urgency reflected his sensitivity to changes in the television business, especially in cable and home video markets. Over one-third of U.S. households now subscribed to cable. Local cable companies produced their own programs; Home Box Office and pay-per-view channels were thriving, as were organizations such as the Appalachian Community Service Network, Black Entertainment Television, Christian Broadcasting Network, Cinemax, C-SPAN, ESPN, Nickelodeon, and Spanish Television Network. Ted Turner's CNN was weaning viewers away from network news shows, and satellite-to-home transmissions were seeming more plausible, potentially altering the playing field with "narrowcasting." Affordable videocassette recorders were changing how and when people experienced visual content in their homes. "The thread that runs through all these developments is that they will give the viewer much more control of what he wants to watch on TV and when"—a situation that would open, Cherkezian argued, a substantial niche market for scientific content.[16] Cherkezian's *Plan of Action for Smithsonian Video Communications* suggested that cable television and home video could "construct electronic bridges" between the institution's "unparalleled resources" and the millions who might benefit from them.[17] The question was not "what do we want to communicate to the public?" Instead, like an open microphone or an unoccupied podium, the availability of a platform had become sufficient rationale to stand up and sing. Speak now and explain your message later.

The breadth of choices, though, threatened to paralyze the process. Which audience or audiences should they seek to reach? Should they mimic National Geographic's format—or pursue something more daring and innovative (and potentially risky)? Should they aim beyond American markets?[18] An internal television committee had proposed four criteria for selecting future projects: programs must have major educational aspects, even if phrased as entertainment; topics should be compatible with the institution's interests and values; Smithsonian staff should be used in development and, as appropriate, within programs themselves; and control of content must remain in Smithsonian hands.[19] After a series of conversations with networks and advertising firms, one Ripley adviser described how millions of dollars were "being spent just for openers in this high stakes poker game."[20] The Smithsonian, despite its intellectual resources and reputation, was not automatically "a shoo-in success" because neither "the public [nor] the interests which are behind this expansion of the world of communications are automatically yearning for increase and diffusion of our brand of knowledge."[21] Merely to sit at the table, the institution needed advice from professional players.

EXCLUSIVITY (ONCE AGAIN)

Contract discussions proceeded in parallel with the search for funding and for an experienced television executive to lead the project. The Smithsonian assumed that an in-house producer would strengthen institutional control over content.[22] The short list of suggested candidates contained such names as Michael Ambrosino (who was involved in creating *NOVA* and unavailable) and Robert Saudek (the creative force behind *Omnibus*); at the end of 1978, WETA hired New York producer Martin Carr.[23]

Of the three people who eventually managed the series, Carr had the most commercial network experience. An intelligent, creative man, then in his late forties, he had worked in the television industry since 1960, producing, directing, and writing award-winning network documentary specials (including *Hunger in America*) and creating segments for network magazine shows like *20/20*, but he had little experience working with scientists. Carr reworked the *Smithsonian World* proposal. WETA and Smithsonian signed a "co-production agreement" and continued the funding search.

For the first few years, Carr maintained offices in New York City and, because he was being paid through WETA, interacted primarily with station president Ward B. Chamberlin. Production estimates spiked. In March 1982, when WETA and the Smithsonian finally secured underwriting from the McDonnell Foundation, Carr estimated that each hour-long show would cost at least $500,000.[24] Moreover, an additional $500,000 "pre-production grant" was essential, he argued, because each program must have "a unifying theme that will explore the subjects occupying the greatest minds on earth": "Many of these people are at the Smithsonian and some are at other institutions throughout the world. The pre-production period will be a chance to meet with those people and find out what they are doing at the present moment."[25] Carr had thus already altered the concept from a series *about* the Smithsonian to a project that treated the institution as one among many potential sources of ideas, as a repository of knowledge but no longer the backbone to the scripts. Moreover, Carr began describing the scientists and curators as hindrances rather than assets.

Both Carr and Cherkezian—as accomplished television professionals— looked at the medium quite differently than did the experts whose research provided ideas for scripts. And all parties found it difficult to abandon arrogance at the door. Cherkezian later described his first meeting with senior scientists at the National Museum of Natural History as "grim, tough": "I didn't give an inch, not an inch. I said to them . . . 'All I want you to do is help us to make it so that more people can understand [your work] than the four people that are your peers. The problem I find is you're only interested in your peers. If you want to do that, do a slide lecture in the basement. Don't come to us.'"[26] Disagreements over who should control access to the scientists' work perturbed the discussions. Carr, like Wolper, insisted on the need for "exclusivity": "We must also be prepared to get commitments from scientists, artists and other experts that they will hold their stories for us. In other words, we must find 'exclusives'—so that we are not chasing after 'Nova,' 'Universe' or any other series."[27] Disputes quickly arose about the extent to which the Smithsonian should allow access to other television groups. The contract with WETA stated that neither organization could produce a related program (presumably, another series about the Smithsonian), but the institution had not relinquished the right to make (or cooperate in) other films or television programs. The institution's Office of Telecommunications had, for example, recently approved development of a pilot for children's specials centered on a Smithsonian puppet theater and exhibitions.[28] When that producer had described the forthcoming "Smithsonian Discovery Theater" project to *Hollywood Variety*, the news story triggered protests from WETA.[29] Carr even challenged films that had been in progress long before the WETA agreement. Outraged by a request that all footage about a Smithsonian coral reef project be embargoed, even though the fully funded film was complete, Cherkezian responded: "How do you rationalize years of work, thousands of dollars from other equally respected sources with SI commitments to them, and plans for a solid, in-depth scientific film being prepared by dedicated scholars to be put on a shelf for three years? All because WORLD tries to make up its mind if and when they might use a piece of eight or nine minutes."[30] Carr had asserted that the *Smithsonian World* project had an obligation to the underwriter "to only run material that hasn't been broadcast before," just as he had dropped plans for a segment on the Golden Gate Bridge after learning that CBS was filming a show about the bridge.[31]

The definition of "exclusivity" was further tested by National Geographic's ongoing efforts to establish a multi-institutional cable, broadcast, and satellite distribution service. In 1981, the society invited over thirty nonprofit organizations (including the British Museum, Audubon Society, World Wildlife Fund, American Museum of Natural History, AAAS, and Smithsonian) to join an "Alliance Network" to produce and distribute educational content. National Geographic's trustees had authorized production of a pilot film with consortium members and commissioned a market analysis.[32] Many senior Smithsonian staff expressed skepticism about the proposed "culture-consortium" project, primarily because cable audiences were not yet well understood ("Access is NOT synonymous with audience—meaning that just because a program has been on cable does not mean that it has been seen by anyone").[33] Nevertheless, discussions continued through early 1985, when the society finally gave up and launched its own weekly series on Nickelodeon.

More galling to the scientists, of course, was the suggestion that they should refrain from public discussion of their research. Carr argued, for example, that the *Smithsonian World* should be given "first refusal" on stories about two particular research projects, citing potential competition from *Discover: The World of Science* and National Geographic programs.[34]

"LUMPING ASTRONOMY AND MOVIE STARS"

In 1981, in the margin of a staff memo about television, Ripley sketched three overlapping circles illustrating his vision for *Smithsonian World* content.[35] He labeled the middle circle "Things that SI [Smithsonian Institution] does"—representing topics suggested by scientists and curators. The circles on either side—"Things that SI is interested in but doesn't do" and "Things that SI should be interested in"—help to explain the project's troubled history. Ripley had encouraged a similar model for the Smithsonian's popular magazine: well-rounded attention to the topics within the center circle and in the overlapping zones. The television professionals, however, never felt bound to confine their choices within what they perceived as arbitrary boundaries.

On the Smithsonian side, the person responsible for overseeing *Smithsonian World* was Thomas H. Wolf, a retired network executive whom Ripley hired in 1981 as the institution's television consultant. Wolf's notes from initial negotiations with the McDonnell Foundation indicate that he stressed the importance of content oversight ("I told them SI had total control of subjects and scripts"). When foundation officials asked if "they could have the same right," WETA's representative explained that PBS policy prohibited allowing underwriters control over a project and assured the Smithsonian that "they would 'reason together.'"[36] Wolf arranged meetings with WETA and Carr "to insure that SI has in fact what we have on paper in our contract: approval of stories, personnel and budget," but, within a few months, the producers were running ideas past McDonnell Foundation officials before, and sometimes instead of, Smithsonian experts or representatives.[37] When a geologist expressed concern about superficial descriptions of his field within treatments submitted to McDonnell ("in a fashion that would be either falsely dramatic or once-over-lightly"), Wolf explained "that much of what was on the paper submitted to McDonnell was there to sell . . . and that before any story was undertaken . . . there would be more than one session with [the expert] to ensure that he was satisfied up front."[38]

Such promises proved unenforceable. Wolf urged the Smithsonian to hire its own writers, and he and Cherkezian pressed for firm production schedules, early drafts, and clear articulation of review procedures, such as how and when rough cuts would be reviewed. Carr, however, insisted that only "extended treatments" could be provided in advance because "it is not possible to script what will be essentially documentary films (you can't put words in the mouths of experts)."[39] The scientists were forced to react piecemeal. At one meeting, a senior producer promised rough cuts of segments in "three to four weeks," but they were not made available until six months later and, even then, shown first to the underwriter.[40]

By summer, Wolf's impatience was undisguised. Initial publicity about the series implied close connection to Smithsonian staff and research, and the regents had been assured that "the Institution has total and final control of both subject matter and script accuracy at all stages."[41] "I know how hard you and all of your associates on Smithsonian World have been working," Wolf wrote to Carr, but "I want to remind you that by contract S.I. must see and approve of all rough and fine cuts and scripts."[42] When Carr continued to disregard the requests, Wolf threatened to notify WETA that they were "in breach of contract."[43] Two months later, the *Smithsonian World* staff finally sent clips of individual segments, but Wolf had yet to receive a tape of a complete show. Wolf warned that Carr should not "underestimate the degree to which the senior SI staff is adamant against forcing stories into arbitrary categories—i.e., lumping astronomy and movie stars. Please don't paint yourself into a corner on this. It clearly won't work."[44]

Lack of communication exacerbated the disputes and contributed to inaccuracies, distortions, and omissions. A script about tigers had, for example, discussed the species' dwindling gene pool but failed to mentioned the Smithsonian's own preeminent research on tiger breeding being done by the National Zoological Park.[45] Scientists at the natural history museum, "badly burned in the mid-1970s when the Smithsonian's TV effort dealt with the Loch Ness Monster, Bigfoot, and other such dubious characters," expressed outrage at a segment exploring the controversial topic of cryptobiology (studying animals purported to exist): "while these subjects may indeed be interesting to the audience out there, they lack sufficient scholarly content to be dealt with officially and so publicly under the Smithsonian banner."[46] As script sections arrived, Wolf often raised questions about relevance, expressing concern that there was no obvious link to the institution: "not every piece has to have a Smithsonian Institution connection . . . but on certain segments there are still questions."[47] When Ripley was shown the first rough cut, he was furious because it ignored relevant collections and experts at the Smithsonian.[48] The rough cut of the third program elicited similar criticism from staff who condemned it as "light weight" and "fluff, almost to the frivolous."[49] In memo after memo, Wolf attempted to persuade Carr to add more objects from or references to the Smithsonian and, at minimum, to consult staff. Finally, in January, Wolf disseminated ground rules for timely notification and review, including approval by all Smithsonian people shown on camera, and screening by a Smithsonian committee before a program could be locked.[50]

CRITICAL REACTION

Uneven critical reception to the first season increased the tension. The publicity had clustered around writer David McCullough, who had been hired as host (for what would be his first television appearance in that role). There were many newspaper photographs of and interviews with the charismatic McCullough, but few of Smithsonian research collections or scientists. The *Louisville Courier-Journal* described the first show as "so loosely threaded that it doesn't hang together well," and the *New York Times* called it "as ambitious, eclectic and chockablock as the great institution itself," full of too many things, with too little connection and insufficient depth, a "serendipiter's journey."[51] George Bullard of the *Detroit News* complained that "the hoped-for cohesiveness doesn't really work. The good parts . . . don't get enough time. The bad parts—like a ho-hum tour through a clock museum—get too much time."[52] He called it "a highbrow *NOVA*" that was "inferior to the best of *NOVA*."[53]

The toughest criticism came from local reporters. *Washington Post* reporter Tom Shales knew the Smithsonian's world well because he had reviewed museum exhibits before concentrating on television. The producers, he wrote, had not yet "mastered the delicate art of making informational programs that don't look like classroom films. . . . the photography is commonplace. . . . The next step for 'Smithsonian World' should be Ye Old Drawing Board."[54] The most scathing reaction came from *Federal Times*. For over a decade, Robert Hilton Simmons had been a dogged critic of the Smithsonian's "glittering scholars, scientists and administrators." *Smithsonian World*, he wrote, fit a long-standing pattern in which desire for publicity overcame good sense: "Years ago, the Smithsonian set forth on another TV series. One show was about the curse of the Hope Diamond. Another about the Abominable Snowman—Secretary S. Dillon Ripley said on screen that he may have seen it up in the Himalayas. . . . The knowledge scam was even more obvious then and the series died in pilothood. . . . "Smithsonian World" may not only survive but may become a winner. Winning is the aim of the game. That's show biz. But is it science and scholarship?"[55] The next programs attracted less press coverage, often only a paragraph or two within stories about the entire PBS lineup. Nevertheless, the series had accomplished that amorphous goal of achieving a television "presence." Critic Shawn McClellan observed that "[i]n a sense, the Smithsonian is doing for itself here what the early 'Disneyland' shows did for Uncle Walt. Just as the various realms of the famed theme park were impressed on the nation's mind even as people were entertained by, for example, Davy Crockett ('Frontierland'! Everyone quickly learned), so various Smithsonian facilities form the backgrounds for these segments."[56]

NEW REGIMES

If Carr assumed that the institution's impending change in leadership created a power vacuum, he was mistaken. In May 1983, the regents had decided to replace Ripley (then in his third term), and by fall, his successor had been named. During February 1984, anthropologist Robert McCormick Adams arrived in Washington to prepare for taking office in June. Adams was unimpressed by the newest rough cuts screened by Wolf. In a carefully phrased letter, Adams urged Carr to pay closer attention to the "differentiated views" and "detailed criticisms'" provided from the "Smithsonian side" and to realize that, without more "Smithsonian involvement," the institution might lose interest (and, it was implied, not cooperate in seeking funding for additional seasons).[57] Wolf now began to play a tougher game. Not every segment in the next season's programs need "relate to a subject in which the Smithsonian itself is directly involved," he wrote, but each must "be of substantial interest to the Institution," and the programs should explain why.[58] Soon thereafter, emphasis shifted from "diffusion of knowledge" to "increase of knowledge," with topics chosen because they were "important aspects of the arts and sciences," rather than simply "attractive" to filmmakers.[59] Wolf actively lobbied for *Smithsonian World* to consider topics related to "the environment," include attention to "U.S. minorities—blacks or Hispanics," and mention research cooperation with Third World countries and scholars.

Review procedures began operating more smoothly, and the series moved toward single themes or topics, away from the first season's hurly-burly, and toward scientific research. "Desk in the Jungle" (January 1985) described the fieldwork of Smithsonian ethnologists, anthropologists, ornithologists, and marine biologists; "Where None Has Gone Before" (October 1985) discussed the first nonstop twenty-five-thousand-mile flight around the world, the discovery of a new class of crustacean, and development of the Hubble Space Telescope. Six of the proposed programs for the third season centered on scientific topics, from geology to botany, archaeology to animal conservation. In 1985, the series was among the ten most-watched PBS programs.[60] Audience size, far less than the average for commercial prime-time shows, remained disappointing, however.

Money dominated internal discussions. Of the first season's $3.5 million grant received from the McDonnell Foundation, Smithsonian received a "fee" of $222,222 and WETA, an overhead payment of $452,604. For the third season, total income from underwriters was $3,182,500. By the sixth season, with project underwriting in jeopardy, the Smithsonian settled for a $75,000 flat fee rather than the $200,000 it had received for the fifth season. There were, nevertheless, other compensations for the partners. During the 1980s, video sales, aftermarket sales of footage, and supplemental rights (such as cable and syndication fees) had become potentially lucrative sources of income for public television science and nature series. By 1988, sales, rentals, and licensing of *Smithsonian World* videos exceeded $20,000 per year.[61] In January 1989, WETA licensed television, cable, and video rights for the first two seasons for a $1,000,000 advance against royalties, money that underwrote part of the fourth season when foundation funding hung in jeopardy.

Reviews for the second season were mixed, with leading television critics still hurling darts. "If seeing scads of money being wasted is your idea of great fun, I urge you not to miss *Smithsonian World*," quipped the *Philadelphia Daily News*: "PBS's all-too-familiar, talking-head approach plodding toward its all-too-predictable result: a boring retelling of a fascinating life."[62] Tom Shales said the program heralded "another drowsy-fusty season of well-meant but tedious public TV specials": the script was "repetitious," the account of the Custer battle "laborious," and narrator David McCullough exhibited "less passion than your average CPA."—"A museum of the air is one thing, a mausoleum of the air is another."[63] By then, a new executive producer had been hired.

CREATIVE SPACE

During initial planning, Cherkezian had met with Adrian Malone (fresh from his success with *The Ascent of Man*). Although Malone had expressed interest in the Smithsonian project, he was engaged in developing *Cosmos*. By late 1985, the forty-eight-year-old Malone was looking for a new challenge and was hired to be chairman of the Smithsonian Television, Film and Radio Communications Council and, through WETA, to oversee *Smithsonian World*.[64] A tall, tweed-coated chain-smoker, Malone was said not to "fit the image of a BBC star producer," yet he could not resist flaunting his British mannerisms and confessed to one reporter that "[y]ou may not believe this but I used to be even more pompous than I am now."[65]

Malone dreamed in creative and economic spaces as large as the cosmos, in tens of millions of dollars, about complex, intellectually challenging topics. Although sensitive to ongoing transitions in television's viewers, money, and power, he was convinced that audiences were not merely indulging in whimsical "channel switching" but were, instead, searching for greater "choice and quality," looking for "imagination, intelligence, accessibility, and professionalism."[66] His assessment of *Smithsonian World*'s first productions apparently matched that of the critics—"divertissements" with insufficient intellectual substance, "well-produced" and possessing "a certain elegance and dignity," reflecting a haphazard view of the institution as "the nation's attic."[67] So he sought to redefine the program's identity: "NOVA is respected for its illumination of science; National

FIGURE 17. *Smithsonian World* producer Adrian Malone and Smithsonian Institution television adviser Thomas H. Wolf, holding the series' 1987 Emmy Award. Courtesy of Smithsonian Institution Archives.

Geographic is respected for its illumination of nature; SMITHSONIAN WORLD must be respected for its illumination of culture."[68]

Fixing the series had become imperative. Despite respectable ratings, survival was uncertain. In March 1986, the McDonnell Foundation reduced its contribution, offering to donate only half the cost of the third and fourth seasons (about $4 million altogether).[69] Contributions from CPB and individual PBS stations helped keep productions on schedule, but almost a year later, a gap of about $1.5 million remained. Finally, in 1988, Southwestern Bell Telephone Company became the main underwriter, on the hope that an "innovative, high-quality series" would attract "educated and influential decision makers" and "enhance" the company's reputation "with this target audience."[70] At that point, half of *Smithsonian World*'s regular viewers were described as "upscale."

REFLECTIONS

The magazine format was abandoned, and *Smithsonian World* settled into themed productions. The tone shifted from celebration to qualified criticism of science, following patterns observable elsewhere on public television. In "The Last Flower" (second season), scientists had been saviors, racing to preserve vanishing treasures in botany, anthropology, biology. "On the Shoulders of Giants" (third season) displayed science as adventure, following vertebrate zoologist David Steadman to the Galápagos and Cook Islands, where his research on exotic creatures sought to shed new light on evolution. Publicity photographs romanticized the handsome Steadman, bearded, suntanned, his hair a bit tousled by the wind, staring off into the distance—almost heroic. By the fourth season, the series was framing science as a cultural activity that provoked moral conflicts and controversy. "Web of Life: Exploring the Genetic Landscape," produced by Sandra Wentworth Bradley, focused on the ethical issues surrounding genetic technologies while projecting "a sense of optimism and wonder at the potentials of genetic research."[71] The film, Bradley emphasized, contained "[n]o Jeremy Rifkin. The 'players' in the film include ethicists, scientists, and a philosopher/poet. . . . the film is about more than scientists' responsibility to society; it's about man the great manipulator and his responsibility to the planet and its future."[72]

Throughout development of "Web of Life," Bradley and her writers had consulted dozens of experts outside the Smithsonian, not all of whom agreed to become involved. After meeting with biologist Maxine Singer,

Bradley explained that "talking experts" would alternate with "exciting visuals."[73] Unpersuaded, Singer declined to be interviewed on camera, citing the proposal's lack of emphasis on technical explanations: "I am deeply troubled by a public consideration of ethical issues regarding subjects that the public does not understand even at a general level. Ethical decisions made in the absence of understanding will always be of questionable merit."[74] Other biologists, such as Nina Federoff and Ruth Hubbard, did appear, but the finished program concentrated, as Singer had feared, more on the ethical and social debate than on the related science. Throughout, a sentence or two from one expert would be interwoven with that of another; no one was allowed to speak at length; every thirty-second statement was chosen carefully to make the producer's case. Negative images (including film of concentration camps and of immigrants being deported from Ellis Island) were interspersed with emotional, romantic scenes of children and even film of the birth of Bradley's own son.

When a senior biologist complained to Secretary Adams about the program's "hand-wringing," "lack of scholarship," errors, bias, and negative messages, Bradley responded that "Smithsonian World is not NOVA. Perhaps on NOVA, you could find the first-class presentation on modern biology you were seeking."[75] *Smithsonian World* was "concerned with ethics and public policy," and she added, "a good film" should be "a reflection for each member of its audience. A relationship develops between the program and the viewer" such that "the viewer sees what (s)he is looking for."

"Tales of the Human Dawn," about changing perceptions of human origins, continued this approach. Despite long-standing extensive Smithsonian expertise, research, and collections related to evolution, the film eschewed any "chronology of the science" and told the story primarily through literature, art, and biography, such as sections based on Kurt Vonnegut's novel *Galapagos* and Elizabeth Marshall Thomas's *Hidden Life of Dogs*.[76]

REALIZATIONS

Malone's heart attack in early 1989 placed Bradley temporarily in charge, and she was eventually named as the third (and what proved to be the last) executive producer. Bradley, a graduate of the UCLA film studies program, had been working in Washington as a professional photographer and filmmaker since 1970 and on the *Smithsonian World* staff since 1983. With far less experience at the senior level than Carr or Malone, Bradley was no more successful in courting internal cooperation or defusing internal criticism, and she began to move the series farther from its original purpose, emphasizing creative rather than pedagogical values. She described *Smithsonian World* as "not a 'documentary' in the informational sense, but a documentary essay," in that "[e]ach program's goal should be to <u>involve</u> the viewer" and "to provoke thought and consideration—to stimulate without being controversial in the political sense."[77]

This perspective, and the difficulty of implementing it, was exemplified in a fifth-season episode about quantum physics. This same program also prompted the Smithsonian to expand the role of its internal advisers to monitoring accuracy, from initial idea and script through rough cut.

Malone had been attempting to make a film about quantum physics since at least 1987, initially as part of Smithsonian Project Discovery and then, when that collaboration with the Discovery Channel collapsed, as a program for *Smithsonian World*. While Malone was recuperating in 1989, Bradley took over production.[78]

"The Quantum Universe" sought to explore the "profound transformations" catalyzed by physics during the twentieth century, and to trace the "intellectual and aesthetic framework . . . the dramatic changes in representation and visualization which occurred within the advent of modern physics, how the invisible has been made visible and our universe been 'reinvented.'"[79] To put flesh on such philosophical bones, Malone's original script included dramatization, a form of "creative fictionalization" that entwined contemporaneity with history. By 1989, the dramatization had been reduced, replaced by extensive interviews with well-known physicists, snatches of art and poetry, and scenes from Tom Stoppard's physics-driven play *Hapgood*.

Locating a balance between technical advisers perceived as "useful" and those dismissed as "interfering" had become problematic, especially for complicated subjects like modern physics. As production began on the fifth season, Malone reminded his producers that the adviser assigned to each program was "not intended as a mere figurehead" but should be consulted at specific stages.[80] Following a presentation by Bradley about that season's programs, Smithsonian researchers raised concerns about the lack of "qualified" scientific consultants, especially for the quantum physics program.[81] Bradley called the scientists' questions about production policies ("how subjects are chosen, how the films are made") intrusive and "a little out of line": "Adrian Malone is the Executive Producer. WETA is the producer, assuming the financial risk, and, for example, we can't shoot the entire African continent on the budget we have, in the time we have, and cram it all into 54 minutes, despite how the scholars feel."[82] Such remarks disturbed Secretary Adams. Bradley argued that *Smithsonian World*'s "major task . . . is to define our culture . . . . [W]e are not NOVA or NATIONAL GEOGRAPHIC. . . . We are not a 'science program' but a 'culture program'."[83] To which Adams replied unequivocally: "The Smithsonian is a major institutional player in the universe business. We, and many who work for us, have a stake in what is said, in how we and the present state of knowledge to which we and they have contributed is represented. Our name will not appear in association with that program unless its scientific content meets our approval."[84] As a result, the Smithsonian demanded detailed film treatments for all proposed shows and assigned Irwin Shapiro, director of the Harvard-Smithsonian Center for Astrophysics, to serve as technical consultant on the physics program.

Adams's deputy, Robert Hoffman, admitted that while the external experts consulted by Bradley might well be useful, they could not be expected to view the "whole elephant." The institution's new technical consultant would "check for accuracy and bias" and flag "presentations that would be potentially embarrassing for the institution." "His or her role is not to take sides, or present a particular point of view, or dictate your approach, or veto your story line," Hoffman wrote the producers, but instead to "work with you in a positive and encouraging way."[85]

Both Bradley and Malone bristled, responding disingenuously that *Smithsonian World* was "a program produced for PBS, funds raised by WETA, risks taken by WETA, which is therefore the organization responsible for content and editorial," whereas the Smithsonian was simply "responsible for verification of fact," not "interpretation" of the content.[86] Malone described the programs as "true documentary films . . . not scripted lesson plans over which visuals are laid," and he explained that, for his team, "the research phase . . . continues throughout production," implying that conclusions revealed themselves in front of the camera and that the interviews scripted the "interpretations" of science (and culture) that would be eventually presented in a completed film, even though the interviewees themselves had been chosen by the producer.[87]

Despite a busy schedule, Shapiro acquitted his advisory duties conscientiously, providing extensive comments over the next six months, detecting errors in script and narration, pointing out where an expert's explanation might be unintelligible to nonphysicists, catching typing errors in transcripts, and noting when interviewers had used such critical terms as "quantum universe" inconsistently (which led to confusing disparities in how the interview subjects responded). Perhaps because of Shapiro's careful criticism, Bradley hired a well-respected physics professor and science writer to draft a new script. By March, Bradley was thanking Shapiro for his help ("more important to me than you probably realize"), and the program became a rich exploration of quantum physics, art, and philosophy, featuring such experts as physicists Sheldon Glashow, Martin Perl, and Burton Richter, Smithsonian astrophysicist Margaret Geller, philosopher Abner Shimony, and historian Peter Galison. The film interleaved intense, difficult exposition of fundamental principles with film of blowing sand, clouds, orreries, blackboards, and billiard balls. From electron clouds to how electrons "regenerate," from Schrödinger's equations to the slow-motion hatching of a luna moth chrysalis, from Albert Einstein to Berenice Abbott describing her photographs of waves and interference ("the scientists' passion to know is paralleled by the artist's passion to see"), "The Quantum Universe" evolved into a tone poem, praised by the *New York Times* as "an elegantly put-together hour," with evocative images and stimulating ideas even though the concepts (that is, the science) remained "elusive."[88]

By fall 1989, as Bradley took charge of planning, Adrian Malone complained about administrative pressure and dwindling funding, and resigned. Southwestern Bell decided not to underwrite a seventh season.[89] The CPB review panel gave exceedingly negative comments about the proposal for a seventh season ("The artistic quality gets in the way of the content"; "Imagery is gorgeous but the series is not cohesive").[90] PBS would not support it, and WETA canceled the series rather than search for a new underwriter.

BASIC COMMUNICATION

The Smithsonian's historic, chartered commitment to diffusion of scientific knowledge should have made the institution the ideal partner for television production. World-renowned staff, collections, and research facilities provided a treasure trove of topics for visualization. So what happened? For sure, individual personalities, fiscal constraints, and political pressures influenced the outcome, but the greatest barriers to success were the attitudes entrenched within two different worlds—television and science. Those same attitudes inhibited many other attempts to create high-quality science content in the United States. Television's economics-driven demand to fence off "exclusive access" to knowledge conflicted with scientists' innate impulse to share insights and information so they could be used. Television's unbending obeisance to audience size as the best measure of quality meant that producers fell (or were pushed) into the trap of choosing to entertain rather than educate, even when the latter might have been more suitable and responsible. And, in the end, the unwillingness of professionals, within both television and science, to compromise in order to serve the audience continually undermined the process of creating the best possible programs and discouraged future popularization ventures. Both groups lost sight of what viewers needed and deserved. The occasional triumphs wrung out of the process at great cost—such as "The Quantum Universe"—offered glimpses of what might have been. They also underlined the opportunities lost during the medium's first half century.

CHAPTER FIFTEEN

All Science, All the Time

[I] wonder if what we have in the way of communication and transport now hasn't supplied society with an infinitely more complex nervous system than it ever had before. . . . [If] you put a human nervous system into an oyster, . . . for the first few weeks the poor oyster is going to have a devil of a time because he has got a great deal more feeling and information rushing in him than he is prepared to handle.

ALAN GREGG, 1938[1]

Science writer Helen Miles Davis brought a color movie camera to the 1952 atomic bomb test at Yucca Flat, Nevada.[2] As her lens swept across "News Nob," the hillside where media representatives were gathered, William L. Laurence and other prominent newspaper reporters smiled and waved. A few frames later, Davis filmed the quintessential symbol of how popularization was changing: a television camera perched atop a KTLA-TV truck, which transmitted images of the test to people all over the country, and allowed viewers in New York City to *see* and *hear* the explosion a day before they *read* Laurence's account in the *New York Times*.[3] No matter how accurate, comprehensive, and insightful the prose, no matter how well-explained the physics, the impact of printed articles could now be eclipsed by the actuality of the visual. To watch Davis's film of the mushroom cloud, roiling and horribly beautiful, like a sky-bound sea anemone, a chrysanthemum of radioactive dust, is to experience what early television viewers found so compelling. We, too, swoon in the gloom, delight in the visualization, our immediate experience of the images absent all context and explanation. At first, we see only the cloud—an amorphous intellectual "black box"—and then, and only then, blinking to rejoin the twenty-first century, remember what we know about that power. We adjust our attention, and impose meaning on the science made visible on the screen.

"Sooner or later," critic Tom Shales has written, "everything becomes television. Or just goes away."[4] The same might be said for what happened to popularized science. Rather than neglecting science, television absorbed it within news, documentaries, and entertainment. Research potential was exaggerated, scientists trivialized into celebrities, and discoveries described with unwarranted drama. Even when nothing seemed worth watching, the sets glowed in the dark and the images rushed past—unmoored from their context. The proportion of content devoted to thoughtful discussions of contemporary scientific research (never all that large) mattered less than the fact that science and scientists, factual and fictional, became part of television, became another "presence" competing for viewers' attention.

No one—neither the media, federal government, scientific community, nor public organizations—stepped up and accepted responsibility for helping viewers assess, order, frame, or evaluate the authenticity and reliability of what was being shown. Perhaps the seductiveness of the illusion, the delights on the screen, or the sheer volume of nightly programming discouraged such efforts. Aside from occasional committees or conferences, no one really considered how this extraordinary technology might be better used to prepare citizens for a modern life intimately engaged with and influenced by scientific knowledge and its applications. Instead, the task of becoming educated via television (about nuclear energy, genetically modified organisms, nanoparticles, or the birds and the bees) was left to viewer discretion. Informing or enlightening audiences about fundamentals—why scientific evidence should be believed, how researchers guard against error and fraud, why some conclusions might be tentative and even later overturned without undermining the reliability of the scientific process—proceeded haphazardly, if at all. Everyone seemed to assume that viewers would fend for themselves. Moreover, like silt stirred from the ocean bottom, the barrage of unsorted content sometimes obscured the very information that viewers needed for evaluation and interfered with intelligent selection.

American television never presented science in its entirety. Instead of attempting to cover the world of science systematically, the industry selected those topics and people that millions of viewers had found interesting in the past. Producers chose the fields (favoring biology and natural history) and the scientists (favoring the charismatic and cooperative) that they assumed audiences would watch, and then marketed those choices to underwriters, sponsors, and networks. The relentless decontextualization of content then further severed connections between evidence and authority, between constructed images of research and the actual standards, practices, and personnel.

What happened to science content on television reflected the natural evolution of a creative medium designed to communicate remotely to mass audiences within a free society. Indeed, in 1953, a University of Chicago Committee on Educational Television had predicted that *how* audiences experienced television would affect what they would gain from it. Viewers received images "simultaneously," that is, many millions of people saw and heard, and therefore could react to, the same program at the same time. To receive was not necessarily to understand, however. Whether a shared experience might stimulate public discussion about major political or social issues was something, the committee emphasized, that would have to occur elsewhere, to be encouraged and enabled by society and its institutions.[5] Other observers during the 1950s bemoaned television's lotuslike lures, the "ambushes of pleasure" that beckoned to "undisciplined" viewers, but those warnings, too, went unheeded. The future, the *Christian Science Monitor* explained, would be determined by viewers' ability to comprehend what was being shown (influenced by their experience and education) and their willingness to demand more sophisticated content.[6] The television industry had no stake in encouraging such skills and no enthusiasm for responding to public pressure to change.

Scientists and their organizations remained mostly on the sidelines, neither launching major television projects nor necessarily supporting the entrepreneurs who did. From time to time, the enthusiasm of creative individuals did make a difference. Whenever accomplished scientists became involved (Roy K. Marshall, Jacques-Yves Cousteau, Jacob Bronowski, Carl Sagan, Philip Morrison, David Suzuki), the results tended to be memorable. Equally important roles were played by intelligent intermediaries (Lynn Poole, Frank C. Baxter, Don Herbert, June Goodfield, David Attenborough) and industry professionals (David L. Wolper, Adrian Malone, Michael Ambrosino) who exploited the medium's potential to the full. The best of television's science fulfilled all parties' high expectations. The majority of television's science, captured within dramatic plots and dialogues, sensational natural history documentaries, or news reports, kept viewers entertained.

The disappointment was that television had initially seemed to offer a remedy for the age-old separation between scientists and society. Viewers could have become voyeurs in the laboratory, peering over researchers' shoulders and learning more about their work. Yet within a decade, television's science had become mediated, transformed into entertainment at a distance, made into spectacle presented by surrogates and celebrities. Perhaps the involvement of scientific associations during those crucial first decades of broadcasting would have made no difference. We will never know, but cannot help but mourn the failure to try. Influential organizations that might have encouraged, enabled, supported, and underwritten programs and popularizers—AAAS, NSF, NAS—hemmed, hawed, and hesitated. From time to time, Johns Hopkins University, the Smithsonian Institution, and other education-oriented organizations supported successful projects, but their commitment to television was erratic. Only the National Geographic Society persisted for decades, taking the necessary financial risks and eventually expanding its productions across the breadth of science and engineering.

TRENDS AND MONEY

Money made the difference. From the 1940s through the 1990s, the marketplace shaped the content made available to viewers, just as had happened with print and lecture popularization.[7] During television's first decade, the number of regularly scheduled nonfictional science series shown on major U.S. networks increased, reached a peak around 1951–1955, and then declined steadily over the next five years, with science never regaining a notable proportion of American television content until the establishment of specialized cable channels.[8] By the 1960s, science-related and educational programs, for all ages, had largely disappeared from broadcast networks, replaced in the premium time slots by drama, comedy, game shows, sports, and similar entertainment. After the mid-1970s, commercial networks sometimes offered no regular science programs at all to their national audiences, leaving local commercial and public stations to fill the gap with syndicated programs (primarily focused on nature). Although the public broadcasting system did eventually develop innovative children's programs, the major commercial networks produced no new children's science shows between 1967 and 1979. By the 1970s, science projects aimed at commercial venues seemed designed to flop, "produced with a highbrow self-righteousness and a humorless reverence for the material."[9] And public television appeared trapped by untested assumptions about what its regular viewers might watch. Comedian Fred Allen once quipped that "[i]mitation is the sincerest form of television."[10] Expensive, beautifully filmed science programs narrated by charismatic Hollywood celebrities followed in well-worn tracks.

By the 1990s, Americans were consuming an average of thirty hours of television every week. By 1993, a typical household could choose from among thirteen free broadcast channels and, with paid subscription, another forty.[11] Over one-third of U.S. households owned three or more sets, and over three-quarters owned a videocassette recorder and were actively taping and rewatching favorite programs.[12] Despite these expanded opportunities, the science content remained limited. In the Washington-Baltimore area in 1992, about 16 percent of programming on the three major public stations could be said to be science-related, yet the region's commercial broadcast stations ignored science unless related to nature or medicine or entwined within fictional drama or comedy.[13] Expansion of channel capacity did not stimulate more science. On Boston-area television in 1996 (thirty-three broadcast and cable channels), under 2 percent of all content (slightly over 10 out of 660 hours analyzed) was devoted to science programming, most focused on health or natural history and most scheduled after midnight or on cable.[14] Two projects typified the PBS response to commercial content. By 1993, *Scientific American Frontiers*, underwritten by GTE Corporation and produced by Graham Chedd and John Angier, was being hosted by actor Alan Alda, who displayed an intelligent science buff's interest in mummies, seashells, and rhinos while cracking the wall of solemnity traditionally surrounding popular science. And in 1995, Adrian Malone's *The Nobel Legacy* miniseries, funded entirely by a pharmaceutical firm, explored science through the lens of postmodern academicism, with commentary from a humanities professor who openly challenged the idea of scientific objectivity.[15]

NOVA's history demonstrates how economic necessity influenced content. Major funding for the first season (1974) came from NSF, and most programs celebrated basic research. By the end of the 1990s, corporate donors dominated the list of underwriters; the most frequent *NOVA* topics were engineering and medical research. For most of its history, *NOVA* had averaged around ten million viewers per broadcast—only one-tenth to one-half of those who watched popular commercial programs. Then, with establishment of the Discovery Channel and the Learning Channel, *NOVA* faced tougher competition. With series like *Raging Planet* (tornadoes, earthquakes, and tidal waves), *Jaws and Claws* (predators), and *The Blast Masters: The Science of Explosion*, the cable operations could attract two and three times more viewers.[16] Pleasing and retaining the core *NOVA* audience (middle-class males) demanded fidelity to certain topics, season after season, clinging to what CNN correspondent Miles O'Brien called an "appeal to Joe 6-Pack" through "sensational visual images."[17] As filmmaker Joe Levine observed, late 1990s science programming frequently pandered to the "'boys' toys crowd" ("relatively affluent males between 20 and 35 years old, who generally have gone to college. Their favorite stuff appears to be sex, violence, machinery, military stuff and sports").[18] Milton Chen has shown that, in an effort to compete for these viewers, *NOVA* increased its sensationalism.[19] *NOVA*'s average audience share for the 1991–1992 season ran about 3.1 percent of all television households (2.85 million out of 92.1 million), while its four highest-rated episodes attracted considerably more: "Hell Fighters from Kuwait" (4.7 percent share, or 4.3 million viewers), "Who Shot Kennedy?" (4.4 percent share, or 4 million), "Submarine!" (4.1 percent share, or 3.8 million), and "Making a Dishonest Buck," about counterfeiting (4 percent share, or 3.7 million). Ever more programs examined topics in archaeology and anthropology, although experts in those fields bemoaned the results: "A viewer lacking previous knowledge about the sites presented or how archaeology works would not necessarily see any distinction between rational deductions drawn from observable evidence, baseless speculations, and ideologically driven pseudoscience."[20]

Nature shows remained audience favorites throughout the 1990s (for example, Discovery Channel's *Wild Sanctuaries, Wildlife International, Wildlife Journeys, Wildlife Tales,* and *Orphans of the Wild*). The Walt Disney Company established its cable channel Animal Planet ("all animals, all the time") in 1996. The National Geographic Society eventually launched its own cable network, National Geographic Network, in 2001, with selections from the society's documentary archives, and *National Geographic Today,* a weeknight science and environmental news show.

There is remarkably little quantitative data about what audiences actually gleaned from any of this content. In the mid-1990s, Milton Chen could locate no assessments of what viewers were learning from *NOVA* or other public broadcasting programming.[21] Studies recorded which programs attracted the most viewers and how many viewers watched each episode, but no one asked whether people really got the information they wanted or just watched the most entertaining shows offered. No comprehensive studies probed what viewers absorbed over the course of a television season, much less a lifetime. Ellen Wartella and her colleagues have concluded that the medium's ubiquitousness and omnipresence have actually confounded obtaining accurate answers to such questions.[22] At some unrecorded moment in American history, exposure to television became as unremarkable as breathing and the long-term impact on viewers difficult, if not impossible, to measure.[23] What we do know is that, from the 1950s through the 1990s, the science content made available to viewers, young and old, grew increasingly more sensationalistic, more entwined within fiction, more preoccupied with moral dilemmas, and more visually inauthentic. Watson Davis once suggested, with characteristic facetiousness, that sensational images (or even words like "gold" and "sex") be sprinkled into science news stories to grab the audience's attention.[24] Many producers of science documentary programming appear to have made similar assumptions about how to ensure good ratings and please loyal viewers.

FICTIONALIZING MEDICINE, SPACE, AND SCIENCE

For subjects like medicine and space, the intermingling of fact and fiction began during television's first seasons. By the late twentieth century, documentary specials about medicine had thoroughly embraced entertainment techniques, focusing on a single patient as if she were a character in a drama or using handheld cameras to follow medical professionals at work just as photographers tracked animals on the hunt. ABC's miniseries *Hopkins 24/7* was filmed at the Johns Hopkins University teaching hospital, and the university's rationale for cooperation echoed that given for *Johns Hopkins Science Review* fifty years earlier: that the series would contribute to public understanding and support of medical research and training.[25] For fictional medical dramas, "realism" continued to be validated by employing high-profile medical consultants, just as James Moser had done for *Medic*.[26] Since the 1970s, physicians' groups had praised daytime soap operas for including plots that sought to communicate reliable information about contemporary personal and public health problems; now, prime-time drama series like *ER* similarly referenced contemporary health policy debates (including access to experimental treatments).[27] Television also created a new type of "celebrity doctor" for the morning and evening news shows. With hair carefully coiffed, dressed in fashionable clothes, and with makeup to erase signs of age or exhaustion, real (and, in some cases, practicing) physicians became regular cast members, often looking as handsome as the actors on *ER*.[28]

Space coverage experienced similar blending of authenticity and fantasy. Independent space heroes disappeared from television drama, rendered less plausible by the international space missions employing thousands of people. Fictional space missions became team efforts. *Star Trek*, *Space: 1999*, *Ark II*, and *Battlestar Galactica* revolved around interdependent groups, militaristic in demeanor and armed with terrible weapons, who achieved victory through cooperative behavior and innovative science and engineering. *The Six Million Dollar Man* and *The Bionic Woman* envisioned fantastic applications of medical technologies to human and social problems; on *V*, *War of the Worlds*, *Alien Nation*, and *The X Files*, aliens exploited scientific savvy in their attempts to conquer Earth.[29]

Drama became such a popular venue for fictional science that physicist and Nobel laureate Leon Lederman decided to campaign for a prime-time drama centered on real scientists' lives and work.[30] The new project (which he named *L.A. Science* in imitation of the popular series *L.A. Law*) would, Lederman explained in 1992, include "the usual sex and drama and car chases, but the hero would be a scientist, and every segment would teach something about science."[31] The idea was not all that radical—twenty years before, an MGM producer had suggested a "weekly action adventure" series based on the work of a Smithsonian scientist.[32] Lederman received grants from NSF and the Department of Energy, and hired producer Adrian Malone to develop a pilot script. Network executives (undoubtedly charmed by the witty, twinkle-eyed physicist with such admirable social goals) listened politely, but said no. Some of the executives reportedly responded that they "didn't do science" (as if it were a sport or hobby).

Science was, of course, already in their network schedules, infused and incorporated within drama, fictional characters, celebrity interviews, documentaries, and news reports, exploited as a plot device for solving crimes or saving lives. Science had become part of television's dramatic content, not set apart from it. Within a decade of the Lederman initiative, on dramas like *CSI* and *Bones*, teams of forensic scientists were combining research with criminal investigation. The characters constructed experiments to test theories about how a crime might have been committed, and employed chemical, biological, and computer analysis techniques, including knowledge of entomology and botany, to identify the culprits: "Rather than alienate viewers who also never majored in science, 'CSI' has become a ratings winner by offering them hope in a troubled time. When everything in life is uncertain, you can always rely on the certainty of science."[33] Episodes of another popular crime series, *Law and Order*, sometimes included fictional scientists who committed murder to protect their flawed or illegal research projects. In all these dramas, whether being used to assist criminality or law enforcement, science was *referenced* as much as represented. Viewers were continually reminded that essential moral and economic choices accompanied the use of scientific knowledge, for right or for wrong.

ARTIFICE, ARTIFICIALITY, AND BLACK BOXES

Did the television era simply offer a routine next stage for science popular-ization, an enhanced version of radio's science shows? Or did the medium's inherent characteristics—its pervasiveness, use of animation and recon-struction, synthetic sense of time, and "simultaneous" mass audiences—distinguish television science content from that disseminated via print, radio, or film? How did the medium alter science's messages?

In 1969, Serena Wade and Wilbur Schramm outlined the unique place that television had already come to occupy in American life, as a dom-inant influence on cultural understanding. "From the parade of events through television," viewers filled in "facts and findings" to supplement their school learning and life experiences, Wade and Schramm wrote.[34] Analysts like Ien Ang argue that television images "pervade the texture of our everyday lives" even when the sets are dark.[35] Raymond Bellour writes that such physical omnipresence "goes hand in hand with the fact that ev-erything, absolutely everything, happens on television (indiscriminately and simultaneously)."[36] Because of this pervasiveness within the culture, television's marginalization, incorporation, and reinterpretation of sci-ence makes a difference, influencing public reception and understanding even as the medium itself now becomes transformed by newer technologies and modes of access, online and off.

Television's "seeming reality" has long posed a special challenge for communication of scientific knowledge.[37] The edges of real life and fantasy blur within even the most well-crafted documentaries. "Science comes to television fully clothed," Roger Silverstone notes; by the time informa-tion about or created by science becomes incorporated into television con-tent, the knowledge has reached the penultimate stage, has shifted from idea (a private process) to publication (where the work is made public).[38] Television then drapes veils of illusion over the science and refashions the resulting narrative into something either entertaining and amazing, in-teresting and amazing, or frightening and amazing—always pushing the account farther from the real research experience. *NOVA* once added the label "based on actual astronomy" to a program, as if no longer relying on viewers' ability to distinguish between fact and fictionalization.

From the beginning, television popularizers had to learn how to engage viewers without boring or confusing them, but that creative impulse in-troduced elements of inauthenticity. Simply filming activities within real laboratories conveyed too much actuality, the result unbearably dull, antiseptic, or inexplicable, and so programs began to include reconstructions of events. Mr. Wizard replaced neighborhood children with actors because the latter could keep the action moving during "experiments" that were, in fact, scripted replicas with predictable outcomes. Directors of natural history programs began to shift points of view with multiple cameras or to combine film footage of animals in one location with sound recordings of similar animals elsewhere. Even David Attenborough's widely praised miniseries *The Life of Birds* adapted sophisticated computer animation to "show" prehistoric birds "flying" and then superimposed those images on film of genuine forests and clouds. Blending reality with imaginary reconstructions and animations, "recreating" astronomical events not visible to the human eye, using time-lapse photography to speed up experiments, all became commonplace techniques in the world of science and nature documentaries.

Broadcasting's synthetic state also continuously distorts time's arrow. There is "real time" and "media time," each with different perspectives.[39] Compare, for example, the experience of watching the 1939 World's Fair parade on television with that of watching the same event in person. The former offered comfortable seating and closer views of the dignitaries; the latter offered a "real" experience. Standing on the sidelines, parade watchers experienced the bad (delays, sore feet) as well as the good. It should not be surprising, then, that people increasingly opted for the synthetic, especially if they had little hope of experiencing the real—or if the real was prohibitively expensive, uncomfortable, or impossible to access (such as the atomic weapons tests). Ray Funkhouser and Eugene Shaw argue that the "prevalence" of synthetic media experiences has also trained viewers to expect an unrealistic level of perfection: "a very high proportion of what we see is far smoother, more expert, and more polished than would be possible for anyone to experience as an on-the-spot observer."[40] Television news producers chose the most articulate and presentable scientists for interviews; experiments on science shows invariably succeeded; and in fictional dramas, scientific testing always proved helpful, if not essential, in preventing disasters or solving crimes. Whenever these synthetic images conformed with real or imagined experience, viewers could more easily accept the mediated world as natural.[41] Constance Penley has suggested that artful synthesis in fact encourages audiences to impose their own "reality" and therefore to "slash" the information, such as merging the real space agency (NASA) in their memories with the fictional organizations of *Star Trek* (into what Penley dubbed "NASA/TREK").[42] Bill McKibben decries similar distortions in how television has represented nature. In an "age of missing information" about the natural world, people accustomed to television's artificial perfection may find unenhanced or untamed reality disappointing.[43] In the wild, real lizards scurry under rocks, and the yellow warblers disappear into the canopy.

Television has tended to present scientific knowledge in isolation (here is what we know) rather than derived from theory and prior effort (here is how we came to know what we know) and to declare everything with equally unqualified assurance. June Goodfield has pointed out that popular narratives often marginalize the significance of hypothesis building, ignoring "the patterns, the limits, the nature of discovery, the balance of certainty and uncertainty . . . the methodology of science and the spirit of science . . . along with their underlying relationship to the basic factual information."[44] Harry Collins and others in the science studies community similarly fault television for describing science as "unambiguous and intractable knowledge" rather than subject to qualifications or amendment.[45] Episodes of magazine-format series like *NOVA*, *Scientific American Frontiers*, and *Nature*, which are produced as separate films, present not only a lack of continuity from week to week but also, at the end of each hour, a false message of completeness, as if no more need be said about black bears, beriberi, or black holes.

By the late 1970s, what television had begun to do increasingly well was to explore science's politics and ethics, aspects otherwise obscured within the laboratory "black box."[46] This trend offers one ray of hope. Through storytelling, biologist Richard D. Klausner points out, popular accounts can explain that "science is part of a long and continuous narrative": "the real stories of science reflect its set of values—the freedom to reject past assumptions; the possibility that the future is different from the present and these differences are made that way by our own work and by our knowledge; that power can come from evidence and not authority. These noble ideas play out in the sweat and grime of real human dramas of competition and failure, of frustration and ambition—but most of all—of amazing curiosity."[47] Such stories have also conformed to a tendency to tell viewers "what to think about" (that is, to set agendas for public debate) rather than tell them what to think.[48] By signaling importance, television encouraged viewers to notice information about science in other sources.[49] By setting science out for display, television primed viewers' interest and declared the topic worthy of attention. By delving into researchers' moral conflicts, documentaries put a human face on science. Fighting to compete in a media environment dominated by sensationalization, controversy, and personalities, television's science adjusted to its own environment, selecting those aspects that intrigued the most viewers.

Moreover, in the television era, the audience was "never really neutral."[50] What appeared on the screen reflected the aggregated, uncoordinated assumptions that networks, stations, producers, writers, advertisers, and underwriters made about the audiences they sought to reach and about what those millions seemed to prefer. It was, after all, the viewers who held in their hands the power to change the channel.

SEEING BEYOND TELEVISION

Even though Frank Baxter had experienced the essential power of television firsthand, he knew its limitations—neither "a vending machine for higher learning" nor a technology for pumping "culture" (or science) into everyone's veins.[51] When the world transitioned into the Internet age, some observers began to suggest that television would become irrelevant or obsolete.[52] Yet, like the vending machine, the technology was adapted to a reconfigured marketplace. Science programs began supplementing their content with online visuals and interactive features, much as telecommunications specialists had, in fact, urged the Smithsonian to prepare to do. Individuals will eventually be able to "access the Smithsonian like a data bank," they predicted in 1972: "The idea is to organize a body of data, graphic and verbal, so that individual packages of material can be assembled by the computer. These would consist of bibliographies, images of materials from the collections, videotaped film, accompanied sometimes by 'real' material from a lending collection. The package could be delivered by CRT display in a part of the museum or at a neighborhood center . . . and eventually it will of course arrive over one's own tube."[53] ARPANET, Internet, and what eventually became graphical interfacing on the Web, along with cable and wireless communications, affordable and portable video recording, and compact, efficient, and inexpensive reception devices were all becoming visible on the horizon. By the first decade of the twenty-first century, individuals could watch television in airports or automobiles or wherever a wireless signal was strong. The knowledge produced by scientific and cultural organizations could, indeed, be accessed like data banks.

The challenges faced by popularizers remained unchanged, however, for no group took on the enormous task of helping consumers make sense of the science information made available. And the expansion of outlets only increased each viewer's dilemma. Faced with overwhelming choices, and increasingly ill-equipped by formal education to be able to discern science from pseudoscience or to unpack complicated technical information,

Americans scurried into comfortable niche sources, too often sampling from favorite programs rather than exploring anything new.

The cultural impact of television, whether you watch it or not, appear on it or not, pay close attention to it or not, has been wide, deep, and complex. Twentieth-century politicization of the public conversation about science, coupled with the mass media's power to construct illusionary images of science and to project them into the kaleidoscopic environment of television entertainment, refashioned the relationship between scientists and the public. Television fractured the artificial formality that had characterized traditional popularization. The medium magnified every mistake, enhanced every bead of sweat on the upper lip, and dropped a veil of fantasy over the whole. Radio, William Albig once wrote, had separated "ideation" from "visual perception," thereby encouraging the mind's eye to supply the vision.[54] On television, those functions seemed initially reunited, but visuality soon reigned supreme. The need to attract the largest possible audiences pushed television's version of science, whether intended as education or fiction, ever more toward sensationalization, politics, celebrities, and representation and away from discussion of ideas, away from the real, away from attention to the thought and reasoning behind scientific conclusions and recommendations.

Was television to blame? As the television age was beginning, CBS educational consultant Lyman Bryson observed that "wisdom and nonsense are found in human beings and not in the machines that they use."[55] We choose how communications technologies are used and, as societies, we make of them what we will. Television has offered humans extraordinary glimpses of our wide, wide world, and of the worlds beyond our own planet, and humanity is richer for those opportunities. But there is more to life than television. Conservationist Aldo Leopold wondered in 1948 whether a higher standard of living, with unlimited access to new technologies, would be "worth its cost in things natural, wild, and free," and he concluded then that the "opportunity to see geese is more important than television."[56] Even if cameras and spectacularly inventive cinematography now allow viewers to share in the illusion that we are gliding on avian wings, with running commentary supplied by famous ornithologists, the reality of the state of the planet supplies a reason to peer beyond the screen and to care where, why, and whether those wild geese fly and about the research attempting to preserve them. There is more to *science* than what appears on television.

Acknowledgments

To EVA, the Tamster, and charter members of the Frank Thone Fan Club, let us keep on dancing and laughing till the cows come home.

Books may be written and read in solitude, but they achieve actuality only through wider interactions. For historians, that social networking begins with archivists and curators, whose skill, ingenuity, and dedication to the preservation of history's raw materials provide the substance on which the research can be built. Throughout this project, I have been fortunate to work with an extraordinary group of people at the Smithsonian Institution Archives (SIA). Thank you to SIA director Anne Van Camp for providing a temporary institutional home to facilitate this research, and to Pamela Henson, *the* Historian of the Smithsonian, for sharing invaluable insights into the institution's people, culture, and evolution and the archival collections. Special thanks, as always, to Ellen Alers, Tammy Peters, Tad Benicoff, Mary Markey, and the SIA Archives Division, and to Marguerite Roby, Michael Barnes, and other SIA photo wizards for so much help through the years. Every member of the SIA staff, however, has assisted this project in some way—from a needed smile on a rainy day to lifting one more box.

Thank you to National Museum of American History curators Ann Seeger and Shannon Perich for assistance in locating images and to Janice Goldblum at the National Academies Archives for help in navigating the IGY records. Thank you also to archivists at the Library of Congress's Manuscript Division and Prints and Photographs Division, Hagley Museum and Library, and Johns Hopkins University's Milton S. Eisenhower Library.

The conceptual phase of this book stretches back decades—from graduate-student debates in a living room under the Kenmore Square CITGO sign to many conversations with television professionals and to observing interactions between the scientists and media representatives with whom I served through the years on the AAAS Committee on Public Understanding of Science and the NAS Committee on Public Understanding of Science. Mary Jane McKinven, Jonathan Cobb, Don Reisman, Donna Gerardi Riordan, and the late Lawrence Badash commented on previous unwieldy manuscripts and, in the process, made this book better. Many friends and colleagues took time to share their "television experiences": Tony Penna recalled watching Edward Teller on television and Wil Lepkowski first mentioned *Mr. I. Magination*, while countless others shouted "Mr. Wizard!" Thank you also to historians Virginia Myhaver and Bruce Lewenstein for providing copies of materials on popularization.

Helen Solorzano shared a copy of her grandmother's film and notes from the 1952 atomic bomb test and thereby inspired several parts of this book. Helen has now donated that and other related films to SIA, where they will be preserved and available to many generations of historians. Helen's uncle, Miles Davis, has been more than kind in describing life with his parents, Watson and Helen Miles Davis, and in generously donating materials to the Smithsonian that will further enrich understanding of science popularization during the twentieth century.

It was University of Chicago Press editorial director Christie Henry who, as I finished a book about popularization via radio, encouraged me to consider tackling the daunting topic of television. Without her support and enthusiasm, this book would not exist. I was also graced with excellent, conscientious peer reviewers, whose anonymous comments and suggestions made this a far better book. The entire crew at the University of Chicago Press, especially Abby Collier and Joann Hoy, has been a pleasure to work with throughout this project.

The support of good-humored friends and neighbors, of course, makes it all worthwhile. I owe so much to Irene and Mickey Schubert, Chris Cherniak, David R. Gessner, Mary Cole, Heather and Raj Sabharwal, and David Walcyzk, and look forward to many more good times and conversations. Little Albert, you chased the song neutrinos through every paragraph.

To my husband Jeffrey Stine, you have made my life more beautiful in every way. Now it is time to turn off the television, take you by the hand, and go for a long walk. Perhaps we will even see some geese flying natural, wild, and free.

Notes

ABBREVIATIONS USED IN MANUSCRIPT CITATIONS

Acc. = Accession
IGY = International Geophysical Year Records
NAA = National Academies Archives
RG = Record Group
RU = Record Unit
SIA = Smithsonian Institution Archives

CHAPTER ONE

1. Quoted in University of Chicago Committee on Educational Television, "Television and the University," *School Review* 61 (April 1953): 225.

2. "Dedication of RCA Seen on Television," *New York Times*, April 21, 1939.

3. H. G. Wells, "World of Tomorrow," *New York Times*, March 5, 1939.

4. Allen B. Du Mont, "Television Now and Tomorrow," *Journal of Marketing* 9 (January 1945): 287.

5. University of Chicago Committee, "Television and the University," 202.

6. Richard Koszarski, "Coming Next Week: Images of Television in Pre-war Motion Pictures," *Film History* 10 (1998): 128–140; and "Radiovision in Homes This Fall," *Science News Letter* 14 (August 25, 1928): 113.

7. "Sights to Be Seen: New York World's Fair, 1939," *New York Times*, March 15, 1939; and Robert W. Rydell, "The Fan Dance of Science: American World's Fairs in the Great Depression," *Isis* 76 (December 1985): 525–542.

8. Helen A. Harrison, "Stuart Davis's 'World of Tomorrow,'" *American Art* 9 (Autumn 1995): 97; and Jody Patterson, "Modernism and Murals at the 1939 New York World's Fair," *American Art* 24 (Summer 2010): 50–73.

9. "Dedication of RCA Seen on Television," *New York Times*, April 21, 1939; Orrin E. Dunlap, Jr., "Today's Eye-Opener," *New York Times*, April 30, 1939; Orrin E. Dunlap, Jr., "Cer-

emony Is Carried by Television as Industry Makes Its Formal Bow," *New York Times*, May 1, 1939; Joseph H. Udelson, *The Great Television Race: A History of the American Television Industry, 1925–1941* (University: University of Alabama Press, 1982), 129–130; Ron Becker, "'Hear-and-See Radio' in the World of Tomorrow: RCA and the Presentation of Television at the World's Fair, 1939–1940," *Historical Journal of Film, Radio, and Television* 21 (October 2001): 370; and Anthony Smith, ed., *Television: An International History* (Oxford: Oxford University Press, 1995), 28–30. See also "RCA Presentation: Television," http://www.archive.org/details/RCAPrese1939 ?start=209.5.

10. "Visitors Take Part in Television Show," *New York Times*, May 4, 1939; and "Notes on Television," *New York Times*, June 11, 1939.

11. Orrin E. Dunlap, Jr., "Act I Reviewed," *New York Times*, May 7, 1939.

12. John Cabot Smith, "Television Reaches Capital and What Seemed Magic Becomes a Fact," *Washington Post*, January 29, 1939; and Orrin E. Dunlap, Jr., "Birth of the News: Scientists and Broadcasters to Open Bag of Tricks at the World's Fair," *New York Times*, March 5, 1939.

13. Dunlap, "Act I Reviewed."

14. Du Mont, "Television Now and Tomorrow," 276–279; William C. Ackerman, "The Dimensions of American Broadcasting," *Public Opinion Quarterly* 9 (Spring 1945): 1–18; and C. V. Newsom, "Radio and Television," *Scientific Monthly* 79 (October 1954): 248–252.

15. Robert N. Farr, "Television, When and How," *Science News Letter* 46 (July 22, 1944): 58–59; and Erik Barnouw, *The Golden Web: A History of Broadcasting in the United States, 1933–1953* (New York: Oxford University Press, 1968), 216–245.

16. David Weinstein, *The Forgotten Network: DuMont and the Birth of American Television* (Philadelphia: Temple University Press, 2004).

17. Du Mont, "Television Now and Tomorrow," 278–279.

18. Dallas W. Smythe, "A National Policy on Television?," *Public Opinion Quarterly* 14 (Autumn 1950): 469.

19. Harry P. Warner, "Television and the Motion Picture Industry," *Hollywood Quarterly* 2 (October 1946): 11–18.

20. Jack Gould, "Television: Boon or Bane?," *Public Opinion Quarterly* 10 (Autumn 1946): 314–320.

21. Ibid., 314–315.

22. Robert T. Bower, *Television and the Public* (New York: Holt, Rinehart and Winston, 1973), 7.

CHAPTER TWO

1. Lynn Poole, "Science on Video," *New York Times*, March 9, 1952.

2. Judy DuPuy, *Television Show Business* (Schenectady, NY: General Electric, 1945); and Robert L. Gibson, "Some Preferences of Television Audiences," *Journal of Marketing* 10 (January 1946): 289–290.

3. The title—identical to the 1945–1946 radio series—was the company's advertising slogan. *Serving through Science: A Series of Talks Delivered by American Scientists on the New York Philharmonic-Symphony Program* (New York: U.S. Rubber Company, 1946); and Tim Brooks and Earle Marsh, *The Complete Directory to Prime Time Network and Cable TV Shows, 1946–Present*, sixth edition (New York: Ballantine Books, 1995), ix, 919.

4. "Writer Hails Use of Teaching Films," *New York Times*, December 4, 1946.

5. Joan Aucourt, "Television: A Double Take," *Hollywood Quarterly* 3 (Spring 1948): 258.

6. Philip Wylie, quoted in Mort Weisinger, "Swooning in the Gloom," *Los Angeles Times*, May 23, 1948. See also Jack Gould, "Television Builds for a Future of a Boundless Promise," *New York Times*, June 13, 1948.

7. Melvin Maddocks, "Revolution in a Living Room," *Christian Science Monitor*, November 22, 1957.

8. "Mars, Saturn, Moon Put on Telecast Act," *Christian Science Monitor*, March 22, 1948; and "Moon over D.C.," *Washington Post*, April 14, 1948.

9. Sonia Stein, "Chevalier Show a Bell-Clanger," *Washington Post*, April 10, 1949.

10. Brooks and Marsh, *Complete Directory*, 732–733. See also Roy K. Marshall, *The Nature of Things* (New York: Henry Holt, 1951).

11. "An Astronomer Clicks Doing TV Commercials," *Chicago Daily Tribune*, October 1, 1950; and John Crosby, "They Do Get Curiousest Items on TV," *Washington Post*, September 26, 1951.

12. Val Adams, "Educational Television: An Opinion," *New York Times*, June 10, 1951.

13. Marshall's television career came to an abrupt end in November 1953, when he was indicted for sending obscene replies to teenage girls who had written fan letters; he was sentenced to probation and treatment. "Judge Reveals Indictment of Commentator," *Chicago Daily Tribune*, January 13, 1954; "Dr. Marshall Confined," *New York Times*, January 16, 1954; and "Scientist Put on Probation for Five Years," *Chicago Daily Tribune*, May 26, 1954.

14. John Crosby, "Nature of Things as Marshall Sees It," *Washington Post*, May 6, 1951.

15. Roy K. Marshall, "Televising Science," *Physics Today* 2 (January 1949): 26; and Roy K. Marshall, "Learning Is Fun—If You Get It via Television," *Washington Post*, September 26, 1948.

16. "Science Programs Go Moon-Gazing," *Washington Post*, April 14, 1948.

17. "A-Bomb Program Alarms Audience," *Washington Post*, September 19, 1948.

18. Don K. Price, *The Scientific Estate* (Cambridge, MA: Belknap Press of Harvard University Press, 1965).

19. Alice Kimball Smith, *A Peril and a Hope: The Scientists' Movement in America, 1945–47*, revised edition (Cambridge, MA: MIT Press, 1970); and David Hill, Eugene Rabinowitch, and John Simpson, "The Atomic Scientists Speak Up," *Life*, October 29, 1945, 45–48.

20. "Lynn Poole, Won Early TV Prizes," *New York Times*, April 16, 1969; Sue De Pasquale, "Live from Baltimore—It's the Johns Hopkins Science Review!," *Johns Hopkins Magazine*, February 1995, http://www.jhu.edu/~jhumag/295web/scirevu.html; Lynn Poole, *Science via Television* (Baltimore: Johns Hopkins University Press, 1950); and Leo Geier, *Ten Years with Television at Johns Hopkins* (Baltimore: Johns Hopkins University, 1958).

21. Johns Hopkins University Archives, JHU News Office Records, RG10.020, Series 10, 1:7.

22. Robert M. Yoder, "TV's Shoestring Surprise," *Saturday Evening Post*, August 21, 1954, 30, 90–92.

23. Irving Stettel, ed., *Top TV Shows of the Year, 1954–1955* (New York: Hastings House, 1955), 6.

24. Larry Wolters, "TV Really Is Educational," *Chicago Daily Tribune*, December 18, 1950; and Larry Wolters, "Lynn Poole Vies with Berle for Video Audience," *Chicago Daily Tribune*, January 9, 1951.

25. Jack Gould, "Science Offered over Television," *New York Times*, October 18, 1950; Leonard D. Pigott, "Biology Reaches a New Horizon: Science on Television," *AIBS Bulletin* 1 (July 1951): 7; Robert Lewis Shayon, "Johns Hopkins Science Review," *Christian Science Monitor*, January 30, 1951; and Jack Gould, "What TV Is—and What It Might Be," *New York Times Sunday Magazine*, June 10, 1951, 22.

26. Lynn Poole, "The Challenge of Television," *College Art Journal* 8 (Summer 1949): 299–304.

27. Wolters, "Lynn Poole Vies with Berle."

28. Ibid.

29. Poole, "Science on Video."

30. Ibid.

31. Geier, *Ten Years with Television*, 21.

32. T. Dale Stewart to J. E. Graf, March 18, 1952, SIA (RU50), 43:2.

33. Poole, *Science via Television*; and "The Hopkins on TV," *Newsweek*, March 17, 1952, 88–89.

34. Pigott, "Biology Reaches a New Horizon," 7; and J. S. Ames to William H. Howell, May 21, 1934, SIA (RU7091), 151:9.

35. John E. McCosker, "I Don't Want to See Any Water: Earl S. Herald, Aquarium Pioneer and Aquatic Dynamo," *California Wild* 56 (Spring 2003): 37–41. See also Graham DuShane, "Science on TV," *Science* 124 (November 16, 1956): 963.

36. Lawrence A. Williams, "'Science in Action' on California Television," *Ward's Natural Science Bulletin* 25 (September 1951): 4.

37. Benjamin C. Draper, "Producing 'Science in Action,'" in Benjamin C. Draper, ed., *The "Science in Action" TV Library*, volume 1 (New York: Merlin Press, 1956), xi, xv.

38. Poole, *Science via Television*, 79.

39. Draper, "Producing 'Science in Action,'" xv.

40. Lawrence E. Davies, "'Science in Action,'" *New York Times*, July 5, 1953; and DuShane, "Science on TV."

41. Gilbert Seldes quoted and interpreted in Carl Beier, Jr., "A New Way of Looking at Things," *Hollywood Quarterly* 2 (October 1946): 1.

42. "Eclipse to Star on TV Tomorrow," *New York Times*, June 29, 1954; and "Final Rehearsals for Eclipse Held," *New York Times*, June 30, 1954.

43. Wayne Thomis, "Big Shots Check In for A-Blast of Same Name," *Chicago Daily Tribune*, April 20, 1952; and William L. Laurence, "Atom Blast Today to Be Strongest of All Except Those at Eniwetok," *New York Times*, April 22, 1952.

44. Helen Miles Davis, "We'll Grope in Dark," *Science News Letter* 61 (May 3, 1952): 275. See also James von Schilling, *The Magic Window: American Television, 1939–1953* (New York: Haworth Press, 2003), 77–78; and Allan M. Winkler, *Life under a Cloud: American Anxiety about the Atom* (New York: Oxford University Press, 1993), 91.

45. Hugh Baillie, "TV Audience Views Atomic Bomb Test for First Time," *Las Vegas Sun*, April 22, 1952.

46. Jack Gould, "TV Brings Atomic Bomb Detonation into Millions of Homes, but Quality of Pictures Is Erratic," *New York Times*, April 23, 1952; "L.A. Sky Watchers Miss Blast Visible over TV," *Los Angles Times*, April 23, 1952; and "History Is Made," *Time*, May 5, 1952.

47. Jack Gould, "Yucca Flat Reflects Danger of Overstressing Atom Destruction at Cost to Civil Defense," *New York Times*, March 18, 1953.

48. Sonia Stein, "Incredibly, 'Bomb' Was Tame on TV," *Washington Post*, March 22, 1953.

49. John Beaufort, "Telecasting in Verbatim Style Weighed," *Christian Science Monitor*, March 24, 1953.

50. Sidney Lohman, "News and Notes from the Studios: The Atom Bomb Tests—Parades, Oddities and Studio Items," *New York Times*, March 15, 1953.

51. William C. Ackerman, "U.S. Radio: Record of a Decade," *Public Opinion Quarterly* 12 (October 1948): 440–454.

52. Flora Rheta Schreiber, "Television's New Idiom in Public Affairs," *Hollywood Quarterly* 5 (Winter 1950): 145.

53. Ibid.

54. Ackerman, "U.S. Radio: Record of a Decade," 442.

55. Charles A. Siepmann and Sidney Reisberg, "'To Secure These Rights': Coverage of a Radio Documentary," *Public Opinion Quarterly* 12 (Winter 1948/1949): 649–658.

56. Irving J. Gitlin, "Radio and Atomic-Energy Education," *Journal of Educational Sociology* 22 (January 1949): 327–328.

57. "ABC Blithely Looks Onward to 1960," *Washington Post*, September 7, 1947.

58. Susan Caudill, "Trying to Harness Atomic Energy, 1946–1951: Albert Einstein's Pub-

licity Campaign for World Government," *Journalism Quarterly* 68 (Spring/Summer 1991): 253–262; Schreiber, "Television's New Idiom," 151; and von Schilling, *Magic Window*, 137–138.

59. During the last season, the series was refocused on medical research and called *Medical Horizons*.

60. Author's analysis of program listings in Daniel Einstein, *Special Edition: A Guide to Network Television Documentary Series and Special News Reports, 1955–1979* (Metuchen, NJ: Scarecrow Press, 1987). See also Jeff Merron, "Murrow on TV: *See It Now, Person to Person*, and the Making of a 'Masscult Personality,'" *Journalism Monographs* 106 (July 1988); and James L. Baughman, *Same Time, Same Station: Creating American Television, 1948–1961* (Baltimore: Johns Hopkins University Press, 2007), 236–249.

61. Author's analysis of program listings in Einstein, *Special Edition*; Baughman, *Same Time, Same Station*, 243. See also Merron, "Murrow on TV."

62. Richard Schickel, *Intimate Strangers: The Culture of Celebrity in America* (Chicago: Ivan R. Dee, 2000), 4.

63. Erik Barnouw, *The Image Empire: A History of Broadcasting in the United States from 1953* (New York: Oxford University Press, 1970), 53–54; and Philip M. Stern, with the collaboration of Harold P. Green, *The Oppenheimer Case: Security on Trial* (New York: Harper and Row, 1969), 450–451.

64. Jack Gould, "Television in Review," *New York Times*, January 9, 1955.

65. Ibid.

66. "Scientists in Debate," *New York Times*, February 21, 1958.

67. Walter Kingson, "The Second New York Television Survey," *Quarterly of Film, Radio, and Television* 6 (Summer 1952): 317–326.

68. Gould, "What TV Is," 10, 22–23.

69. Gould, "Television in Review," *New York Times*, May 21, 1954.

70. Martin Grams, Jr., *The Official Guide to the History of the Cavalcade of America* (self-published by author, 1998), Hagley Museum and Library, Cavalcade of America Collection.

71. Larry Wolters, "Quiz and Forum Responses Go Over Big on TV," *Chicago Daily Tribune*, May 4, 1949.

72. Evan Z. Vogt, "Anthropology in the Public Consciousness," *Yearbook of Anthropology* (1955): 365; Brooks and Marsh, *Complete Directory*, 1122; Alex McNeil, *Total Television*, fourth edition (New York: Penguin Books, 1996), 905; "Profs as TV Pros," *Newsweek*, February 4, 1952, 47; and "Experts," *New Yorker*, May 3, 1952, 68, 70. In the beginning, objects were taken from a fake "pirate's chest"; later, they emerged in a puff of smoke. Froelich Rainey, "The Origins and History of *What in the World?*," *Pennsylvania Triangle* (November 1953).

73. "The University Museum TV Program," n.d., 2, SIA (RU50), 43:2.

74. "Experts," 68.

75. Baughman, *Same Time, Same Station*, 9–10, 56–81.

76. House Committee on the Judiciary, Antitrust Subcommittee, *The Television Broadcasting Industry*, 85th Cong., 1st Sess., June 24, 1957, H. R. Report No. 607, 1–2.

77. Thomas Streeter, *Selling the Air: A Critique of the Policy of Commercial Broadcasting in the United States* (Chicago: University of Chicago Press, 1996).

78. James Schwoch, *Global TV: New Media and the Cold War, 1946–69* (Urbana: University of Illinois Press, 2009).

79. "Competition and TV Program Content," *University of Chicago Law Review* 19 (Spring 1952): 556–573.

80. Roger Manvell, "Experiments in Broadcasting and Television," *Hollywood Quarterly* 2 (July 1947): 388; and "BBC Announces Television Plan Resumes in June," *Christian Science Monitor*, April 9, 1946.

81. Manvell, "Experiments in Broadcasting," 388.

82. "Science on Television," *Discovery* 14 (April 1953): 103.

83. Robert J. Williams, "The Politics of American Broadcasting: Public Purposes and Private Interests," *Journal of American Studies* 10 (December 1976): 331.

84. Paul C. Adams, "Television as Gathering Place," *Annals of the Association of American Geographers* 82 (March 1992): 117–135.

85. Robert Lewis Shayon, "Johns Hopkins Science Review," *Christian Science Monitor*, January 30, 1951.

CHAPTER THREE

1. Frank Baxter quoted in "The Wide, Wide World," *Time*, April 11, 1955.

2. Sidney Lohman, "Radio Row: One Thing and Another," *New York Times*, February 3, 1946.

3. Telford Taylor, "Finding a Place for Education on TV," *New York Times Sunday Magazine*, January 28, 1951, 9, 14–15; Charles A. Siepmann, "The Case for TV in Education," *New York Times Sunday Magazine*, June 2, 1957, 13, 42, 44, 47; and Parker Wheatley, "Radio and Television as Instruments of Education," *Bulletin of the American Academy of Arts and Sciences* 3 (October 1949): 2–4.

4. Walter Kingson, "The Second New York Television Survey," *Quarterly of Film, Radio, and Television* 6 (Summer 1952): 317–326.

5. University of Chicago Committee on Educational Television, "Television and the University," *School Review* 61 (April 1953): 202–225.

6. Kingson, "Second New York Television Survey."

7. Jack Gould, "What TV Is—and What It Might Be," *New York Times Sunday Magazine*, June 10, 1951, 22.

8. Taylor, "Finding a Place," 9.

9. Ibid., 14.

10. R. D. Heldenfels, *Television's Greatest Year—1954* (New York: Continuum, 1994), 182.

11. J. M. Hutzel, "AAAS Centenary: A Preliminary Report," *Science* 108 (October 22, 1948): 428–429.

12. Warren Weaver, "AAAS Policy," *Science* 114 (November 2, 1951): 471–472; and John A. Behnke, "Television Takes Education to the People," *Science* 118 (October 9, 1953): 3.

13. Bruce V. Lewenstein, "The AAAS and Scientific Perspectives on Public Understanding, 1945–1980," paper presented at the 1991 annual meeting of the American Association for the Advancement of Science; and Bruce V. Lewenstein, "'Public Understanding of Science' in America, 1945–1965" (PhD dissertation, University of Pennsylvania, 1987); Weaver, "AAAS Policy."

14. Lewenstein emphasizes this point in "'Public Understanding of Science' in America," and "AAAS and Scientific Perspectives."

15. Lewenstein, "'Public Understanding of Science' in America," chapter 3; and Lewenstein, "AAAS and Scientific Perspectives," 22. See also Sidney S. Negus, "Public Information at Philadelphia," *Science* 115 (February 15, 1952): 177–178.

16. Remarks by Warren Weaver in *Conference on the Interpretation of the Natural Sciences for a General Public*, Rye, NY, June 15–16, 1938, volume 1, 21–22 (SIA [RU7091], 381).

17. Remarks by W. F. Ogburn, ibid., 29.

18. Paul F. Lazarsfeld and Patricia L. Kendall, *Radio Listening in America: The People Look at Radio—Again* (New York: Prentice-Hall, 1948), 4–8; and Sydney W. Head, *Broadcasting in America: Survey of Television and Radio* (Cambridge, MA: Riverside Press, 1956), 183.

19. John R. Zaller, *The Nature and Origins of Mass Opinion* (Cambridge: Cambridge University Press, 1992), 20.

20. Ivan Karp, "High and Low Revisited," *American Art* 5 (Summer 1991): 14.

21. Head, *Broadcasting in America*, 420.

22. Bernard Rosenberg, "Mass Culture in America," in Bernard Rosenberg and David Manning White, eds., *Mass Culture: The Popular Arts in America* (Glencoe, IL: Free Press, 1957), 5; and Paul R. Gorman, *Left Intellectuals and Popular Culture in Twentieth-Century America* (Chapel Hill: University of North Carolina Press, 1996), 165, 114.

23. Quoted by Paul Lazarsfeld in Norman Jacobs, ed., *Culture for the Millions? Mass Media in Modern Society* (Boston: Beacon Press, 1961), xii.

24. Head, *Broadcasting in America*, 420–421.

25. Frank Stanton in Jacobs, *Culture for the Millions?*, 348.

26. Stanton in ibid., 352.

27. Neal Gabler, *Life the Movie: How Entertainment Conquered Reality* (New York: Knopf, 1999), 20–21.

28. Ibid., 20–21.

29. Gorman, *Left Intellectuals*, 114, 121, 165.

30. Karp, "High and Low Revisited," 15.

31. Gorman, *Left Intellectuals*, 165; and Patrick Hazard in Jacobs, *Culture for the Millions?*, 157.

32. Jeremy Green, "Media Sensationalism and Science: The Case of the Criminal Chromosome," in Terry Shinn and Richard Whitley, eds., *Expository Science: Forms and Functions of Popularisation*, special issue of *Sociology of the Sciences* 9 (1985): 139.

33. Ibid., 139–161.

34. Anthony Smith, "Technology, Identity, and the Information Machine," *Daedalus* 115 (Summer 1986): 160. Smith was paraphrasing Plato.

35. "The Wide, Wide World," *Time*, April 11, 1955.

36. Evan Z. Vogt, "Anthropology in the Public Consciousness," *Yearbook of Anthropology* (1955): 366.

37. Triangle Publications, Radio and Television Division, *The University of the Air: Commercial Television's Pioneer Effort in Education* (Philadelphia: Triangle Publications, 1959).

38. Marguerite Lehr, "An Experiment with Television," *American Mathematical Monthly* 62 (January 1955): 15–21.

39. Triangle Publications, *University of the Air*. See also Paul L. Chessin, "Spies, Electric Chairs, and Housewives," *American Mathematical Monthly* 65 (June/July 1958): 416–421.

40. "Education: Eye Opener," *Time*, February 9, 1959.

41. Brendan Gill, "E/M Etc.," *New Yorker*, February 7, 1959, 24–25.

42. "Education: Eye Opener."

43. For example, Frederick Mosteller, "Continental Classroom's Television Course in Probability and Statistics," *Revue de l'Institut International de Statistique* 31, no. 2 (1963): 153–162.

44. Vogt, "Anthropology in the Public Consciousness," 366–368; and Ayrlene McGahey Jones, "Television Activity, Department of Mathematics, University of Alabama," *American Mathematical Monthly* 65 (June/July 1958): 421. MIT graduate Val Fitch hosted *Science Reporter* on WGBH-TV during the 1960s.

45. "Report of the Committee on the Role and Opportunities in Broadcasting—To President Edward H. Levi, August 15, 1972," *University of Chicago Record* 7 (April 21, 1973): 161.

46. Frederic A. Leigh, "Educational and Cultural Programming," in Brian G. Rose, ed., *TV Genres: A Handbook and Reference Guide* (New York: Greenwood Press, 1985), 367.

47. Gould, "What TV Is," 22.

48. *Report to Annual Meeting of Board of Trustees of Science Service*, April 24, 1949, 7, Library of Congress, Robert A. Millikan Collection (microfilm), reel 16, folder 15.4, frame 373.

49. Fellowship brochure, Johns Hopkins University Archives, JHU News Office Records, RG10.020, Series 10, 1:5.

50. Reprint of article "From Mousetraps to the Moon with Johns Hopkins File 7" (n.d.,

published between 1956 and 1960), Johns Hopkins University Archives, JHU News Office Records, RG10.020, Series 10, 1:3.

51. Leo Geier, *Ten Years with Television at Johns Hopkins* (Baltimore: Johns Hopkins University, 1958), 17.

52. Gary Schultz, "Is Video a Good Education Medium?," *Wisconsin Alumnus* 53 (June 1952): 8–10; and Heldenfels, *Television's Greatest Year*, 179.

53. Schultz, "Is Video a Good Education Medium?," 8–10.

54. John Crosby (1950) quoted in Heldenfels, *Television's Greatest Year*, 178.

55. House Committee on the Judiciary, Antitrust Subcommittee, *The Television Broadcast Industry*, 85th Cong., 1st Sess., June 24, 1957, H. R. Report No. 607, 41.

56. Heldenfels, *Television's Greatest Year*, 183.

57. Gordon Hubbel to Alexander Wetmore, June 15, 1948, SIA (RU50), 43:2.

58. Alexander Wetmore to Gordon Hubbel, July 7, 1948, SIA (Acc. T91027), 1:1.

59. For example, J. E. Graf to Alexander Wetmore, November 19, 1945; and Selden Menefee to Webster True, October 5, 1946, and October 21, 1946, SIA (Acc. T91027), 1:1.

60. Theodore F. Koop to Alexander Wetmore, March 7, 1951, SIA (RU50), 43:2.

61. "To the Planning Committee," May 14, 1951, SIA (RU50), 43:2.

62. Alexander Wetmore, "Memo to Files," November 3, 1951, SIA (RU50), 43:2.

63. Frank A. Taylor, memorandum, February 26, 1952, SIA (RU50), 43:2.

64. Martin Took, "Senator Asks TV Show for Smithsonian," *Minneapolis Tribune*, April 25, 1953; and Jack Wilson, "Case Would Save Shoes—Let TV Tour Smithsonian," *Des Moines Register*, May 10, 1953.

65. Francis Case to Leonard Carmichael, April 13, 1953; and Leonard Carmichael to Francis Case, April 15, 1953, SIA (RU50), 215:6.

66. J. Hugh E. Davis to Leonard Carmichael, October 2, 1953, SIA (RU50), 43:2.

67. Paul H. Oehser to Leonard Carmichael, June 12, 1953, SIA (RU50), 215:4.

68. Correspondence and draft contract, SIA (RU50), 215:10.

69. Smithsonian regent Vannevar Bush brokered these meetings. Vannevar Bush to Cleo F. Craig, March 9, 1954, SIA (RU50), 43:2; and various material in SIA (RU50), 214:8.

70. William M. Jones and Andrew Walworth, "Saudek's *Omnibus*: Ambitious Forerunner of Public TV," 1999, http://www.current.org/coop/coopomni.html.

71. Robert Saudek to Leonard Carmichael, August 31, 1954, SIA (RU50), 215:2.

72. Robert Saudek to Leonard Carmichael, September 15, 1954, SIA (RU50), 215:2.

73. Leonard Carmichael to Robert Saudek, September 30, 1954, SIA (RU50), 215:2.

74. Leonard Carmichael to Paul A. Rosen, December 13, 1954, SIA (RU50), 214:10.

75. Correspondence in SIA (RU50), box 43.

76. Jack Gould, "Television: Boon or Bane?," *Public Opinion Quarterly* 10 (Autumn 1946): 319–320.

77. Dallas W. Smythe, "A National Policy on Television?," *Public Opinion Quarterly* 14 (Autumn 1950): 462.

78. Gould quoted in Smythe, "National Policy," 462.

79. Lawrence W. Lichty, "Success Story," *Wilson Quarterly* 5 (Winter 1981): 63–64. For a succinct history of the devolution of cooperation between educators and commercial broadcasters, see Eugene Leach, "Tuning Out Education: The Cooperation Doctrine in Radio," *Current*, December 13, 1999 (originally published as "Snookered 50 Years Ago," 1983) http://www.current.org/coop/.

80. Head, *Broadcasting in America*, 408.

81. John Crosby, "The Full Flowering of Wally Cox," *Washington Post*, August 6, 1952. See also Val Adams, "The Amiable 'Mr. Peepers' in the Flesh," *New York Times*, July 27, 1952.

82. Jack Gould, "Radio and Television," *New York Times*, August 13, 1952, 29.

83. Walt Disney, "Mickey as Professor," *Public Opinion Quarterly* 9 (Summer 1945): 119-125.
84. Ibid., 120.

CHAPTER FOUR

1. David Sarnoff, "Probable Influence of Television on Society," *Journal of Applied Physics* 10 (July 1939): 428, quoted in Joseph H. Udelson, *The Great Television Race: A History of the American Television Industry, 1925-1941* (University: University of Alabama Press, 1982), 95.
2. Bernard Lightman, *Victorian Popularizers of Science: Designing Nature for New Audiences* (Chicago: University of Chicago Press, 2007).
3. Irving Lerner, "Director's Notes," *Hollywood Quarterly* 1 (January 1946): 183.
4. Alan O'Connor, ed., *Raymond Williams on Television: Selected Writings* (London: Routledge, 1989), 3-4.
5. William Hawes, *American Television Drama: The Experimental Years* (University: University of Alabama Press, 1986), xix.
6. Raymond Williams, *Television: Technology and Cultural Form* (New York: Schocken Books, 1975), 55-60.
7. Roger Silverstone, "Science and the Media: The Case of Television," in S. J. Doorman, ed., *Images of Science: Scientific Practice and the Public* (Aldershot, UK: Gower, 1989), 195; and Gary Edgerton, "The American Made-for-TV Movie"; and Thomas W. Hoffer, Robert Musburger, and Richard Alan Nelson, "Docudrama," in Brian G. Rose, ed., *TV Genres: A Handbook and Reference Guide* (New York: Greenwood Press, 1985), 151-211.
8. Sybil DelGaudio, "If Truth Be Told, Can 'Toons Tell It? Documentary and Animation," *Film History* 9 (1997): 189-199.
9. Hoffer, Musburger, and Nelson, "Docudrama."
10. Ibid.
11. Carl Beier, Jr., "A New Way of Looking at Things," *Hollywood Quarterly* 2 (October 1946): 4.
12. Anthony R. Michaelis, *Research Films in Biology, Anthropology, Psychology, and Medicine* (New York: Academic Press, 1955); and Gregg Mitman, "Cinematic Nature: Hollywood Technology, Popular Culture, and the American Museum of Natural History," *Isis* 84 (December 1993): 636-661.
13. Gregg Mitman, *Reel Nature: America's Romance with Wildlife on Film* (Cambridge, MA: Harvard University Press, 1999), esp. 26-35. See also Garth Jowett, *Film: The Democratic Art* (Boston: Focal Press, 1985).
14. "Video Brings Up Problem for Families," *Chicago Daily Tribune*, January 2, 1948.
15. Ernest P. Walker to William M. Mann, December 22, 1947, SIA (RU74), 213:1.
16. Freeman M. Shelly to William M. Mann, March 18, 1953, SIA (RU74), 214:4.
17. Gil Hodges to William Mann, May 6, 1953; and Leonard Carmichael to Gil Hodges, May 8, 1953, SIA (RU74), 214:4.
18. Mario DePrato to Theodore Reed, April 16, 1960, SIA (RU380), 12:49.
19. Theodore Reed to Thomas Willette, April 25, 1960, SIA (RU380), 12:49.
20. Marlin Perkins, *My Wild Kingdom* (New York: E. P. Dutton, 1982).
21. Marlin Perkins, *Zooparade* (Chicago: Rand McNally, 1954), 10; and Perkins, *My Wild Kingdom*, 113. See also Mitman, *Reel Nature*, 132-133; Marcia Winn, "Enticing Show Packs Them In for 78 Years," *Chicago Daily Tribune*, March 24, 1946; Charles and Jean Komaiko, "The Animal Kingdom of Marlin Perkins," *Pageant*, April 1952, 98-101; and Ralph Knight, "Chicago's Sunday Jungle," *Saturday Evening Post*, October 25, 1952, 32-33, 64, 68, 70.
22. Larry Wolters, "Obeler Drama about Chicago W-G-N Feature," *Chicago Daily Tribune*, July 26, 1945.

23. Larry Wolters, "Radio and Video Whirligig Has Its High Spots," *Chicago Daily Tribune*, August 3, 1949.

24. A. C. Nielsen (now the Nielsen Company) measures the audiences for television programming, providing estimates of the number of viewers and the share of the potential audience during the time slot.

25. Mitman, "Cinematic Nature," 652. See also Donna J. Haraway, *Primate Visions: Gender, Race, and Nature in the World of Modern Science* (New York: Routledge, 1989).

26. Alexander Wilson, *The Culture of Nature: North American Landscape from Disney to the Exxon Valdez* (Cambridge: Blackwell Publishers, 1992), 134.

27. Chris Palmer, *Shooting in the Wild: An Insider's Account of Making Movies in the Animal Kingdom* (San Francisco: Sierra Club Books, 2010).

28. Mitman, *Reel Nature*; Wilson, *Culture of Nature*.

29. Scott MacDonald, "Up Close and Political: Three Short Ruminations on Ideology in the Nature Film," *Film Quarterly* 59 (Spring, 2006): 5.

30. Mitman, "Cinematic Nature," 653.

31. Mitman, *Reel Nature*, 110; Mitman, "Cinematic Nature," 660.

32. Steven Watts, *The Magic Kingdom: Walt Disney and the American Way of Life* (Boston: Houghton Mifflin, 1997), 305.

33. Ibid., 304–306; Wilson, *Culture of Nature*, 118–119.

34. Watts, *Magic Kingdom*, 305.

35. Elizabeth Walker Mechling and Jay Mechling, "The Atom According to Disney," *Quarterly Journal of Speech* 81 (November 1995): 437.

36. Author's analysis of segments on Disney Studios television series, seasons 1 (1954–1955) through 16 (1969–1970), based on descriptions in Larry James Gianakos, *Television Drama Series Programming: A Comprehensive Chronicle, 1947–59* (Metuchen, NJ: Scarecrow Press, 1980); and Larry James Gianakos, *Television Drama Series Programming: A Comprehensive Chronicle, 1959–75* (Metuchen, NJ: Scarecrow Press, 1978). Disney series on broadcast television included *Disneyland* (1954–1958), *Walt Disney Presents* (1958–1961), *Walt Disney's Wonderful World of Color* (1961–1969), *The Wonderful World of Disney* (1969–1979), and *Disney's Wonderful World* (1979–1983).

37. Douglas L. Gilbert, "Television as a Wildlife Education Medium," *Journal of Wildlife Management* 20 (October 1956): 456–458.

38. Evan Z. Vogt, "Anthropology in the Public Consciousness," *Yearbook of Anthropology* (1955): 365–366. See also Mitman, *Reel Nature*, 142–144.

39. Anton Remenih, "New Adventure Series to Have Help of Science," *Chicago Daily Tribune*, May 2, 1953.

40. "Museum Signed Up for New Telecast," *New York Times*, March 26, 1953; and Vogt, "Anthropology in the Public Consciousness," 366.

41. Sig Mickelson, "One Approach to Educational TV," *New York Times*, May 10, 1953. When CBS approached the Smithsonian with a similar program idea, the network was rebuffed. Handwritten note on *Adventure* announcement in *Museum News* (April 15, 1953), SIA (RU50), 214:10.

42. Mickelson, "One Approach to Educational TV."

43. Jack Gould, "Television in Review," *New York Times*, May 15, 1953.

44. Publicity sheet for "Science Series Bell System TV Program," 1956, SIA (Acc. T91027), 1:8.

45. *The Restless Sea* starred a different host and was distributed in an abbreviated version.

46. Sydney W. Head, *Broadcasting in America: Survey of Television and Radio* (Cambridge, MA: Riverside Press, 1956), 410n12.

47. David Templeton, "Weird Science," *Sonoma County Independent*, September 23–29, 1999.

48. "The Wide, Wide World," *Time*, April 11, 1955.

49. James Burkhardt Gilbert, *Redeeming Culture: American Religion in an Age of Science* (Chicago: University of Chicago Press, 1997). Other prominent scientists served as technical advisers, including astronomer Donald Menzel, astrophysicist Walter Orr Roberts, biologist George Wald, and physicists Carl D. Anderson and Bruno Rossi.

50. Press publicity for *Our Mr. Sun*, November 18, 1956, SIA (Acc. T91027), 1:8.

51. Ira M. Freeman, "Interpreting Science," *Science* 136 (June 8, 1962): 902–903.

52. Jack Gould, "Television: 'Our Mr. Sun,'" *New York Times*, November 30, 1956; and John Fink, "TV Film on Science Is Unusual," *Chicago Daily Tribune*, November 20, 1956.

53. John Crosby, "Capra's 'Hemo' Aimed at Thirst for Knowledge," *Washington Post*, March 25, 1957; and Jack Gould, "TV: The Story of Blood," *New York Times*, March 21, 1957.

54. Malvin Maddocks, "Natural Science Series Aims to Be Entertaining," *Christian Science Monitor*, October 29, 1957.

55. Gilbert, *Redeeming Culture*, 221.

56. Jack Gould, "TV: A Study of Genetics," *New York Times*, December 10, 1960.

57. Joseph Turner, "A Mystery Story without Mystery," *Science* 135 (February 23, 1962): 635.

58. John Crosby, "'The Unchanged Goddess' Is Pretty Sticky Weather," *Washington Post*, February 17, 1958.

CHAPTER FIVE

1. Leslie Gelb, quoted in Herbert I. Schiller, *Information Inequality: The Deepening Social Crisis in America* (New York: Routledge, 1996), 18.

2. W. W. Bauer and Thomas G. Hall, *Health Education of the Public: A Practical Manual of Technic* (Philadelphia: W. B. Saunders, 1937); and W. W. Bauer and Leslie Edgley, *Your Health Dramatized: Selected Radio Scripts* (New York: E. P. Dutton, 1939).

3. *Radio in Health Education, Prepared under the Auspices of the New York Academy of Medicine* (New York: Columbia University Press, 1945).

4. Ibid., 26, 101–102.

5. Roy Gibbons, "Surgeons See Television as Training Boon," *Chicago Daily Tribune*, April 4, 1948; and "Operations to Be Televised at D.C. Medical Assembly," *Washington Post*, September 16, 1947.

6. "Caesarean Birth Is Broadcast in Television Test," *Chicago Daily Tribune*, June 19, 1948; "2000 on TV See Birth of Baby," *Washington Post*, June 15, 1951; and Jack Goodman, "Television Course in Post-graduate Medicine," *New York Times*, November 15, 1953.

7. "A.M.A. Telecast Shows How TB Attacks Body," *Chicago Daily Tribune*, August 21, 1948.

8. William L. Laurence, "TV Takes Surgery across Continent," *New York Times*, December 8 1951.

9. Sidney Lohman, "News of TV and Radio," *New York Times*, June 8, 1952; and "Medical Meeting Televised," *Washington Post*, June 8, 1952.

10. William L. Laurence, "Major Operation to Save a Life Put on Network TV for First Time," *New York Times*, June 11, 1952.

11. "TV Debut Is Scheduled for Artificial Kidney," *Washington Post and Times Herald*, October 9, 1955.

12. "Man's Heartbeat Seen and Heard over Television," *Los Angeles Times*, December 5, 1952; and "TV Medical Telecasts Scheduled," *Washington Post*, November 30, 1952.

13. Larry Wolters, "Critics Argue over Caesarian [*sic*] Birth on Television," *Chicago Daily Tribune*, December 13, 1952. See also Jack Gould, "Succession of Medical Programs Rivals Horror Shows in Elements of Shock," *New York Times*, December 6, 1954.

14. See, for example, Truman J. Keesey, "Three TV Series in Washington, D.C.," *Public*

Health Reports 69 (June 1954): 599–605; Charles E. Pinckney, "'Your Lease on Life' in Denver," *Public Health Reports* 69 (June 1954): 606–608; and Romayne Wicks Spangler, "We're on TV Every Week," *American Journal of Nursing* 55 (May 1955): 592–593.

15. Quote from a script in Frank Warren, *Television in Medical Education: An Illustrated Handbook* (Chicago: American Medical Association, 1955), 43. See also Harriet H. Hester, H. L. Fishel, and Martin Magner, *Television in Health Education* (Chicago: American Medical Association, 1955).

16. Hester, Fishel, and Magner, *Television in Health Education*, 6–9.

17. Ibid.

18. *Journal of the American Medical Association* editorial, as quoted in "A.M.A. Assails Use by TV of Profession's Symbols," *New York Times*, June 27, 1953. See also "Medical Garb on Commercials Hit," *Los Angeles Times*, June 28, 1953.

19. Jack Gould, "Dubiousness of Medical Commercial Is Raised in Wake of Complaint by the A.M.A.," *New York Times*, November 11, 1953.

20. Robert K. Plumb, "Medical Society Hits TV 'Doctors,'" *New York Times*, May 14, 1958.

21. Gould, "Dubiousness."

22. Joseph Turow and Rachel Gans-Boriskin, "From Expert in Action to Existential Angst: A Half-Century of Television Doctors," in Leslie J. Reagan, Nancy Tomes, and Paula A. Treichler, eds., *Medicine's Moving Pictures: Medicine, Health, and Bodies in American Film and Television* (Rochester, NY: University of Rochester Press, 2007), 263.

23. Ibid., 264–265.

24. Robert S. Alley, "Medical Melodrama," in Brian G. Rose, ed., *TV Genres: A Handbook and Reference Guide* (New York: Greenwood Press, 1985), 74.

25. Robert S. Alley, "Media Medicine and Morality," in Richard Adler and Douglass Cater, eds., *Television as a Cultural Force* (New York: Praeger, 1976), 96.

26. Gould, "Succession."

27. Turow and Gans-Boriskin, "From Expert in Action," 265–266.

28. Joseph Turow, *Playing Doctor* (Oxford: Oxford University Press, 1989), 12, quoting a Paramount publicity release.

29. Ibid., 30; and Alley, "Medical Melodrama," 74.

30. Turow, *Playing Doctor*, 37.

31. Ann Hudson Jones, "Medicine and the Physician in Popular Culture," in M. Thomas Inge, ed., *Handbook of American Popular Culture* 3 (Westport, CT: Greenwood Press, 1981), 183.

32. Lawrence Laurent, "Is the AMA Bad Medicine for Practitioners on TV?," *Los Angeles Times*, January 15, 1963; and Turow, *Playing Doctor*, 61–62.

33. Quoted in Turow, *Playing Doctor*, 28–29.

34. LACMA official quoted in Larry Wolters, "Doctors Check Medic's Pulse, Find It's Strong," *Chicago Daily Tribune*, March 6, 1955.

35. Larry Wolters, "The Medic Attracts Host of TV Patients," *Chicago Daily Tribune*, January 21, 1956.

36. Alley, "Media Medicine and Morality," 74; and Wolters, "Doctors Check Medic's Pulse."

37. Turow, *Playing Doctor*, 37.

38. Ibid., 25.

39. Alley, "Medical Melodrama," 74.

40. Turow, *Playing Doctor*, 271.

41. Alley, "Media Medicine and Morality," 96; and Alley, "Medical Melodrama," 75. Turow (*Playing Doctor*, 42–43) states that the network did little to defuse these, possibly because it wanted an excuse to cancel the series.

42. Gould, "Succession."

43. "Smith, Kline & French Plan TV Medical Series," *Wall Street Journal*, November 7, 1955; and Robert Lee Bailey, "An Examination of Prime Time Network Television Special Programs, 1948 to 1966" (Ph.D. dissertation, University of Wisconsin, 1967).

44. Sonia Stein, "Televised Surgery Promoting Apples," *Washington Post*, December 14, 1952.

45. Irving Stettel, ed., *Top TV Shows of the Year, 1954–1955* (New York: Hastings House, 1955), 77.

46. Dael Wolfle, "Editorial Responsibility," *Science* 122 (November 25, 1955): 1001.

47. Alexander Mark Soranno, "A Descriptive Study of Television Network Prime Time Programming, 1958–59 through 1962–63" (master's thesis, San Francisco State College, 1966). Comparison of the Soranno data to Dimmick's research on television westerns demonstrates that, when the popularity of westerns increased during the late 1950s, the number of medical and science fiction dramas declined briefly. John Dimmick, "The TV Western Program Cycle: Decision Uncertainty and Audience Habituation," *Mass Comm. Review* 4 (Spring 1977): 13, figure 1.

48. John Keats, "Rx for an M.D. on TV," *New York Times*, May 27, 1962; and Philip A. Kalisch and Beatrice J. Kalisch, "Nurses on Prime-Time Television," *American Journal of Nursing* 82 (February 1982): 264–270.

49. Tim Brooks and Earle Marsh, *The Complete Directory to Prime Time Network and Cable TV Shows, 1946–Present*, sixth edition (New York: Ballantine Books, 1995), 90.

50. Turow, *Playing Doctor*, 62–63. Turow says that AMA committee members also "saw it as their duty to consider each script's possible effect on the public and the image of medicine" (62).

51. Keats, "Rx for an M.D. on TV."

52. LACMA executive quoted in *Time*, February 21, 1955, as cited in Alley, "Medical Melodrama," 75.

53. Laurent, "Is the AMA Bad Medicine."

54. Erik Barnouw, *The Image Empire: A History of Broadcasting in the United States from 1953* (New York: Oxford University Press, 1970), 204.

55. Michael R. Real, *Mass-Mediated Culture* (Englewood Cliffs, NJ: Prentice Hall, 1977), 120–126; Alley, "Media Medicine and Morality," 100; Jones, "Medicine and the Physician," 183; and Turow and Gans-Boriskin, "From Expert in Action," 269–270.

56. Alley, "Media Medicine and Morality."

57. Turow, *Playing Doctor*, 292.

58. David A. Kirby, *Lab Coats in Hollywood: Science, Scientists, and Cinema* (Cambridge, MA: MIT Press, 2011).

59. Alley, "Media Medicine and Morality," 100.

60. Alex McNeil, *Total Television*, fourth edition (New York: Penguin Books, 1996), 226–228.

61. Jack Gould, "TV: N.B.C. and C.B.S. Offer Medical Melodrama," *New York Times*, October 5, 1962.

62. "Eleventh Hour—Never Too Late," *Los Angeles Times*, October 10, 1962.

63. Jack Gould, "Disturbed Television," *New York Times*, October 28, 1962.

64. Michael Amrine, "Psychology in the News," *American Psychologist* 19 (1959): 74–78. See also Emma Harrison, "TV Show Assailed by Psychologists," *New York Times*, December 2, 1962; "Psychiatrists Assail TV Presentation," *Christian Science Monitor*, December 3, 1962; and "Psychologists Join Protest on TV Show," *New York Times*, December 8, 1962.

65. Vernon Scott, "Corey and the Analysts," *Chicago Daily Tribune*, February 3, 1963.

66. Conclusion based on data in Brooks and Marsh, *Complete Directory*; Harry Castleman and Walter J. Podrazik, *Watching TV: Four Decades of American Television* (New York: McGraw-Hill, 1982); Daniel Einstein, *Special Edition: A Guide to Network Television Documentary Series and Special News Reports, 1955–1979* (Metuchen, NJ: Scarecrow Press, 1987); McNeil, *Total Television*; Stettel, *Top TV Shows*; Vincent Terrace, *The Complete Encyclopedia of Television Programs, 1947–1976* (South Brunswick, NJ: A. S. Barnes, 1976); and Erwin K. Thomas and Brown H. Carpenter, eds., *Handbook on Mass Media in the United States* (Westport, CT: Greenwood Press, 1994).

67. Robert M. Weitman, "Television Is Coming of Age," in Stettel, *Top TV Shows*, xv.

68. Todd Gitlin, "Prime Time Ideology: The Hegemonic Process in Television Entertainment," *Social Problems* 26 (February 1979): 266.

CHAPTER SIX

1. John Crosby, "'The Unchained Goddess' Is Pretty Sticky Weather," *Washington Post*, February 17, 1958.

2. "What TV Is Doing to America," *U.S. News & World Report*, September 2, 1955, 36.

3. University of Michigan Survey Research Center, *The Public Impact of Science in the Mass Media: A Report on a Nation-Wide Survey for the National Association of Science Writers* (Ann Arbor: University of Michigan, 1958), 29; and U.S. Bureau of the Census, *Social Indicators III* (Washington, DC: GPO, 1980), table 11/3. See also Erik Barnouw, *The Golden Web: A History of Broadcasting in the United States, 1933-1953* (New York: Oxford University Press, 1968), 210-211, 269.

4. University of Michigan Survey Research Center, *Public Impact of Science*, 113, table IV-20. About 4.8 percent of the respondents named General Electric Company advertisements starring Don Herbert in his role as Mr. Wizard.

5. Ibid.

6. National Association of Science Writers, *Science, the News, and the Public: Who Gets What Science News, Where They Get It, and What They Think about It* (New York: New York University Press, 1958), 13.

7. Russel Nye, *The Unembarrassed Muse: The Popular Arts in America* (New York: Dial Press, 1970), esp. 273-274; William Sims Bainbridge, *Dimensions of Science Fiction* (Cambridge, MA: Harvard University Press, 1986); Robert Lambourne, Michael Shallis, and Michael Shortland, *Close Encounters? Science and Science Fiction* (Bristol, UK: Adam Hilger, 1990); David Seed, *American Science Fiction and the Cold War: Literature and Film* (Edinburgh: Edinburgh University Press, 1999); Richard Hodgens, "A Brief, Tragical History of the Science Fiction Film," *Film Quarterly* 13 (Winter 1959): 30-39; and William M. Tsutsui, "Looking Straight at *Them!* Understanding the Big Bug Movies of the 1950s," *Environmental History* 12 (April 2007): 237-253.

8. Donald F. Glut and Jim Harmon, *The Great Television Heroes* (New York: Doubleday, 1975).

9. Mark Siegel, "Science Fiction and Fantasy TV," in Brian G. Rose, ed., *TV Genres: A Handbook and Reference Guide* (New York: Greenwood Press, 1985), 92.

10. Tim Brooks and Earle Marsh, *The Complete Directory to Prime Time Network and Cable TV Shows, 1946-Present*, sixth edition (New York: Ballantine Books, 1995), 165; David Weinstein, *The Forgotten Network: DuMont and the Birth of American Television* (Philadelphia: Temple University Press, 2004), 69-90.

11. James Burkhardt Gilbert, *Redeeming Culture: American Religion in an Age of Science* (Chicago: University of Chicago Press, 1997); and Siegel, "Science Fiction and Fantasy TV."

12. Siegel, "Science Fiction and Fantasy TV," 93.

13. Patrick Luciano and Gary Colville, *American Science Fiction Television Series of the 1950s: Episode Guides for Casts and Credits for Twenty Shows* (Jefferson, NC: McFarland, 1998), 10.

14. Anton Remenih, "Talking Worms, Beetle Face No Problem for TV," *Chicago Daily Tribune*, April 12, 1953.

15. Luciano and Colville, *American Science Fiction*, 12, 178.

16. Script dialogue reprinted in ibid., 12. See also Brooks and Marsh, *Complete Directory*, 908.

17. Luciano and Colville, *American Science Fiction*, 178.

18. Ibid., 180.

19. Transcript for *Face the Nation*, August 14, 1955, NAA (IGY), Office of Information: Radio & TV: 1954-1955.

20. "News of Science: Sputnik," *Science* 126 (October 18, 1957): 739-740.

21. Jack Lule, "Roots of the Space Race: Sputnik and the Language of U.S. News in 1957," *Journalism Quarterly* 68 (Spring/Summer 1991): 76-86; and Alfred Robert Hogan, "Televising the Space Age: A Descriptive Chronology of CBS News Special Coverage of Space Exploration from 1957 to 2003" (master's thesis, University of Maryland, 2005).

22. Edwin Diamond, *The Rise and Fall of the Space Age* (Garden City, NY: Doubleday, 1964), 99.

23. Fae L. Korsmo, "Shaping Up Planet Earth: The International Geophysical Year (1957–1958) and Communicating Science through Print and Film Media," *Science Communication* 26 (December 2004): 165.

24. See various correspondence in NAA (IGY), Office of Information: Radio & TV: 1956–1957; and Donald C. Thompson to Hugh Odishaw, August 7, 1957, and September 16, 1957, NAA (IGY), Office of Information: Radio & TV: NBC, 1955–1957.

25. Reuven Frank to Director, Office of Public Information, Department of Defense, July 31, 1956, NAA (IGY), Office of Information: Radio & TV: NBC, 1957.

26. Robert Emmett Ginna to Hugh Odishaw, March 11, 1957, NAA (IGY), Office of Information: Radio & TV: NBC, 1957.

27. NAA (IGY), Office of Information: Radio & TV: NBC, 1955–1957.

28. John C. Truesdale to John Goetz, November 8, 1957, NAA (IGY), Office of Information: Radio & TV: NBC, 1955–1957.

29. Arnold W. Frutkin to Charles R. Denny, September 1957 (draft), NAA (IGY), Office of Information: Radio & TV: NBC, 1955–1957.

30. NAA (IGY), Office of Information: Radio & TV: Educational TV & Radio Center with NBC.

31. Evans G. Valens to Joseph Kaplan, August 30, 1957, NAA (IGY), Office of Information: Radio & TV: Educational TV & Radio Center with NBC.

32. Daniel Einstein, *Special Edition: A Guide to Network Television Documentary Series and Special News Reports, 1955–1979* (Metuchen, NJ: Scarecrow Press, 1987), 188.

33. John S. Coleman to Irving Gitlin, August 29, 1957, NAA (Administration), Public Relations: General, 1957.

34. See summary of reaction to first *Conquest* episode in R. E. Paulson to R. M Wheatley, December 9, 1957, NAA (Central Policy Files, 1957–1961), Governing Board: Advisory Board on Education, Film and Television Programs, 1957.

35. R. E. Paulson memo to R. M. Wheatley, December 9, 1957, NAA (Central Policy Files, 1957–1961), Governing Board: Advisory Board on Education, Film and Television Programs, 1957.

36. See, especially, Detlev V. Bronk to Charles Allen Thomas, May 24, 1958, NAA (Central Policy Files, 1957–1961), Public Relations: General, 1958.

37. Stafford Clark to Mr. Thorman, December 7, 1957, NAA (IGY), Office of Information: Commercial Radio & TV, 1958.

38. Arnold W. Frutkin to Stafford Clark, December 26, 1957, NAA (IGY), Office of Information: Commercial Radio & TV, 1958.

39. Steven Watts, *The Magic Kingdom: Walt Disney and the American Way of Life* (Boston: Houghton Mifflin, 1997), 309, 311.

40. Elizabeth Walker Mechling and Jay Mechling, "The Atom According to Disney," *Quarterly Journal of Speech* 81 (November 1995): 437; Watts, *Magic Kingdom*, 312, 372; Spencer R. Weart, *Nuclear Fear: A History of Images* (Cambridge: Harvard University Press, 1988), 169n42; Heinz Haber, *Our Friend the Atom* (New York: Simon and Schuster, 1957), 13; and Allan M. Winkler, *Life under a Cloud: American Anxiety about the Atom* (New York: Oxford University Press, 1993), 140–141.

41. Watts, *Magic Kingdom*, 312; Haber, *Our Friend the Atom*, 13; and Winkler, *Life under a Cloud*, 140–141.

42. Paul S. Boyer, "From Activism to Apathy: The American People and Nuclear Weapons, 1963–1980," *Journal of American History* 70 (March 1984): 823.

43. Paul S. Boyer, *By the Bomb's Early Light: American Thought and Culture at the Dawn of the Atomic Age* (New York: Pantheon, 1985); and Allan M. Winkler, "The 'Atom' and American Life," *History Teacher* 26 (May 1993): 317–337. See also Paul S. Boyer, *Fallout: A Historian Reflects on America's Half-Century Encounter with Nuclear Weapons* (Columbus: Ohio State University Press, 1998).

44. Boyer, "From Activism to Apathy," 824.

45. Howard E. McCurdy, *Space and the American Imagination* (Washington, DC: Smithsonian Institution Press, 1997).

46. Erik Barnouw, *The Image Empire: A History of Broadcasting in the United States from 1953* (New York: Oxford University Press, 1970), 196, 207.

47. E. G. Sherburne, Jr., "Television Coverage of the Gemini Program," *Science* 149 (September 17, 1965): 1329.

48. Robert Lewis Shayon, *Saturday Review* (August 9, 1969), as quoted in Barnouw, *Image Empire*, 331.

49. Barnouw, *Image Empire*, 266.

50. Vivian Carol Sobchack, *Screening Space: The American Science Fiction Film*, second edition (New Brunswick, NJ: Rutgers University Press, 1997), 29.

51. Barnouw, *Image Empire*, 326.

52. Hogan, "Televising the Space Age," 32, 35.

53. Siegel, "Science Fiction and Fantasy TV," 94–95.

54. Barnouw, *Image Empire*, 264.

55. Robert Poole, *Earthrise: How Man First Saw the Earth* (New Haven, CT: Yale University Press, 2008); and Benjamin Lazier, "Earthrise; or, The Globalization of the World Picture," *American Historical Review* 116 (June 2011): 602–630.

56. Frederic M. Philips to S. Dillon Ripley, July 18, 1969, SIA (RU145), 7:16; "Instant History," *Washington Post*, July 21, 1969; and Michael Kernan, "Earth Lore: Always Traditional (A Party)," *Washington Post*, July 19, 1969.

57. James Schwoch, *Global TV: New Media and the Cold War, 1946–69* (Urbana: University of Illinois Press, 2009), 154.

CHAPTER SEVEN

1. John K. Mackenzie, "Educating the Public." *Science* 150 (October 1, 1965): 7.

2. Jay G. Blumler and Michael Gurevitch, *The Crisis of Public Communication* (London: Routledge, 1995), 99.

3. James Lawrence Fly, "Regulation of Radio Broadcasting in the Public Interest," *Annals of the American Academy of Political and Social Science* 213 (January 1941): 102–105.

4. Commission on Freedom of the Press, *A Free and Responsible Press* (Chicago: University of Chicago Press, 1947).

5. Peter L. Walsh, "This Invisible Screen: Television and American Art," *American Art* 18 (Summer 2004): 8.

6. A. William Bluem, *Documentary in American Television: Form, Function, Method* (New York: Hastings House, 1965), 7.

7. Bruce V. Lewenstein, "The AAAS and Scientific Perspectives on Public Understanding, 1945–1980," paper presented at the 1991 annual meeting of the American Association for the Advancement of Science.

8. Michael Curtin, "The Discourse of 'Scientific Anti-Communism' in the 'Golden Age' of Documentary," *Cinema Journal* 32 (Autumn 1992): 3–25.

9. Ibid. See also Erik Barnouw, *Tube of Plenty: The Evolution of American Television*, revised edition (New York: Oxford University Press, 1982); and James L. Baughman, *The Republic of Mass Culture: Journalism, Filmmaking, and Broadcasting in America since 1941* (Baltimore: Johns Hopkins University Press, 1992).

10. Curtin, "Discourse of 'Scientific Anti-Communism,'" 3–25.

11. Andrew Jamison and Ron Eyerman, *The Seeds of the Sixties* (Berkeley: University of California Press, 1994), 66–68, 92–100.

12. Mark Hamilton Lytle, *The Gentle Subversive: Rachel Carson, "Silent Spring," and the Rise of the Environmental Movement* (New York: Oxford University Press, 2007), 113.

13. Zuoyue Wang, "Responding to *Silent Spring*: Scientists, Popular Science Communication, and Environmental Policy in the Kennedy Years," *Science Communication* 19 (December 1997): 141-163; and Zuoyue Wang, *In Sputnik's Shadow: The President's Science Advisory Committee and Cold War America* (New Brunswick, NJ: Rutgers University Press, 2008), 208-214. See also Frank Graham, Jr., *Since Silent Spring* (Boston: Houghton Mifflin, 1970); and Linda Lear, *Rachel Carson: Witness for Nature* (New York: Henry Holt, 1997).

14. Author's reanalysis of data in the appendix to Robert Lee Bailey, "An Examination of Prime Time Network Television Special Programs, 1948 to 1966" (Ph.D. dissertation, University of Wisconsin, 1967).

15. Author's analysis of program listings in Daniel Einstein, *Special Edition: A Guide to Network Television Documentary Series and Special News Reports, 1955-1979* (Metuchen, NJ: Scarecrow Press, 1987), entries 880-1148.

16. See discussion of *60 Minutes* in Anthony Smith, ed., *Television: An International History* (Oxford: Oxford University Press, 1995). See also William C. Spragens, *Electronic Magazines: Soft News Programs on Network Television* (Westport, CT: Praeger, 1995), chapter 2.

17. Thomas W. Moore, "ABC Network Cites Reality of TV in Color," *Chicago Tribune*, November 12, 1967. See also Louise Sweeney, "'Event TV' Joins the Language," *Christian Science Monitor*, June 9, 1967.

18. Moore, "ABC Network Cites Reality."

19. Frank Luther Mott, *A History of American Magazines* (Cambridge, MA: Harvard University Press, 1957), 4:620-632; and Catherine A. Lutz and Jane L. Collins, *Reading National Geographic* (Chicago: University of Chicago Press, 1993).

20. "National Geographic's Newest Adventure: A Color Television Series," *National Geographic Magazine*, September 1965, 448.

21. Ibid., 451.

22. Marion Purcelli, "Chimp vs. Man Highlighted on Video Special," *Chicago Tribune*, December 19, 1965.

23. Chris Palmer, *Shooting in the Wild: An Insider's Account of Making Movies in the Animal Kingdom* (San Francisco: Sierra Club Books, 2010), xviii.

24. Alexander Wilson, *The Culture of Nature: North American Landscape from Disney to the Exxon Valdez* (Cambridge: Blackwell Publishers, 1992), 141.

25. "2 to Sponsor TV Series by Geographic Society," *Washington Post*, May 26, 1965.

26. Bill McKibben, *The Age of Missing Information* (New York: Plume Books, 1993), 217.

27. Einstein, *Special Edition*, 60.

28. Tim Brooks and Earle Marsh, *The Complete Directory to Prime Time Network and Cable TV Shows, 1946-Present*, sixth edition (New York: Ballantine Books, 1995), 768, 1037, 1084-1085; Einstein, *Special Edition*, 53; Alex McNeil, *Total Television*, fourth edition (New York: Penguin Books, 1996), 616, 834, 960, 1058; and Wilson, *Culture of Nature*, 137-140.

29. McKibben, *Age of Missing Information*, 217.

30. As quoted in Wilson, *Culture of Nature*, 137.

31. Paul H. Oehser to Leonard Carmichael, May 29, 1956, SIA (Acc. T91027), 1:8.

32. Jack Warner, Jr., to Robert V. Fleming, June 9, 1962, SIA (RU50), 43:2.

33. See correspondence in SIA (RU137), 51:2; and SIA (RU99), 80:7, 79:4. By September 1965, Smithsonian official William C. Grayson was discussing a possible series with all three networks and the producer David L. Wolper.

34. "The Museum as Enigma," address by S. Dillon Ripley at the bicentennial celebration commemorating the birth of James Smithson, September 18, 1965, SIA (RU99), 79:3.

35. NBC press release, Fall 1966, SIA (RU99), 80:3.

36. Contracts and correspondence in SIA (RU99), 80:3, 6. Publisher McGraw-Hill

marketed the film versions to schools. NBC probably charged around $4,500 per minute for commercials during the first season. In 1972, the institution was still earning about $11,000 a year in royalties from *The Smithsonian*.

37. NBC-Smithsonian agreement, August 23, 1966, 1–2, SIA (RU99), 80:6.

38. For example, Frank Taylor to William C. Grayson, August 2, 1966, SIA (RU99), 80:3.

39. Scripts in SIA (RU99), 80:5, 6.

40. Announcement by S. Dillon Ripley, July 6, 1966, SIA (RU276), 53:1.

41. See materials in SIA (RU99), 80:3.

42. "NBC Network Series 'The Smithsonian,'" 1968, SIA (RU145), 15:17.

43. Statement by S. Dillon Ripley, January 6, 1967, SIA (RU99), 79:3.

44. Richard D. Heffner to Donald F. Squires, November 28, 1966; Donald F. Squires to Richard D. Heffner, December 14, 1966; Donald F. Squires to Walter Orr Roberts, December 14, 1966; and Donald F. Squires to S. Dillon Ripley, May 24, 1967, and October 19, 1967, SIA (RU99), 80:3, 4. See also "NBC Network Series 'The Smithsonian'" (undated), SIA (RU145), 15:19.

45. Philip Ritterbush to S. Dillon Ripley, March 11, 1967, SIA (RU99), 80:4.

46. Craig B. Fisher to S. Dillon Ripley, March 8, 1967, SIA (RU99), 80:2.

47. Robert W. Mason (memo) to William C. Grayson, March 28, 1967, SIA (RU99), 80:2.

48. William C. Grayson to S. Dillon Ripley, June 30, 1967, SIA (RU102), 5:8.

49. Ripley had hired Lee M. Talbot in 1965 to manage the institution's international conservation activities, and appointed Talbot's wife and research partner Marty Talbot to the research staff. See correspondence in SIA (RU218), 13:16.

50. Lee M. Talbot to S. Dillon Ripley, June 14, 1968, SIA (RU145), 7:16. In 1968, a rating of 33.5 translated to about twenty-nine million viewers.

51. Frederic M. Philips to Mr. [Robert] Mason, December 12, 1967, SIA (RU99), 80:4.

52. Donald V. Meaney to S. Dillon Ripley, February 21, 1968, and February 23, 1968, SIA (RU145), 7:16.

53. Frederic M. Philips to S. Dillon Ripley, February 26, 1968, SIA (RU145), 7:16.

54. W. W. Warner, "Note to Files," March 14, 1968, SIA (RU145), 7:16.

55. Frederic M. Philips memos, SIA (RU145), 7:16.

56. Craig B. Fisher to S. Dillon Ripley, April 20, 1979, SIA (RU145), 7:16.

57. SIA (RU145), 6:3. "British Couple Begin Harvard Science Study," *Chicago Tribune*, October 10, 1964; and David Wade Chambers, "History of Science on the Silver Screen," *Isis* 57 (Winter 1966): 494–497. In 1967, Goodfield was Treves Professor in the History and Philosophy of Science at Wellesley College, and her husband, Stephen Toulmin, was professor of philosophy and history of ideas at Brandeis University.

58. William W. Warner, "Memo to Files after Visit from Richard Crewdson," December 6, 1968; and June Goodfield and Stephen Toulmin, "Memorandum Proposal for a Comprehensive Program on Ecology and the Biosphere," SIA (RU145), 6:5.

59. See correspondence in SIA (RU145), 6:5.

60. William W. Warner to S. Dillon Ripley, March 3, 1969, SIA (RU367), 9:10. See also Richard Crewdson to June Goodfield Toulmin, September 21, 1967, SIA (RU145), 6:5.

61. S. Dillon Ripley to Fred W. Friendly, March 27, 1969, SIA (RU99), 312:1.

62. David M. Davis (Ford Foundation) to William W. Warner, November 7, 1969, SIA (RU145), 6:5.

CHAPTER EIGHT

1. "Report of the Committee on the Role and Opportunities in Broadcasting—To President Edward H. Levi, August 15, 1972," *University of Chicago Record* 7 (April 21, 1973): 167.

2. Ibid., 158.

3. Correspondence in SIA (RU137), 51:2; William W. Warner and Fred Philips to S. Dillon Ripley, July 6, 1970, SIA (RU367), 8:5.

4. Contracts and related correspondence, SIA (RU247), 7:3.

5. William W. Warner and Fred Philips to S. Dillon Ripley, July 6, 1970, SIA (RU367), 8:5.

6. William W. Warner to S. Dillon Ripley, November 30, 1970, "Some Potentially Difficult Problems concerning Staff Reactions to CBS Television Series," SIA (RU367), 8:5.

7. Elizabeth Stevens, "An Archeological Find Named Iris Love," *New York Times Sunday Magazine*, March 7, 1971; and Meryle Secrest, "Condescending 'Goddess of Love,'" *Washington Post*, June 12, 1971.

8. Clifford Evans to Frederic Philips, May 14, 1971, SIA (RU367), 8:5.

9. S. Dillon Ripley to John Schneider, August 4, 1971; William W. Warner to S. Dillon Ripley, May 28, 1971; and Frederic M. Philips to S. Dillon Ripley, July 29, 1971, SIA (Acc. T90110), 2:2.

10. William W. Warner and Carl Larsen to S. Dillon Ripley, "Television and the Smithsonian," April 18, 1972; William W. Warner to Burton Benjamin, May 2, 1972, and June 14, 1972; Robert Burstein to William W. Warner, June 28, 1972; and Julian T. Euell to Robert Burstein, July 5, 1972, SIA (RU145), 15:20.

11. Carl W. Larsen to S. Dillon Ripley, February 9, 1972, SIA (RU367), 8:2; and Warner, "Television and the Smithsonian."

12. Carl W. Larsen to S. Dillon Ripley et al., April 17, 1972; and Carl W. Larsen to S. Dillon Ripley, May 15, 1972, SIA (RU145), 15:23.

13. Ralph Lee Smith, "The Wired Nation," *Nation*, May 18, 1970, 582–606.

14. "Report of the Committee," 161.

15. Ibid., 183.

16. Ibid., 182.

17. Max Dawson, "Home Video and the 'TV Problem,'" *Technology & Culture* 48 (July 2007): 524–549.

18. Julian T. Euell to S. Dillon Ripley, November 8, 1972, SIA (RU367), 8:3.

19. Julian T. Euell to S. Dillon Ripley, October 5, 1972; and Carl W. Larsen to Julian Euell, October 4, 1972, SIA (RU145), 15:14.

20. Estimates in Carl W. Larsen to Robert A. Brooks, January 18, 1973, SIA (RU137), 90:1; and *Proceedings of the Meeting of the Board of Regents, Smithsonian Institution, January 24, 1973*, appendix 2, p. 8, SIA (RU1), 10:2.

21. Carl W. Larsen to Robert A. Brooks, January 18, 1973, SIA (RU137), 90:1; and *Proceedings . . . January 24, 1973*, appendix 2, p. 8.

22. R. S. Cowen to Julian T. Euell, August 10, 1972, SIA (RU137), 82:16.

23. William C. Grayson to Julian T. Euell and Carl W. Larsen, September 25, 1972, SIA (RU145): 15:14; Carl W. Larsen to Julian T. Euell, December 22, 1972, SIA (RU145), 15:23.

24. *Proceedings . . . January 24, 1973*, appendix 2, p. 13.

25. *Smithsonian Telecommunications Study Notes Number 1*, December 1972, SIA (RU367), 8:3.

26. Richard Lusher to Julian T. Euell, August 16, 1972, SIA (RU145), 15:14.

27. Carl W. Larsen to William W. Warner, April 24, 1972, SIA (RU367), 8:8; Edward R. Trapnell memorandum, June 7, 1973, SIA (RU337), 8:24; and related documents in SIA (RU276), 22:3, 4, 5.

28. "MHT Preliminary Proposal," November 12, 1973, and other documents, SIA (RU276), 22:3, 4, 5.

29. National Science Foundation grant GM-38587, awarded May 1973.

30. See correspondence in SIA (RU367), 4:11. Smithsonian donor Paul L. Davies brokered the Wolper meeting; Daniel Boorstin, head of the MHT, had also recently advised on a Wolper series. S. Dillon Ripley to Paul L. Davies, SIA (RU613), 170:1; Carl Larsen to S. Dillon Ripley,

July 20, 1972, SIA (RU367), 4:11. Thank you to Virginia Myhaver for bringing the Davies correspondence to my attention.

31. Rick Du Brow, "Outsider Enters TV Documentary Field," *Washington Post*, January 1, 1968; and "Mr. TV Documentary Eyes Lush Film Pastures," *Los Angeles Times*, January 7, 1968.

32. SIA (RU367), box 4, folder "Wolper Contract 1972–74" and folder "Executive Comm. 1/9 -Wolper Contract 1973."

33. Virginia Myhaver, "Funding the 'Folk': New Federalism and Public-Private Partnerships during the American Bicentennial," presentation at National Museum of American History, December 15, 2009.

34. William L. Eilers to James Billington, September 27, 1973, SIA (RU218), 13:21.

35. Carl W. Larsen to Julian T. Euell, September 25, 1972; and Ruth Frazier, "Conversations of Larsen and Euell in New York on September 20," SIA (RU145), 15:14, 20.

36. Julian T. Euell to S. Dillon Ripley, draft memorandum, January 5, 1973, SIA (RU145), 15:14.

37. *Proceedings . . . January 24, 1973*, appendix 2, p. 13.

38. Ibid.

39. Christian C. Hohenlohe, memo to Executive Committee, March 12, 1973, SIA (RU367), 8:1.

40. S. Dillon Ripley to Edmund D. Andrews, December 13, 1973, SIA (RU367), 4:5.

41. S. Dillon Ripley to Thomas J. Watson, Jr., April 5, 1973, SIA (RU137), 90:1; and S. Dillon Ripley to Paul L. Davies, SIA (RU613), 170:1. Two regents—James E. Webb and Caryl Haskins—raised questions about the television deal. SIA (RU367), 8:1.

42. *Proceedings . . . January 24, 1973*.

43. Julian T. Euell to S. Dillon Ripley and Robert A. Brooks, March 5, 1973, SIA (RU137), 90:1.

44. Jim Stingley, "Saving of All Wildlife: Idea Good, But . . . ," *Los Angeles Times*, January 16, 1972; David L. Wolper, "'No Rabbit up My Sleeve,'" *New York Times*, February 6, 1972; "Defect Notification," *Washington Post*, May 12, 1973; and Dian Olson Belanger, *Managing American Wildlife: A History of the International Association of Fish and Wildlife Agencies* (Amherst: University of Massachusetts Press, 1988), 122–123. Belanger points out that the controversy prompted NBC to develop guidelines for reenactments and dramatization of animal behavior (123).

45. *Hollywood Reporter*, January 16, 1973.

46. The press praised the project as the Smithsonian "reaching out to the masses." Jean White, "Reaching Out," *Washington Post*, July 9, 1973. See also "Smithsonian, Wolper Sign Agreement for Series of Television Specials," Smithsonian press release, June 17, 1973, SIA (RU586), 3:25.

47. S. Dillon Ripley to Heads of Organization Units, etc., August 14, 1973, SIA (RU276), 53:1.

48. Cecil Smith, "David Wolper Sees Comedies as Way to Express Views," *Canton (OH) Repository*, July 16, 1973.

49. *Proceedings . . . January 24, 1973*, appendix 2, p. 11; *Proceedings of the Meeting of the Board of Regents, Smithsonian Institution, May 9, 1973*, 16, SIA (RU1), 10:3; and Carl W. Larsen to Robert A. Brooks, May 8, 1973, SIA (RU367), 4:9.

50. Charles DeVault to Julian T. Euell, March 11, 1974, SIA (RU367), 4:11.

51. Julian T. Euell to S. Dillon Ripley, March 18, 1974, SIA (RU586), 3:25; and Julian T. Euell to Brooke Hindle, March 6, 1975, SIA (RU276), 22:7.

52. S. Dillon Ripley to Julian Euell, March 15, 1974, SIA (RU367), 4:11.

53. Edward F. Rivinius to S. Dillon Ripley, March 15, 1974, SIA (RU367), 4:11.

54. S. Dillon Ripley to Carl Larsen, June 15, 1974, SIA (RU416), 20:10.

55. Julian T. Euell to S. Dillon Ripley, October 21, 1974, SIA (RU367), 4:11.

56. Isaac Asimov, "They Don't Make Monsters like They Used To," *TV Guide*, November 23–29, 1974, 13–15.

57. "'Monsters' Scores Big in Nielsens," *Los Angeles Times*, December 7, 1974. Previous record for a documentary had been "We Live with Elephants" (1973), with a 19.6 rating and a 30 share. Wolper later broke his own record with his miniseries *Roots*. Nazaret Cherkezian to Dorothy Rosenberg, November 27, 1974, SIA (RU367), 4:13; and oral history interview with Nazaret Cherkezian, 1986, SIA (RU9541).

58. Joseph N. Bell, "On the Trail of 'Bigfoot,'" *Christian Science Monitor*, November 22, 1974; Joel Dreyfuss, "Monster Myths?," *Washington Post*, November 25, 1974; and Cecil Smith, "Hotfooting It to Bigfoot in Oregon," *Los Angeles Times*, November 25, 1974.

59. Terence O'Flaherty, "Views TV," *San Francisco Chronicle*, November 25, 1974.

60. William C. Grayson to George Lefferts, July 31, 1974, SIA (RU367), 4:13.

61. David L. Wolper to William C. Grayson, August 2, 1974, SIA (RU367), 4:13.

62. Ibid.

63. "'Monsters! Mysteries or Myths?' Explores Nature's Puzzles in Smithsonian Special to Be Presented Monday, Nov. 25 on the CBS Television Network," press release dated October 21, 1974, SIA (RU416), 20:10.

64. John J. O'Connor, "TV: Special on U.F.O.'s," *New York Times*, December 13, 1974.

65. David L. Wolper to S. Dillon Ripley, December 13, 1974, SIA (RU367), 4:5.

66. S. Dillon Ripley to David L. Wolper, December 23, 1974, SIA (RU367), 4:5.

67. Mel Stuart (Wolper) to Nazaret Cherkezian, December 2, 1974, SIA (RU367), 4:12.

68. Jay Sharbutt, "Special Explores Aviation History," *Los Angeles Times*, January 31, 1975.

69. Jay Sharbutt, "'Flight' an Uplifting Subject for Tonight," *Independent-Journal*, January 31, 1975.

70. Charles DeVault, memo, April 10, 1974, SIA (RU416), 20:10; and Nazaret Cherkezian to David L. Wolper, January 14, 1975, SIA (RU586), 6:16.

71. Carl W. Larsen to Warren V. Bush, May 16, 1974, SIA (RU416), 20:10; and Julian T. Euell to David L. Wolper, December 13, 1974, SIA (RU367), 4:14.

72. Richard Kurin, *Hope Diamond: The True Story of a Legendary and Cursed Gem* (New York: HarperCollins/Smithsonian Books, 2006).

73. William C. Grayson to Carl Larson, May 9, 1974, SIA (RU586), 6:16.

74. Porter M. Kier to Nazaret Cherkezian, January 8, 1975, SIA (RU586), 6:16.

75. William C. Grayson to Julian T. Euell, September 25, 1974, SIA (RU367), 4:14; Porter M. Kier to S. Dillon Ripley, January 3, 1975, SIA (RU416), 20:10; George Lefferts to Carl W. Larsen, May 20, 1974; and Paul E. Desautels to Charles DeVault, May 28, 1974, SIA (RU586), 6:16.

76. Julian T. Euell to S. Dillon Ripley, January 17, 1975, SIA (RU367), 8:9.

77. S. Dillon Ripley to Porter M. Kier, January 23, 1975, SIA (RU586), 6:15.

78. Julian T. Euell to S. Dillon Ripley, March 24, 1975, SIA (RU367), 8:9.

79. Nazaret Cherkezian to David L. Wolper, April 11, 1975, SIA (RU586), 3:25.

80. Correspondence in SIA (RU586), 2:9 and SIA (RU276), 22:5. Examples included refrigeration, telegraphy, movies, Bakelite, cyclotrons, antibiotics, and ATMs.

81. See correspondence in SIA (RU367), 4:8; David L. Wolper to Julian T. Euell, December 27, 1973, SIA (RU367), 4:5; and notes dated January 8, 1974, SIA (RU367), 4:8.

82. Charles DeVault to Julian T. Euell, February 1, 1974, SIA (RU367), 4:5.

83. David A. Hounshell to Silvio A. Bedini, May 10, 1974; and Eugene S. Ferguson to Silvio A. Bedini, May 9, 1974, SIA (RU276), 22:6.

84. Eugene S. Ferguson to Silvio A. Bedini, December 11, 1973, SIA (RU367), 4:8.

85. Eugene S. Ferguson to Silvio A. Bedini, March 1974, SIA (RU276), 22:6.

86. Brooke Hindle to Eugene S. Ferguson, March 25, 1974, SIA (RU276), 22:6.

87. Brooke Hindle to David Wolper, October 31, 1974; and "Feasibility Study for 'The American Dream,'" October 25, 1974, SIA (RU367), 4:8; David L. Wolper to Nazaret Cherkezian, January 16, 1975; and Nazaret Cherkezian to Brooke Hindle, February 5, 1975, SIA (RU276), 63:7.

88. Eugene S. Ferguson to Brooke Hindle, March 26, 1974, SIA (RU276), 22:6.

89. Nazaret Cherkezian to Julian T. Euell and Carl W. Larsen, June 5, 1975, SIA (RU367), 8:9.

90. In 1976, Wolper unsuccessfully sought Smithsonian cooperation on a CBS special "Pyramids," for which he had already secured sponsorship. David L. Wolper to S. Dillon Ripley, September 17, 1976, SIA (RU367), 18:7.

91. "Report of the Committee," 159.

92. Julian T. Euell to S. Dillon Ripley, January 17, 1975, SIA (RU367), 8:9.

93. Ibid.

CHAPTER NINE

1. Huw Wheldon, "Creativity and Collaboration in Television Programmes," Frank Nelson Doubleday Lecture at Museum of History and Technology, March 7, 1974, SIA (RU586), 2:9.

2. Roger Manvell, "Experiments in Broadcasting and Television," *Hollywood Quarterly* 2 (July 1947): 388-392.

3. David Paterson, untitled essay in *Science on Television* (Washington, DC: AAAS, 1975), 29.

4. Wheldon, "Creativity and Collaboration in Television Programmes."

5. Paterson in *Science on Television*, 31; Paul Saltman, "The Softest Hard Sell: Bronowski's Approach to Communicating Science," *Leonardo* 18 (1985): 243-244; and Christine Russell, "The Man behind 'Ascent of Man': Jacob Bronowski," *BioScience* 25 (January 1975): 9-12.

6. "Science on Television," *Discovery* 14 (April 1953): 104.

7. Charlotte Hays, "Big Crowd, Few Seats at 'Ascent,'" *Washington Post*, November 29, 1973; and Jeanette Smyth, "Crowds for 'Ascent' Films," *Washington Post*, November 30, 1973.

8. See materials in SIA (RU367), 9:9. During spring 1974, the National Academy of Sciences showed all thirteen films in its auditorium.

9. Kenneth Clark, "Television," in R. B. McConnell, ed., *Art, Science and Human Progress* (New York: Universe Books, 1983), 38-39.

10. James L. Baughman, *The Republic of Mass Culture: Journalism, Filmmaking, and Broadcasting in America since 1941* (Baltimore: Johns Hopkins University Press, 1992); Robert G. Finney, "Television," in Erwin K. Thomas and Brown H. Carpenter, eds., *Handbook on Mass Media in the United States* (Westport, CT: Greenwood Press, 1994), 171-172; and Frederic A. Leigh, "Educational and Cultural Programming," in Brian G. Rose, ed., *TV Genres: A Handbook and Reference Guide* (New York: Greenwood Press, 1985), 365-380.

11. Robert T. Bower, *Television and the Public* (New York: Holt, Rinehart and Winston, 1973), 50-52.

12. Alan O'Connor, ed., *Raymond Williams on Television: Selected Writings* (London: Routledge, 1989), 76-77.

13. Virginia Myhaver, "Funding the 'Folk': New Federalism and Public-Private Partnerships during the American Bicentennial," presentation at National Museum of American History, December 15, 2009.

14. WGBH Educational Foundation owned and operated the station, working with the Lowell Institute Cooperative Broadcasting Council, which included Boston College, Boston University, MIT, and other area universities.

15. Michael Ambrosino, "The Science Program Group for Public Television in the United States," ca. 1973, SIA (RU145), 15.

16. Lynde McCormick, "British TV Series, 'Nova,' Makes Natural History Breathtaking," *Christian Science Monitor*, March 6, 1974, 7. A complete list of *NOVA* episodes may be found at http://www.pbs.org/wgbh/nova.

17. John J. O'Connor, "TV: Science on 'Nova,'" *New York Times*, May 10, 1974.

18. Richard P. Adler, "Making Science Come Alive on TV," *Wall Street Journal*, August 2, 1977.

19. Ambrosino, "Science Program Group," 15.

20. Throughout this chapter, calculations are based on *NOVA*'s own lists of its season productions and do not include repeat broadcasts (which varied from station to station).

21. John Mansfield, "The Making of '*NOVA*,'" in *NOVA: Adventures in Science* (Reading, MA: Addison-Wesley, 1983), 8–13.

22. "NOVA Production Contracts, etc.," September 26, 1985, SIA (Acc. 03-022), 13:20.

23. Henry R. Cassirer, "Educational Television: World-Wide," *Quarterly of Film, Radio, and Television* 8 (Summer 1954): 373–374.

24. Most seasons of twenty-six programs included repeat broadcasts from previous years. Graham Chedd in *Science on Television*, 24; and John Henahan in *Science on Television*, 21. The familial relationship between *NOVA* and British television continued for years; in 2002, a *NOVA* producer referred to "our sister series BBC's *Horizon*." Remarks by Evan Hadingham, annual meeting of the American Association for the Advancement of Science, 2002.

25. Georgine M. Pion and Mark Lipsey, "Public Attitudes toward Science and Technology: What Have the Surveys Told Us?," *Public Opinion Quarterly* 145 (1981): 313.

26. Robert C. Cowen, "The Uneasy Marriage between Science and Society," *Christian Science Monitor*, January 3, 1980.

27. Robert C. Cowen, untitled essay in *Science in the Newspaper* (Washington, DC: AAAS, 1974), 29.

28. Peter Conrad, *Television: The Medium and Its Manners* (London: Routledge , 1982), 144.

29. Mansfield, "Making of '*NOVA*,'" 12.

30. Ibid., 12–13.

31. This analysis uses *NOVA*'s own subject classifications, as published on its website.

32. Daniel Q. Haney, "'Nova' Gives Viewers New Look on Special Science Test Tonight," *Paducah Sun-Democrat*, October 16, 1984, SIA (RU372), 25:33.

33. Jonathan Weiner, "Prime Time Science," *The Sciences*, September 1980, 9.

34. John J. O'Connor, "'Plutonium Connection' Proves a Dud," *New York Times*, March 9, 1975.

35. *Congressional Record*, March 11, 1975.

36. Stephen Jay Gould to John Mansfield, November 13, 1981, photocopy in author's collection. For discussion of the 1980 creationism controversy, see Marcel Chotkowski LaFollette, ed., *Creationism, Science, and the Law: The Arkansas Case* (Cambridge, MA: MIT Press, 1983).

37. Jon Palfreman, "Bringing Science to a Television Audience," *Nieman Reports* (Fall 2002): 2–3.

38. Mansfield, "Making of '*NOVA*,'" 11.

39. Barbara Isenberg, "Backers Polish the Corporate Image," *Los Angeles Times*, December 10, 1979.

40. Rose, *TV Genres*, 7; Isenberg, "Backers Polish the Corporate Image."

41. Isenberg, "Backers Polish the Corporate Image."

42. Tony Schwartz, "Public TV Faces Financial Crisis: Cuts Due in Programs and Staffs," *New York Times*, November 13, 1980; "Will Public TV Fade Away?," *Washington Post*, November 21, 1980; Arthur Unger, "Why It's a Good Idea to Underwrite PBS: The Companies' View," *Christian Science Monitor*, November 7, 1983; and Irvin Molotsky, "For PBS's New President, The Challenge Is Still Spelled M-o-n-e-y," *New York Times*, July 1, 1984.

43. Oscar H. Gandy, *Beyond Agenda Setting: Information Subsidies and Public Policy* (Norwood, NJ: Ablex, 1982).

44. Paula Apsell, "Sex, Lies, and Science Television," in Gail Porter, ed., *Communicating the Future: Best Practices for Communication of Science and Technology to the Public* (Gaithersburg, MD: National Institute of Standards and Technology, 2002).

45. Arthur Unger, "PBS Emerges as the 'Network of Science,'" *Christian Science Monitor*, August 23, 1979.

46. Peter J. Schuyten, "PBS's 'Connections': The Many Sides of Technology," *New York Times*, October 21, 1979.

47. David Bianculli, *Teleliteracy: Taking Television Seriously* (New York: Continuum, 1992), 190.

48. Schuyten, "PBS's 'Connections.'"

49. Todd Gitlin, "Prime Time Ideology: The Hegemonic Process in Television Entertainment," *Social Problems* 26 (February 1979): 254.

50. Wheldon, "Creativity and Collaboration in Television Programmes."

51. Apsell, "Sex, Lies, and Science Television."

52. Ibid.

53. Palfreman, "Bringing Science to a Television Audience," 2–4.

CHAPTER TEN

1. Anthony Smith, *The Shadow in the Cave: The Broadcaster, His Audience, and the State* (Urbana: University of Illinois Press, 1973), 46.

2. "Science on TV: Problems and Promise," *SIPIscope* 16 (Spring 1988): 8.

3. Anthony Smith, *The Newspaper: An International History* (London: Thames and Hudson, 1979), 7.

4. *The Spectator*, no. 452 (August 8, 1712).

5. John Tebbel, *The Compact History of the American Newspaper*, revised edition (New York: Hawthorn Books, 1963), 186; and Edwin Emery, *The Press and America*, third edition (Englewood Cliffs, NJ: Prentice Hall, 1972), 432.

6. Deems Taylor, "The Very Human Magazine," *Century Magazine*, July 1917, 425.

7. J. Herbert Altschull, "What Is News?," *Mass Comm. Review* 2 (December 1974): 17–18.

8. Boyce Rensberger, "What Makes Science News?," *The Sciences*, September 1978, 10–13; and Dorothy Nelkin, "An Uneasy Relationship: The Tensions between Medicine and the Media," *The Lancet* 374 (June 8, 1996): 1600–1603.

9. *Conference on the Interpretation of the Natural Sciences for a General Public*, Rye, NY, June 15–16, 1938, 2:159 (SIA [RU7091], 381:20, 21).

10. Vince Kiernan, *Embargoed Science* (Urbana: University of Illinois Press, 2006).

11. Samuel Goudsmit, "Publicity," *Physical Review Letters* 4 (1960): 1–2; and J. H. Crawford, "Editorial," *Applied Physics Letters* 1 (1962): 51.

12. Malcolm Peltu, "The Role of Communications Media," in Harry Otway and Malcolm Peltu, eds., *Regulating Industrial Risks: Science, Hazards and Public Protection* (London: Butterworths, 1985), 137–138; Stephen Klaidman, *Health in the Headlines: The Stories behind the Stories* (New York: Oxford University Press, 1991), 15; Roger Wallis and Stanley J. Baran, *The Known World of Broadcast News: International News and Electronic Media* (London: Routledge, 1990); and Michael R. Hawthorne, "The Media, Economic Development, and Agenda-Setting," in Robert J. Spitzer, ed., *Media and Public Policy* (New York: Praeger, 1993), 81–99.

13. For example, Kris Wilson, "Television Weathercasters as Science Communicators," *Public Understanding of Science* 17 (2008): 73–87; and Mary L. Nucci and Robert Kubey, "'We Begin Tonight with Fruits and Vegetables': Genetically Modified Food on the Evening News, 1980–2003," *Science Communication* 29 (December 2007): 147–176.

14. Project for Excellence in Journalism and Medill News Service Washington Bureau, *Changing Definitions of News* (Washington, DC: Project for Excellence in Journalism, 1998).

15. Henry J. Kaiser Family Foundation, *Assessing Local Television News Coverage of Health Issues* (Menlo Park, CA: Kaiser Family Foundation, 1998).

16. Vanderbilt University began videotaping the *ABC Evening News*, *CBS Evening News*, and *NBC Nightly News* broadcasts in 1968 and eventually began to publish broadcast summaries online. Application of a sampling scheme (every twenty-sixth day from August 1968 through December 1989) yielded slightly over 3 percent of summaries then available online. Conclusions about coverage combine data on all networks. On some nights, especially during the first decade of the Vanderbilt project, not every broadcast was videotaped (and thus there was no online summary).

17. This proportion is comparable to that in other media. Between 1955 and 1985, slightly under 12 percent of editorials in three mainstream U.S. newspapers addressed science and invention topics, with the proportions declining over the thirty-year period. Ernest C. Hynds, "Changes in Editorials: A Study of Three Newspapers, 1955–1985," *Journalism Quarterly* 67 (Summer 1990): 305, table 1. Five to fifteen percent of newspaper front-page stories and news magazine covers in the late 1990s focused on science, technology, health, or medicine. Project for Excellence in Journalism and Medill News Service, *Changing Definitions of News*.

18. Arthur Asa Berger, *Essentials of Mass Communications Theory* (Thousand Oaks, CA: Sage Publications, 1995), 65; and Herbert J. Gans, *Deciding What's News: A Study of CBS Evening News, NBC Nightly News, Newsweek, and Time* (New York: Pantheon Books, 1979).

19. Martin W. Bauer, Kristina Petkova, Pepka Boyadjieva, and Galin Gornev, "Long-Term Trends in the Public Representation of Science across the 'Iron Curtain,' 1946–1995," *Social Studies of Science* 36 (February 2006): 101.

20. James L. Baughman, *The Republic of Mass Culture: Journalism, Filmmaking, and Broadcasting in America since 1941* (Baltimore: Johns Hopkins University Press, 1992), 9. Data on soft news from David K. Scott and Robert H. Gobetz, "Hard News/Soft News Content of National Broadcast Networks, 1972–1987," *Journalism Quarterly* 69 (Summer 1992): 406–412; and William C. Spragens, *Electronic Magazines: Soft News Programs on Network Television* (New York: Praeger, 1995), 13. See also Ellen Hume, *Tabloids, Talk Radio, and the Future of News: Technology's Impact on Journalism* (Washington, DC: Annenberg Washington Program in Communications Policy Studies of Northwestern University, 1995), 12; and Project for Excellence in Journalism and Medill News Service, *Changing Definitions of News*.

21. This trend in biomedical coverage conforms to what Louis Nathe found in three mainstream U.S. newspapers in 1963 and 1973. Untitled essay in *Science in the Newspaper* (Washington, DC: AAAS, 1974), 18, table 3. Bauer, Petkova, Boyadjieva, and Gornev detected cycles in the intensity of coverage over fifty-year periods ("Long-Term Trends").

22. *CBS Evening News*, August 23, 1984.

23. Doris A. Graber, "Seeing Is Remembering: How Visuals Contribute to Learning from Television News," *Journal of Communication* 40 (Summer 1990): 139–140; and Hal Himmelstein, *Television Myth and the American Mind* (New York: Praeger, 1984), 253.

24. Shanto Iyengar and Donald R. Kinder, *News That Matters: Television and American Opinion* (Chicago: University of Chicago Press, 1987), 16–46.

25. William A. Gamson and Andre Modigliani, "Media Discourse and Public Opinion on Nuclear Power: A Constructionist Approach," *American Journal of Sociology* 95 (July 1989): 1–37; Allan Mazur, "The Journalists and Technology: Reporting about Three Mile Island and Love Canal," *Minerva* 22 (Spring 1984): 45–66; Allan Mazur, "Nuclear Power, Chemical Hazards, and the Quantity of Reporting," *Minerva* 28 (Autumn 1990): 294–323; and Thomas H. Moss and David L. Sills, eds., *The Three Mile Island Nuclear Accident: Lessons and Implications* (New York: New York Academy of Sciences, 1981).

26. Gamson and Modigliani, "Media Discourse."

27. Mazur, "Journalists and Technology," 45–46.

28. The existence of film or video of an event affects coverage. Carole Gorney, "Numbers versus Pictures: Did Network Television Sensationalize Chernobyl Coverage?," *Journalism Quarterly* 69 (Summer 1992): 460–463, citing data from Michael R. Greenberg, David B. Sachsman, and Peter M. Sandman, "Risk, Drama, and Geography in Coverage of Environmental Risk by Network TV," *Journalism Quarterly* 66 (Summer 1989): 267–276.

29. Doris A. Graber, "Failures in News Transmission: Reasons and Remedies," in Philip Gaunt, ed., *Beyond Agendas: New Directions in Communication Research* (Westport, CT: Greenwood Press, 1994), 77.

30. Story lengths on January 25, 1978, were 2 minutes, 50 seconds for *ABC Evening News*; 4 minutes, 40 seconds for *CBS Evening News*; and 1 minute, 30 seconds for *NBC Nightly News*.

31. *NBC Nightly News*, July 26, 1978.

32. *ABC Evening News*, July 26, 1978.

33. *CBS Evening News*, July 26, 1978.

34. Quoted in Jonathan Weiner, "Prime Time Science," *The Sciences*, September 1980, 10.

35. Ibid., 11.

36. Ibid.

37. Robert Sahr, "Credentialing Experts: The Climate of Opinion and Journalist Selection of Sources in Domestic and Foreign Policy," in Spitzer, *Media and Public Policy*, 153–169.

38. Ibid., 154.

39. Robert Karl Manoff, "Modes of War and Modes of Social Address: The Text of SDI," *Journal of Communication* 39 (Winter 1989): 68.

40. Statement by Michael Crew, news director of WJKS-TV, Jacksonville, Florida, quoted in "Report from the Cutting Edge," *SIPIscope* 17 (Spring 1989): 21.

41. Shepherd found that 77 percent of news reports on marijuana research made no explicit reference to research studies but "merely reported the statements of individuals presumed to have scientific or medical expertise." R. Gordon Shepherd, "Science News of Controversy: The Case of Marijuana," *Journalism Monographs* 62 (August 1979): 15, 17. See also Jonathan Alter, "Just the Facts?," *Washington Monthly* 31 (January/February 1999): 23.

42. Richard Streckfuss, "Objectivity in Journalism: A Search and a Reassessment," *Journalism Quarterly* 67 (Winter 1990): 973–983.

43. Rudy M. Baum, "Overrated Objectivity," *Chemical & Engineering News*, October 6, 1997, 5.

44. Ibid.

45. John C. Burnham, "Of Science and Superstition: The Media and Biopolitics," *Gannett Center Journal* 4 (Summer 1990): 34.

46. *CBS Evening News*, October 15, 1971.

47. Eagles: *ABC Evening News* and *CBS Evening News*, June 2, 1971.

48. Polar bears: *ABC Evening News* and *CBS Evening News*, June 2, 1971.

49. "Science, Technology, and the Press: Must the 'Age of Innocence' End?," *Technology Review*, March/April 1980, 54–55.

50. *NBC Nightly News*, April 20, 1977.

51. "New Trends in Interpreting Science to the Public" session, annual meeting of the American Association for the Advancement of Science, 1978; Marcel Chotkowski LaFollette, "Observations on Science, Media, and the Public at the 1978 Meeting of the AAAS," *Science, Technology, & Human Values* 3 (April 1978): 25–29.

52. Michael Mulkay, "Frankenstein and the Debate over Embryo Research," *Science, Technology, & Human Values* 21 (Spring 1996): 158.

53. R. Lance Holbert, "A Typology for the Study of Entertainment Television and Politics," *American Behavioral Scientist* 49 (November 2005): 436–453.

54. See Richard Malmsheimer, *"Doctors Only": The Evolving Image of the American Physician* (New York: Greenwood Press, 1988).

55. Rae Goodell, "How to Kill a Controversy: The Case of Recombinant DNA," in Sharon M. Friedman, Sharon Dunwoody, and Carol L. Rogers, eds., *Scientists and Journalists: Reporting Science as News* (New York: Free Press, 1986), 173; Rae Goodell, "Problems with the Press: Who's Responsible?," *BioScience* 35 (March 1985): 154; and June Goodfield, *Reflections on Science and the Media* (Washington, DC: AAAS, 1981).

56. Clifford Grobstein, *A Double Image of the Double Helix: The Recombinant-DNA Debate* (San Francisco: W. H. Freeman, 1979), 104.

57. Goodell, "Problems with the Press," 154.

58. "An Interview with Dr. David Baltimore," *SIPIscope* 10 (March/April 1982): 16; and David Baltimore, "Baltimore's Travels," *Issues in Science and Technology* 5 (Summer 1989): 41.

59. Issues raised in Charles R. Eisendrath, "The Press as Guilty Bystander," in David A. Jackson and Stephen P. Stich, eds., *The Recombinant DNA Debate* (Englewood Cliffs, NJ: Prentice Hall, 1979), 279–299.

60. J. Samuel Walker, *Three Mile Island: A Nuclear Crisis in Historical Perspective* (Berkeley: University of California Press, 2004), 1–3; Spencer R. Weart, *Nuclear Fear: A History of Images* (Cambridge, MA: Harvard University Press, 1988), 351, 363; and David A. Kirby, *Lab Coats in Hollywood: Science, Scientists, and Cinema* (Cambridge, MA: MIT Press, 2011), 173–176.

61. Moss and Sills, *Three Mile Island Nuclear Accident*; Sharon M. Friedman, "A Case of Benign Neglect: Coverage of Three Mile Island before the Accident" and Anne Marie Cunningham, "Not Just Another Day in the Newsroom: The Accident at TMI," in Friedman, Dunwoody, and Rogers, *Scientists and Journalists*, 182–212.

62. Walker, *Three Mile Island*.

63. Quoted in David M. Rubin, "What the President's Commission Learned about the Media," in Moss and Sills, *Three Mile Island Nuclear Accident*, 98.

64. Alan McGowan, "Editor's Note," *SIPIscope* 7 (March–June 1979): 1; and David M. Rubin, "How the News Media Reported on Three Mile Island and Chernobyl," *Journal of Communication* 37 (Summer 1987): 42–57.

65. Mazur, "Journalists and Technology," 56.

66. Peggy Girshman, "Beyond the Basics: A Roundtable Discussion," in Sharon M. Friedman, Sharon Dunwoody, and Carol L. Rogers, eds., *Communicating Uncertainty: Media Coverage of New and Controversial Science* (Hillsdale, NJ: Lawrence Erlbaum, 1999), 253.

67. SIPI, established in 1963, had absorbed scientific and activist organizations such as the Committee for Environmental Information founded by biologist Barry Commoner. *SIPIscope* 7 (March–June 1979).

68. Clarence Page, "Where Is Mr. Wizard When We Need Him?," *Chicago Tribune*, April 10, 1979.

69. Cecil Smith, "CBS Specials: Of Men, Monsters," *Los Angeles Times*, May 18, 1980; Jerry Buck, "Jason Miller's Latest Acting Outing Finds Him Creating 'Monster' Movie," *Chicago Tribune*, May 23, 1980; Cecil Smith, "On the Cover," *Los Angeles Times*, May 25, 1980; Cecil Smith, "Town vs. Gown in 'Monster,'" *Los Angeles Times*, May 27, 1980; John J. O'Connor, "TV: Science Amok and the Vietnam War," *New York Times*, May 27, 1980; Tom Shales, "'The Henderson Monster,'" *Washington Post*, May 27, 1980; and Jo Gladstone, "Letter to the Editor," *Science, Technology, & Human Values* 6 (Fall 1980): 78–79.

70. Smith, "CBS Specials."

71. Buck, "Jason Miller's Latest."

72. Ibid.

73. Smith, "CBS Specials."

74. As quoted in Guy E. Lometti, "Broadcast Preparations for and Consequences of *The Day After*," in J. M. Wober, ed., *Television and Nuclear Power: Making the Public Mind* (Norwood, NJ: Ablex, 1992), 4. See also Lawrence Badash, *A Nuclear Winter's Tale: Science and Politics in the 1980s*

(Cambridge, MA: MIT Press, 2009); and Lawrence Badash, "Nuclear Winter: Scientists in the Political Arena," *Physics in Perspective* 3 (2001): 76–105.

75. Lometti, "Broadcast Preparations," 5.

76. David M. Rubin and Constance Cummings, "Nuclear War and Its Consequences on Television News," *Journal of Communication* 39 (Winter 1989): 42; and Badash, *Nuclear Winter's Tale*, 63–76.

77. Allan M. Winkler, *Life under a Cloud: American Anxiety about the Atom* (New York: Oxford University Press, 1993), 195.

78. Lometti, "Broadcast Preparations," 5–6; Ronald Bailey, *Eco-Scam: The False Prophets of Ecological Apocalypse* (New York: St. Martin's Press, 1993), 114; and Kenneth R. Clark, "'Day After' Fallout: Critics Accuse ABC of Irresponsibility," *Chicago Tribune*, April 28, 1984, 14.

79. Lometti, "Broadcast Preparations," 6.

80. Stanley Feldman and Lee Sigelman, "The Political Impact of Prime-Time Television: 'The Day After,'" *Journal of Politics* 47 (June 1985): 556–578; Lometti, "Broadcast Preparations," 12–17; and Robert W. Kubey, "U.S. Opinion and Politics before and after *The Day After*: Television Movie as Rorschach," in Wober, *Television and Nuclear Power*, 19–30.

81. "Europe/'The Day After,'" *NBC Nightly News*, Wednesday, December 7, 1983; and "Hiroshima/'The Day After,'" *NBC Nightly News*, Saturday, January 28, 1984.

82. George Gerbner, Larry Gross, Michael Morgan, and Nancy Signorielli, "Health and Medicine on Television," *New England Journal of Medicine* 305 (October 8, 1981): 901–904; George Gerbner, Larry Gross, Michael Morgan, and Nancy Signorielli, "Scientists on the TV Screen," *Society*, May/June 1981, 41–44.

83. John Chancellor remarks excerpted in "Science and the People: Television's Role in Making the Connection," *SIPIscope* 16 (Spring 1988): 18.

84. Ibid.

85. Ibid.

86. Ibid., 19.

CHAPTER ELEVEN

1. James Burke, *Connections* (Boston: Little, Brown, 1978), 5.

2. U.S. Bureau of the Census, *Social Indicators III* (Washington, DC: GPO, 1980), 561.

3. Carl Sagan, "There's No Hint of the Joys of Science," *TV Guide*, February 4, 1978, 6. See also Carl Sagan, "In Praise of Science and Technology," *New Republic*, January 22, 1977, 24; and "Sagan Says Scientists Need a Better Shake on Television," *Broadcasting*, April 4, 1977, 72–73.

4. Daniel S. Greenberg, "Scientific Magazines Bursting Out All Over," *Science & Government Report* 9 (January 15, 1979): 1–2; William Bennett, "Science Goes Glossy," *Sciences*, September 1979, 10–15, 22; William Bennett, "Science Hits the Newsstand," *Columbia Journalism Review*, January/February 1981, 53, 55–57; and William J. Broad, "Science Magazines: The Second Wave Rolls In," *Science* 215 (January 15, 1982): 272–273.

5. Mark Dowie, as quoted in "Science, Technology, and the Press: Must the 'Age of Innocence' End?" *Technology Review*, March/April 1980, 51.

6. Ien Ang, *Desperately Seeking the Audience* (London: Routledge, 1991), 18.

7. Lawrence W. Lichty, "Success Story," *Wilson Quarterly* 5 (Winter 1981): 62.

8. William Poundstone, *Carl Sagan: A Life in the Cosmos* (New York: Henry Holt, 1999), 140; and Keay Davidson, *Carl Sagan: A Life* (New York: John Wiley, 1999), 253, 319.

9. Adrian Malone to Stanley Weiss, April 13, 1987, SIA (RU594), 1:3; and Les Brown, "Two British TV Directors Set Up Shop in U.S.," *New York Times*, January 19, 1978.

10. Davidson, *Carl Sagan*.

11. Ibid., 140; and Poundstone, *Carl Sagan*, 253–254. *The Dragons of Eden* (1977) won a Pulitzer Prize, and Sagan's novel *Contact* (1985) was made into a Hollywood movie.

12. Rae Goodell, *The Visible Scientists* (Boston: Little, Brown, 1977), 171; and Davidson, *Carl Sagan*, 262–264.

13. Poundstone, *Carl Sagan*, 178.

14. Timothy Ferris, "The Risks and Rewards of Popularizing Science," *Chronicle of Higher Education*, April 4, 1997, B6; and Clark Chapman, "Two Views of a Star's Life," *Science* 287 (January 7, 2000): 47.

15. Davidson, *Carl Sagan*, 330.

16. Arthur Unger, "Carl Sagan: Cosmic Storyteller," *Christian Science Monitor*, May 8, 1979.

17. Davidson estimates the total audience for original broadcast and reruns as over 400 million. Davidson, *Carl Sagan*, 318, 331–333.

18. Revenue data in Peter M. Robeck to Nazaret Cherkezian, August 18, 1977, SIA (RU367), 18:11.

19. Burke, *Connections*; John G. Burke and Marshall C. Eakin, eds., *Technology and Change: A Courses by Newspaper Reader* (San Francisco: Boyd and Fraser, 1979).

20. Sales were so good that, in 1981, Sagan demanded a record-breaking $2 million advance for his novel *Contact*. Poundstone, *Carl Sagan*, 262, 268; and Davidson, *Carl Sagan*.

21. Peter M. Robeck to Nazaret Cherkezian, August 18, 1977, SIA (RU367), 18:11. Note that the companion books for nonscience blockbuster miniseries grossed even greater sums.

22. Quoted in Richard Zoglin, "Science on TV—How Sharp Is the Focus?," *New York Times*, April 26, 1981.

23. David Perlman, "An Unscientific Science Show," *San Francisco Chronicle*, September 19, 1981.

24. Fred Jerome, "Editor's Note," *SIPIscope* 10 (Winter 1982/1983): 1.

25. Barbara J. Culliton, "Science in the Media," *SIPIscope* 10 (Winter 1982/1983): 5.

26. John Angier, "No Optimism for TV Science Series," *SIPIscope* 11 (September/October 1983): 15.

27. *I, Leonardo* promotional kit, author's collection.

28. "Newspaper Science Sections: Testing the Mass Audience," *SIPIscope* 12 (Autumn 1984): 1.

29. "Are Science Sections Such a Swell Idea?," *SIPIscope* 14 (Autumn 1986): 15.

30. "Newspaper Science Sections Spreading Nationwide," *SIPIscope* 14 (Autumn 1986): 1, 9.

31. "Newspaper Science Sections Still on the Rise," *SIPIscope* 18 (Spring 1990): 26–27; and Renate G. Bader, "How Science News Sections Influence Newspaper Science Coverage: A Case Study," *Journalism Quarterly* 67 (Spring 1990): 88–96.

32. Jack Horkheimer, director of the Miami Science Museum planetarium, appeared in a syndicated five-minute program, *Star Gazer*, from 1976 until his death in 2010.

33. The miniseries, shown in England as *From the Face of the Earth*, was a coproduction of WGBH/Boston and Video Arts Television in London.

34. Quoted in Jonathan Meades, "Medical Mysteries: Quest for the Killers," *Dial*, September 1985, 35–36.

35. Korsmo says that the 1960 IGY film series inspired *Planet Earth*. Fae L. Korsmo, "Shaping Up Planet Earth: The International Geophysical Year (1957–1958) and Communicating Science through Print and Film Media," *Science Communication* 26 (December 2004): 180.

36. James Day, *The Vanishing Vision: The Inside Story of Public Television* (Berkeley: University of California Press, 1995), 287–289.

37. Press release for "The Ring of Truth," August 1987, author's collection.

38. Michael Arlen, quoted in Jonathan Weiner, "Prime Time Science," *The Sciences*, September 1980, 9.

39. Ibid., 6.

40. Ibid.

41. Neil Postman, "The Las Vegasizing of America," *National Forum* 62 (Summer 1982): 8.

42. Orrin E. Dunlap, Jr., "Two Arts as One?," *New York Times*, January 8, 1939, 134; and James L. Baughman, *Same Time, Same Station: Creating American Television, 1948–1961* (Baltimore: Johns Hopkins University Press, 2007), 304–307.

43. National Geographic Society press release, "National Geographic Society Announces Plans for New Cable TV Series," n.d. (received February 14, 1985), author's collection.

44. David T. Suzuki, "Information Overload: More TV Science Could Add to the Confusion," *SIPIscope* 14 (January/February 1986): 4.

45. Ibid.

46. David Bianculli, *Teleliteracy: Taking Television Seriously* (New York: Continuum, 1992), 190–191. See also "Origin of Species," *Times* (London), August 25, 1992.

47. Donna J. Haraway, *Primate Visions: Gender, Race, and Nature in the World of Modern Science* (New York: Routledge, 1989), 401n13.

48. Jan-Christopher Horak, "Wildlife Documentaries: From Classical Forms to Reality TV," *Film History* 18 (2006): 472.

49. David Attenborough, *Life on Air: Memoirs of a Broadcaster* (Princeton, NJ: Princeton University Press, 2002).

50. Gail Davies, "Science, Observation and Entertainment: Competing Visions of Post-war British Natural History Television, 1946–1967," *Ecumene* 7 (2000): 432–459.

51. Attenborough, *Life on Air*.

52. David Attenborough, as quoted in Bianculli, *Teleliteracy*, 191.

53. Gregg Mitman, *Reel Nature: America's Romance with Wildlife on Film* (Cambridge, MA: Harvard University Press, 1999), 205.

54. George Page, director of WNET arts and science programming, served as host and narrator. Arthur Unger, "George Page Wants Television to Peer Deeper into Science," *Christian Science Monitor*, October 26, 1984.

55. Remarks by Thomas E. Lovejoy at CBE/EASE annual meeting, Washington, DC, 1998.

56. Russell W. Peterson, *Rebel with a Conscience* (Newark: University of Delaware Press, 1999), 303.

57. Ibid., 312.

58. Ibid., 306. By 2000, the society had produced over forty new specials, cocreated *All Bird TV* for the Animal Planet channel, and was coproducing *Audubon's Animal Adventures* with Disney.

59. *Living Wild* publicity, February 1984, author's collection.

60. See, for example, Benjamin DeMott, "The Trouble with Public Television," *Atlantic Monthly*, February 1979, 42–47; Stephen Chapman, "Down with Public Television," *Harper's*, August 1979, 77–84; and Robert W. Fleming, "The Future of Public Broadcasting," *National Forum* 61 (Summer 1981): 23.

61. Eric Konigsberg, "Stocks, Bonds, and Barney," *Washington Monthly*, September 1993, 12–15.

62. Vincent Campbell, "The Extinct Animal Show: The Paleoimagery Tradition and Computer Generated Imagery in Factual Television Programs," *Public Understanding of Science* 18 (2009): 199–213; Anneke M. Metz, "A Fantasy Made Real: The Evolution of the Subjunctive Documentary on U.S. Cable Science Channels," *Television & New Media* 9 (July 2008): 333–348; and Horak, "Wildlife Documentaries."

63. Postman, "Las Vegasizing," 8.

64. Ibid.

CHAPTER TWELVE

1. William Melody and Wendy Ehrlich, "The History of Children's Television," in William Melody, *Children's Television: The Economics of Exploitation* (New Haven, CT: Yale University Press, 1973), 36.

2. Margaret Midas, "Without TV," *American Quarterly* 3 (Summer 1951): 152–166.

3. Ibid., 166.

4. The eight-part *Christian Science Monitor* series was reprinted as Robert Lewis Shayon, *Television and Our Children* (New York: Longmans, Green and Co., 1951).

5. Jack Gould, "Something Is Amiss," *New York Times*, December 2, 1951.

6. Melody and Ehrlich, "History of Children's Television," 36.

7. Hal Erikson, *Syndicated Television: The First Forty Years, 1947–1987* (Jefferson, NC: McFarland, 1989), 75.

8. Jack Gould, "Programs in Review," *New York Times*, May 27, 1951.

9. Alex McNeil, *Total Television*, fourth edition (New York: Penguin Books, 1996), 627.

10. Tim Brooks and Earle Marsh, *The Complete Directory to Prime Time Network and Cable TV Shows, 1946–Present*, sixth edition (New York: Ballantine Books, 1995), 1111; McNeil, *Total Television*, 562; Vincent Terrace, *The Complete Encyclopedia of Television Programs, 1947–1976* (South Brunswick, NJ: A. S. Barnes, 1976), 120; and Jeffery Davis, *Children's Television, 1947–1990* (Jefferson, NC: McFarland, 1995), 151–153.

11. Donald Jeffry Herbert Kemske (1917–2007) was born in Minnesota and attended high school and college in Wisconsin. Richard Goldstein, "Don Herbert, 'Mr. Wizard' to Science Buffs, Dies at 89," *New York Times*, June 13, 2007; Martin Weil, "Don Herbert; Mr. Wizard of Children's Television," *Washington Post*, June 13, 2007; and Terry Rindfleisch, "A Life Remembered: 'Mr. Wizard' Lets Kids Experience Science," *La Crosse Tribune*, June 14, 2007.

12. Diane Dismuke, "Don 'Mr. Wizard' Herbert: Still a Science Trailblazer," *NEA Today* 12 (April 1994): 9.

13. Willie Schatz, "Welcome Back, Mr. Wizard," *Washington Post Television Magazine*, November 9–15, 1980, 5.

14. Larry Wolters, "Mr. Wizard Has Skilful [*sic*] Touch with Science," *Chicago Daily Tribune*, June 30, 1951; and "Scientific Interest Stirred by TV Show," *Washington Post*, December 20, 1951.

15. George Alexander, "Don Herbert: Gee Wiz!," *SciQuest*, May/June 1981, 21–28.

16. Anton Remenih, "TV's Mr. Wizard Lures Kids with Science Magic," *Chicago Daily Tribune*, January 25, 1953.

17. Walter Carlson, "Space Age 'Wizard,'" *New York Times*, April 4, 1965.

18. Val Adams, "N.B.C. Will Cancel 'Mr. Wizard' Show," *New York Times*, April 17, 1965.

19. Jane E. Brody, "See the Monkey—and Learn," *New York Times*, September 18, 1966.

20. Ibid.

21. George Gent, "TV Networks Move to Upgrade Much-Criticized Children's Fare," *New York Times*, May 31, 1971; and Carol Kramer, "His Wizardry Makes Aerodynamics Snappy," *Chicago Tribune*, October 31, 1971.

22. Cleveland Amory, "Mr. Wizard," *TV Guide*, 22 April 1972.

23. "How a Television Commercial Is Made," *General Electric Review* (September 1956), in Wilbur Schramm, ed., *Mass Communications*, second edition (Urbana: University of Illinois Press, 1960), 161–174; and Lawrence Laurent, "'Mr. Wizard' Is a Product of Mating Two Careers," *Washington Post*, September 30, 1954.

24. Robert Young, "Nuclear Energy Plan 'Breeds' Lobby Alliance," *Chicago Tribune*, June 23, 1975; and Tom Bronzini, "Mr. Wizard Still Working Science Magic," *Los Angeles Times*, March 22, 1976.

25. Lee Margulies, "Dipping into Mr. Wizard's Bag," *Los Angeles Times*, February 7, 1980; Schatz, "Welcome Back, Mr. Wizard," 5; and Alexander, "Don Herbert."

26. James Quinn, "Mr. Wizard's Pop-Science TV Show Updated, Returned to Airways after 19-Year Absence," *Los Angeles Times*, March 25, 1984; Dismuke, "Don 'Mr. Wizard' Herbert," 9; George Maksian, "Mr. Wizard's Coming Back," *New York Daily News*, April 1, 1976; and K. C. Cole, "'Mr. Wizard' Is Still Up to His Old Tricks," *Chicago Tribune*, June 27, 1984.

27. Quinn, "Mr. Wizard's Pop-Science TV Show."

28. Ben H. Bagdikian, "Out of the Can and into the Bank," *New York Times Magazine*, October 21, 1973, 110, quoting Clay Whitehead and Vice President Spiro T. Agnew.

29. Barnouw says the networks created these series in response to complaints of excessive violence on television. Erik Barnouw, *The Image Empire: A History of Broadcasting in the United States from 1953* (New York: Oxford University Press, 1970), 186. See also Melody and Ehrlich, "History of Children's Television," 49.

30. Clara Jeffrey, "The Less You Know: From Einstein to Aniston," *Mother Jones*, January/February 1997, 16.

31. Tom Yohe, "History and Kid's T.V.," *Public History News* 13 (Spring 1993): 1–2.

32. For example, Surgeon General's Advisory Committee on Television and Social Behavior, *Television and Social Behavior: A Technical Report*, five volumes (Washington, DC: GPO, 1972).

33. *Notice of Inquiry and Notice of Proposed Rulemaking in re Children's Television Programs*, 28 FCC 2d 369–70 (1971), as cited in U.S. Commission on Civil Rights, *Window Dressing on the Set: Women and Minorities in Television* (Washington, DC: GPO, 1977), 70.

34. Lloyd N. Morrisett, "The Age of Television and the Television Age," *Peabody Journal of Education* 48 (January 1971): 112–121.

35. *The Television Code of the National Association of Broadcasters* (June 1975), as quoted in U.S. Commission on Civil Rights, *Window Dressing on the Set*, 67.

36. Sonia Livingstone, "Half a Century of Television in the Lives of Our Children," *Annals of the American Academy of Political and Social Science* 625 (September 2009): 155–156.

37. Ibid., 156.

38. During the mid-1970s, two-to-five-year-olds in the United States consumed 42 percent of their daily amount of television during network prime time (7:00 to 11:00 p.m.); more than half of children ages seven through eleven were allowed to watch television whenever they wanted, and one-third of those could watch whatever programs they wanted. National Television Institute data cited in William A. Henry III, "Some Sense of Violence," *Boston Globe*, August 1, 1977. See also Margita White, "Mom, Why's the TV Set Sweating?," *New York Times*, March 29, 1978; and Frank Mankiewicz and Joel Swerdlow, *Remote Control: Television and the Manipulation of American Life* (New York: Times Books, 1978), 17.

39. In 1977, the Boston area had three network-affiliated commercial, two public, and four independent stations; Massachusetts Educational Television broadcast weekdays during the school year via public UHF and VHF channels.

40. Conclusions in this section are based on the author's analysis, described in Marcel Chotkowski LaFollette, "Wizards, Villains, and Other Scientists: The Science Content of Television for Children," survey conducted for Action for Children's Television (Harvard University, 1978); and Marcel Chotkowski LaFollette, "Science on Television: Influences and Strategies," *Daedalus* 111 (Fall 1982): 183–198.

41. A. C. Nielsen data cited in National Science Foundation, *Research on the Effects of Television Advertising on Children*, RANN Report NSF/RA 770115 (Washington, DC: NSF, 1977), 15.

42. NBC press release, November 16, 1976.

43. In the series *Monster Squad*, the name Frankenstein Jr. was given to the monster. In Mary Shelley's *Frankenstein*, the name was that of the monster's creator.

44. Henry, "Some Sense of Violence."

45. Melvin DeFleur and Lois DeFleur, "The Relative Contribution of Television as a Learning Source for Children's Occupational Knowledge," *American Sociological Review* 32 (1967): 789; and Sharon M. Friedman, "Women in Engineering: Influential Factors for Career Choice," *Newsletter on Science, Technology & Human Values*, no. 20 (June 1977): 14–16.

46. Constance Holden, "Science Show for Children Being Developed for TV," *Science* 202 (November 17, 1978): 731.

47. Jonathan Weiner, "Prime Time Science," *The Sciences*, September 1980, 10.

48. Ibid.

49. "Science on Television" segment, *CBS Sunday Morning*, March 9, 1986. Thanks to Carol Rogers for bringing this quotation to my attention.

50. After cancellation of *Newton's Apple*, its producers created *Dragonfly TV*, which combined basic science with discussion of sports and similar topics.

51. Data cited in Reed Hundt, "Reading the First Amendment in Favor of Children: Implementing the Children's Television Act of 1990," speech to Brooklyn Law School, Brooklyn, NY, December 4, 1995.

52. Kit Boss, "On Seattle TV, It's a Comic, It's an Engineer—It's Science Guy," *Washington Post*, October 4, 1989.

53. Harry F. Waters, "Wacky Science and True Grit," *Newsweek*, October 11, 1993, 62; and Mark S. Goodman, *People Weekly* 41 (April 11, 1994): 87–88.

54. Waters, "Wacky Science and True Grit," 62.

55. ABC canceled *Cro* after two seasons, despite loud protests from the educational and scientific communities. *The Magic School Bus* lasted four seasons on PBS and continued to rerun on cable for over a decade.

CHAPTER THIRTEEN

1. Richard Leacock, "Technology and Reality at the Movies," *Technology Review*, February 1973, 26.

2. Marilee Long, Greg Boiarsky, and Greg Thayer, "Gender and Racial Counterstereotypes in Science Education Television: A Content Analysis," *Public Understanding of Science* 10 (2001): 255–269; and Marilee Long, Jocelyn Steinke, Brooks Applegate, Maria Knight Lapinski, Marne J. Johnson, and Sayani Ghosh, "Portrayals of Male and Female Scientists in Television Programs Popular among Middle School-Age Children," *Science Communication* 32 (2010): 356–382.

3. Marcel Chotkowski LaFollette, "Eyes on the Stars: Images of Women Scientists in Popular Culture," *Science, Technology, & Human Values* 13 (Fall 1988): 262–275.

4. John R. Zaller, *The Nature and Origins of Mass Opinion* (Cambridge: Cambridge University Press, 1992), 1–9.

5. Steve Barkin, quoted in Doris A. Graber, "Media Impact on the Political Status Quo: What Is the Evidence?," in Robert J. Spitzer, ed., *Media and Public Policy* (New York: Praeger, 1993), 21.

6. *Conference on the Interpretation of the Natural Sciences for a General Public*, Rye, NY, June 15–16, 1938, 1:55 (SIA [RU7091], 381:20, 21).

7. Erik Barnouw, *The Image Empire: A History of Broadcasting in the United States from 1953* (New York: Oxford University Press, 1970), 31.

8. Alberto Elena, "Skirts in the Lab: Madame Curie and the Image of the Woman Scientist in the Feature Film," *Public Understanding of Science* 6 (1997): 276.

9. Sydney W. Head, "Content Analysis of Television Drama Programs," *Quarterly of Film, Radio, and Television* 9 (1954): 181.

10. U.S. Commission on Civil Rights, *Window Dressing on the Set: Women and Minorities in Television* (Washington, DC: GPO, 1977).

11. Ibid.; George Gerbner, Larry Gross, Michael Morgan, and Nancy Signorielli, "Health and Medicine on Television," *New England Journal of Medicine* 305 (October 8, 1981): 901–904; Tim Brooks and Earle Marsh, *The Complete Directory to Prime Time Network and Cable TV Shows, 1946-Present*, sixth edition (New York: Ballantine Books, 1995); Alex McNeil, *Total Television*, fourth edition (New York: Penguin Books, 1996); and Vincent Terrace, *The Complete Encyclopedia of Television Programs, 1947-1976*, volumes 1 and 2 (South Brunswick, NJ: A. S. Barnes, 1976).

12. Anthony Dudo, Dominique Brossard, James Shanahan, Dietram A. Scheufele, Michael Morgan, and Nancy Signorielli, "Science on Television in the 21st Century: Recent Trends in Portrayals and Their Contributions to Public Attitudes toward Science," *Communication Research* 20 (December 2010): 8–10.

13. Raymond Stedman, *The Serials: Suspense and Drama by Installment*, second edition (Norman: University of Oklahoma Press, 1977).

14. Eva Flicker, "Between Brains and Breasts—Women Scientists in Fiction Film: On the Marginalization and Sexualization of Scientific Competence," *Public Understanding of Science* 12 (2003): 316.

15. Jocelyn Steinke, "Women Scientist Role Models on Screen: A Case Study of *Contact*," *Science Communication* 21 (December 1999): 111-136.

16. Jocelyn Steinke, "Cultural Representations of Gender and Science: Portrayals of Female Scientists and Engineers in Popular Films," *Science Communication* 27 (September 2005): 53.

17. Peter Conrad, *Television: The Medium and Its Manners* (London: Routledge, 1982), 145.

18. Robert Lambourne, Michael Shallis, and Michael Shortland, *Close Encounters? Science and Science Fiction* (Bristol, UK: Adam Hilger, 1990), 101.

19. Gilbert Seldes, "The Nature of Television Programs," *Annals of the American Academy of Political and Social Science* 213 (January 1941): 142.

20. Joshua Gamson, *Claims to Fame: Celebrity in Contemporary America* (Berkeley: University of California Press, 1994), 42, 98.

21. Ibid., 172; Richard Schickel, *Intimate Strangers: The Culture of Celebrity in America* (Chicago: Ivan R. Dee, 2000), ix.

22. Schickel, *Intimate Strangers*, 4.

23. Rae Goodell, *The Visible Scientists* (Boston: Little, Brown, 1977); Anita Rae Simpson Goodell, "The Visible Scientists" (Ph.D. dissertation, Stanford University, 1975).

24. Goodell, "Visible Scientists" (dissertation), 276-311.

25. Goodell, *Visible Scientists*, 21-22; and Donna J. Haraway, *Primate Visions: Gender, Race, and Nature in the World of Modern Science* (New York: Routledge, 1989), 158.

26. Goodell, *Visible Scientists*, 21.

27. E-mail to author from Donna Gerardi, February 17, 1999.

28. Maxine Singer, "Heroines and Role Models," *Science* 253 (July 19, 1991): 249.

29. Flora Rheta Schreiber, "Television's New Idiom in Public Affairs," *Hollywood Quarterly* 5 (Winter 1950): 144-145.

30. Ibid., 146.

31. Elena, "Skirts in the Lab," 276.

32. U.S. Commission on Civil Rights, *Window Dressing on the Set*, 1.

33. Ibid.

34. LaFollette, "Eyes on the Stars"; and Marcel Chotkowski LaFollette, *Science on the Air: Popularizers and Personalities on Radio and Early Television* (Chicago: University of Chicago Press, 2008), 221-222.

35. Author's analysis of "Guests on Johns Hopkins Television Series, 1948-60," n.d., Johns

Hopkins University Archives, JHU News Office, RG10.020, Series 10, 1:1. Analysis based on descriptions of 425 of 491 programs in the four Johns Hopkins series.

36. Ibid.

37. Author's analysis of new *NOVA* programs, as described on the series website. The analysis did not count repeat broadcasts, but did include programs produced originally for *NOVA* as well as those purchased from or coproduced with another network. *NOVA* did include a program about female astronauts, one of whom was trained as a scientist.

38. Author's analysis of online summaries from the Vanderbilt University News Archive. For a discussion of that analysis, see chapter 10.

39. "Women, Men and Media" study sponsored by the Gannett Foundation and University of Southern California (1989), as described in Maureen H. Beasley, "Women Audiences," in Erwin K. Thomas and Brown H. Carpenter, eds., *Handbook on Mass Media in the United States: The Industry and Its Audiences* (Westport, CT: Greenwood Press, 1994), 216–217.

40. Study by D. C. Whitney and colleagues, as cited by Lana F. Rakow and Kimberlie Kranich, "Woman as Sign in Television News," *Journal of Communication* 41 (Winter 1991): 12–13.

41. "Women, Men and Media," 216–217. See also Marlene Sanders and Marcia Rock, *Waiting for Prime Time: The Women of Television News* (New York: Harper and Row, 1990).

42. NOW data cited in "The Viewers' Bill of Rights," *Digital Beat*, May 14, 1999, http://www.benton.org.

43. Mwenya Chimba and Jenny Kitzinger, "Bimbo or Boffin? Women in Science: An Analysis of Media Representations and How Female Scientists Negotiate Cultural Contradictions," *Public Understanding of Science* 19 (2010): 621.

44. The average number of scientists shown on camera in network news stories about science rose from one every two stories in the mid-1970s to almost one per story in the 1990s. Author's analysis of online summaries from the Vanderbilt University News Archive. For more extensive discussion of that analysis, see chapter 10.

45. Chimba and Kitzinger, "Bimbo or Boffin?," 621.

46. R. Gordon Shepherd, "Science News of Controversy: The Case of Marijuana," *Journalism Monographs* 62 (August 1979): 5.

47. Alfred Robert Hogan, "Televising the Space Age: A Descriptive Chronology of CBS News Special Coverage of Space Exploration from 1957 to 2003" (master's thesis, University of Maryland, 2005).

48. Jocelyn Steinke and Marilee Long, "A Lab of Her Own? Portrayals of Female Characters on Children's Educational Science Programs," *Science Communication* 18 (December 1996): 97–98.

49. Ibid., 107.

50. Jocelyn Steinke, "A Portrait of a Woman as a Scientist: Breaking Down Barriers Created by Gender-Role Stereotypes," *Public Understanding of Science* 6 (1997): 409–428.

CHAPTER FOURTEEN

1. S. Dillon Ripley, "The View from the Castle," *Smithsonian Magazine*, October 1982, 12. Ripley used a similar description of television in his 1969 report to the Smithsonian regents. *Smithsonian Year 1969, Annual Report of the Smithsonian Institution for the Year Ended 30 June 1969* (Washington, DC: Smithsonian Institution Press, 1969), 28.

2. Wil Lepkowski, letter to the author, August 16, 1990.

3. For a good summary of this change, see Christopher P. Tuomey, *Conjuring Science: Science as Meaning in American Culture* (New Brunswick, NJ: Rutgers University Press, 1996).

4. Larry Taylor to Sam Hughes, July 25, 1980, SIA (RU392), 14:9.

5. S. Dillon Ripley, "If the Masses Won't Come to Museums . . . ," *Los Angeles Times*, August 23, 1976.

6. Nazaret Cherkezian said that he "brought" the *Smithsonian World* concept with him when he joined the staff in 1972, assembled the "concept" at the Smithsonian, and hired "a writer I knew." SIA (RU367), 18:5; and "Oral History Interview with Nazaret Cherkezian, 1986," SIA (RU9541). The January 1976 proposal listed Nazaret Cherkezian and Michael de Guzman as authors; a third draft, dated September 29, 1976, was titled "Smithsonian World—A film perspective of the Smithsonian Institution with Secretary Dillon Ripley," with de Guzman as writer, David Vassar as director, and Cherkezian as executive producer; by March 1977, the proposal read "created by Nazaret Cherkezian, Office of Telecommunications, and written by Michael de Guzman." SIA (Acc. 01–203), 3:21. Martin Carr later told journalists that he and Cherkezian had developed the idea during the 1970s.

7. Carl W. Larsen to Nazaret Cherkezian, November 5, 1975, SIA (RU586), 4:10.

8. Correspondence in SIA (RU367), 18:5.

9. Ibid.

10. G. B. Butler to S. Dillon Ripley, August 17, 1977, SIA (RU367), 18:5.

11. Ward B. Chamberlin, Jr., to James Symington, January 7, 1980; James Symington to Ward B. Chamberlin, Jr., January 8, 1980; and S. Dillon Ripley to John A. Schneider, December 15, 1980, SIA (RU367), 26:4.

12. Tony Schwartz, "Public TV Faces Financial Crisis: Cuts Due in Programs and Staff," *New York Times*, November 13, 1980.

13. James Symington to Ward B. Chamberlin, Jr., January 8, 1980, SIA (RU367), 26:4.

14. Nazaret Cherkezian to S. Dillon Ripley, February 4, 1981, SIA (RU586), 6:1.

15. Ibid.

16. Nazaret Cherkezian to S. Dillon Ripley, January 18, 1980, SIA (RU367), 26:4.

17. Nazaret Cherkezian to S. Dillon Ripley, December 10, 1980, *A Plan of Action for Smithsonian Video Communications*, SIA (RU367), 26:4; and Nazaret Cherkezian to S. Dillon Ripley, February 4, 1981, SIA (RU586), 6:1.

18. Edward K. Thompson to S. Dillon Ripley, March 21, 1981, SIA (RU392), 14:30.

19. *Report to Smithsonian TV Committee*, March 10, 1981, SIA (RU392), 14:30. See also Nazaret Cherkezian to Executive Committee of the Smithsonian, December 22, 1981, SIA (RU392), 14:29.

20. Edward K. Thompson to S. Dillon Ripley, March 21, 1981, SIA (RU392), 14:30.

21. Ibid.

22. See, for example, Larry Taylor to Sam Hughes, July 25, 1980, SIA (RU392), 14:9.

23. Nazaret Cherkezian to S. Dillon Ripley, December 28, 1977, SIA (RU367), 18:5; and Nazaret Cherkezian to S. Dillon Ripley, January 25, 1978, SIA (RU586), 4:2.

24. Martin Carr to Ward B. Chamberlin, Jr., March 2, 1982, SIA (Acc. 03–022), 13:27. In 1984, the organization was renamed the James S. McDonnell Foundation.

25. Ibid.

26. "Oral History Interview with Nazaret Cherkezian, 1986," 24, SIA (RU9541).

27. Martin Carr to Ward B. Chamberlin, Jr., March 2, 1982, SIA (Acc. 03–022), 13:27.

28. Correspondence in SIA (RU372), 20:5.

29. WETA "reminded" the Smithsonian that neither party may "engage in a large-scale project . . . with a similar format or title prior to initial broadcast of the entire series." Gerald Slater to Tom Wolf, February 18, 1983; and Thomas H. Wolf to Elizabeth Brownstein, February 24, 1983, SIA (Acc. 03–022), 10:40.

30. Nazaret Cherkezian to Thomas H. Wolf, March 18, 1983, SIA (Acc. 01–203), 1:36.

31. Correspondence in SIA (Acc. 03–022), 10:35.

32. Gilbert Grosvenor to S. Dillon Ripley, November 30, 1982; Gilbert Grosvenor to

S. Dillon Ripley, August 8, 1983; and Tim Cowling and Tim Kelly (NGS) to Nazaret Cherkezian, May 6, 1983, SIA (RU586), 1:15.

33. Thomas H. Wolf to Philip Hughes, March 14, 1984, SIA (RU586), 1:15.

34. Martin Carr to Thomas H. Wolf, July 17, 1984; David Clark to Martin Carr, July 16, 1984; and Thomas H. Wolf to Martin Carr, July 23, 1984, SIA (Acc. 03-022), 10:39.

35. Jim Hobbins to S. Dillon Ripley, March 11, 1981, SIA (RU392), 14:30.

36. Thomas H. Wolf to P. S. Hughes, November 8, 1982, SIA (Acc. 03-022), 10:40.

37. Ibid.

38. Thomas H. Wolf to Martin Carr, n.d. (ca. November 1982), SIA (Acc. 03-022), 10:40.

39. Martin Carr to Ward B. Chamberlin, Jr., March 2, 1982, SIA (Acc. 03-022), 13:27.

40. Thomas H. Wolf to S. Dillon Ripley, October 11, 1983, SIA (Acc. 03-022), 10:40.

41. "Series Overview," February 1983, author's personal collection; and "Television," report to Smithsonian Institution Board of Regents, Audit and Review Committee, May 19, 1983, SIA (Acc. 03-022), 12:31.

42. Thomas H. Wolf to Martin Carr, July 6, 1983, SIA (Acc. 03-022), 10:40.

43. Thomas H. Wolf to Martin Carr, August 4, 1983, SIA (Acc. 03-022), 10:40.

44. Thomas H. Wolf to Martin Carr, January 12, 1983, SIA (Acc. 03-022), 10:40.

45. Thomas H. Wolf to Martin Carr, October 13, 1983, SIA (Acc. 03-022), 10:40.

46. Richard S. Fiske to Thomas H. Wolf, March 31, 1983, SIA (Acc. 03-022), 10:40.

47. Minutes, Smithsonian World Staff meeting, August 11, 1983, SIA (Acc. 03-022), 13:38.

48. S. Dillon Ripley to Thomas H. Wolf, October 7, 1983, SIA (Acc. 03-022), 10:46; Tom Simon to Ward Chamberlin, November 9, 1983, SIA (Acc. 03-022), 13:38; and Thomas H. Wolf to Gerald Slater, December 14, 1983, SIA (Acc. 03-022), 10:40. Carr and WETA claimed that Smithsonian officials had not complained about these problems, but the records show otherwise. Minutes of Smithsonian World staff meeting, August 11, 1983, SIA (Acc. 03-022), 13:38.

49. Undated and unsigned memo from Ripley staff member to "Tom Wolf and SI World Committee Members," SIA (Acc. 03-022), 10:39.

50. Thomas H. Wolf to Martin Carr and Gerald Slater, January 23, 1984; and Tom Wolf to Tom Simon, January 23, 1984, SIA (Acc. 03-022), 10:39.

51. Tom Dorsey, *Louisville-Courier Journal*, ca. January 1984 (no date on clipping; SIA [Acc. 01-203]); and John Corry, "TV: 'Smithsonian World' Starting," *New York Times*, January 18, 1984.

52. George Bullard, *Detroit News*, ca. January 1984 (no date on clipping; SIA [Acc. 03-022]).

53. George Bullard, "Still Defying the Ratings," *Detroit News*, January 10, 1984.

54. Tom Shales, "'World' Weary," *Washington Post*, April 11, 1984. See also Tom Shales, "Smithsonian on View," *Washington Post*, January 18, 1984.

55. Robert Hilton Simmons, "'Smithsonian World': It's Rich, It's Glamorous, but Is It Truth?," *Federal Times*, April 2, 1984.

56. Shawn McClellan, "Ties to the Past," *New Orleans Times-Picayune–The States-Item*, December 9, 1984.

57. Robert McCormick Adams to Martin Carr, February 20, 1985, SIA (Acc. 03-022), 10:43.

58. Thomas H. Wolf to Martin Carr, June 11, 1984, SIA (Acc. 03-022), 10:39; and Thomas H. Wolf to Martin Carr, November 15, 1984, SIA (Acc. 03-022), 10:39. Wolf asked David McCullough to pressure Carr to increase attention to environment, minorities, and Third World issues. Thomas H. Wolf to David McCullough, November 14, 1984, SIA (Acc. 03-022), 10:39.

59. Thomas H. Wolf to Martin Carr, September 24, 1984, and October 4, 1984, SIA (Acc. 03-022), 10:39.

60. Joyce Campbell to Ward Chamberlin and Tammy Robinson, February 14, 1985, SIA (Acc. 03-022), 11:2.

61. Elizabeth Brownstein to Sandra W. Bradley, October 29, 1985, SIA (Acc. 03–022), 9:22.

62. David Friedman, "No Time for 'Heroes,'" *Philadelphia Daily News*, April 24, 1985.

63. Tom Shales, "'World' without End," *Washington Post*, April 24, 1985.

64. Mike Hughes, "'Smithsonian' Producer Shedding Pompous Image," *Lansing (MI) State Journal*, November 19, 1986.

65. Robin Marantz Henig, "Malone in Charge," *Dial*, November 1986, 30–31, 35; and Hughes, "'Smithsonian' Producer Shedding Pompous Image."

66. Discussion paper, Adrian Malone Productions Inc., 1983, SIA (RU594), 5:9.

67. Adrian Malone to Ward Chamberlin et al., January 7, 1986, SIA (RU594), 5:65.

68. Ibid.

69. "PBS's 'Smithsonian World' Looks for Backing," *New York Times*, February 2, 1987.

70. *Enterprise* (Southwestern Bell corporate publication), November 2, 1988, 8 (SIA [RU372], 35:52).

71. Sandra W. Bradley to Maxine Singer, April 7, 1988, SIA (Acc. 03–022), 5:48.

72. Sandra W. Bradley to Ben Wattenberg, January 17, 1989, SIA (Acc. 03–022), 15:1.

73. Sandra W. Bradley to Maxine Singer, April 7, 1988, SIA (Acc. 03–022), 5:48.

74. Maxine F. Singer to Sandra W. Bradley, April 20, 1988, SIA (Acc. 03–022), 5:48.

75. Steven P. Briggs to Robert McCormick Adams, January 26, 1989, SIA (Acc. 03–022), 18:2; and Sandra W. Bradley to Steven P. Briggs, February 13, 1989, SIA (Acc. 03–022), 15:27.

76. See letters from Meryl Loonin, May 2–10, 1989, SIA (Acc. 03–022), 6:18.

77. Sandra W. Bradley to Tammy Robinson and Bob Dierker, memo, Feb 22, 1990, SIA (Acc. 03–022), 12:1.

78. Since her company, Wentworth Films, had the $300,000 contract to produce the film, Bradley had an interest in concluding the project successfully.

79. Description used in Sandra W. Bradley's letters to potential interviewees; for example, Sandra W. Bradley to Freeman Dyson, August 3, 1989, SIA (Acc. 03–022), 7:63.

80. Adrian Malone to Sandra W. Bradley, April 13, 1989, SIA (Acc. 03–022), 7:63.

81. Howard Smith to Sandra W. Bradley, July 20, 1989, SIA (Acc. 03–022), 20:9.

82. Sandra W. Bradley to Adrian Malone and Tammy Robinson, July 27, 1989, SIA (Acc. 03–022), 20:9.

83. Sandra W. Bradley to Howard Smith, July 27, 1989, SIA (Acc. 03–022), 20:9.

84. Robert McCormick Adams to Robert Hoffman, August 3, 1989, SIA (Acc. 03–022), 20:9.

85. Robert Hoffman to Sandra W. Bradley, August 9, 1989, SIA (Acc. 03–022), 20:9.

86. Adrian Malone to Robert Hoffman, August 16, 1989, SIA (Acc. 03–022), 20:9.

87. Ibid.

88. Walter Goodman, "Examining God's Dice with Quantum Mechanics," *New York Times*, June 6, 1990.

89. The corporation's new chief executive reportedly wanted to emphasize sports promotion rather than television "cultural programming." Correspondence in SIA (Acc. 03–022), 15:25.

90. John Carmody, "The TV Column," *Washington Post*, February 5, 1991; and Ann Pincus, "Statement of Loss of Smithsonian World Funding by SBC," January 22, 1991, SIA (Acc. 03–022), 14:48. See also various correspondence in SIA (Acc. 03–022),15:18.

CHAPTER FIFTEEN

1. *Conference on the Interpretation of the Natural Sciences for a General Public*, Rye, NY, June 15–16, 1938, 2:199 (SIA [RU7091], 381).

2. Film by Helen Miles Davis, April 1952, SIA.

3. William L. Laurence, "Atom Bomb Fired with Troops Near; Chutists Join Tests," *New York Times*, April 23, 1952. Laurence's story contained a box in its first column calling attention to television coverage (Jack Gould, "TV Brings Atomic Bomb Detonation into Millions of Homes, but Quality of Pictures Is Erratic," *New York Times*, April 23, 1952).

4. Tom Shales, "All the World's a Critic," *Washington Post*, July 1, 2009.

5. University of Chicago Committee on Educational Television, "Television and the University," *School Review* 61 (April 1953): 211.

6. "Revolution in the Living Room: 10 Years of Television," *Christian Science Monitor*, November 22, 1957.

7. Aileen Fyfe and Bernard Lightman, eds., *Science in the Marketplace: Nineteenth-Century Sites and Experiences* (Chicago: University of Chicago Press, 2007); and Peter J. Bowler, *Science for All: The Popularization of Science in Early Twentieth-Century Britain* (Chicago: University of Chicago Press, 2009).

8. Conclusions based on program lists in Tim Brooks and Earle Marsh, *The Complete Directory to Prime Time Network and Cable TV Shows, 1946–Present*, sixth edition (New York: Ballantine Books, 1995); Harry Castleman and Walter J. Podrazik, *Watching TV: Four Decades of American Television* (New York: McGraw-Hill, 1982); Daniel Einstein, *Special Edition: A Guide to Network Television Documentary Series and Special News Reports, 1955–1979* (Metuchen, NJ: Scarecrow Press, 1987); Sydney W. Head, *Broadcasting in America: Survey of Television and Radio* (Cambridge, MA: Riverside Press, 1956); R. D. Heldenfels, *Television's Greatest Year—1954* (New York: Continuum, 1994); Alex McNeil, *Total Television*, fourth edition (New York: Penguin Books, 1996); Irving Stettel, ed., *Top TV Shows of the Year, 1954–1955* (New York: Hastings House, 1955); Vincent Terrace, *The Complete Encyclopedia of Television Programs, 1947–1976*, volumes 1 and 2 (South Brunswick, NJ: A. S. Barnes, 1976); and Erwin K. Thomas and Brown H. Carpenter, eds., *Handbook on Mass Media in the United States* (Westport, CT: Greenwood Press, 1994).

9. Hilde T. Himmelweit and Betty Swift, "Adolescent and Adult Media Use and Taste: A Longitudinal Study," quoted in *Harvard University Program on Technology and Society, 1964–72: A Final Review* (Cambridge, MA: Harvard University, 1972), 181.

10. Quoted in George Will, "TV for Voyeurs," *Washington Post*, June 21, 2001.

11. Milton Chen, "Television and Informal Science Education: Assessing the Past, Present, and Future of Research," in Valerie Crane, ed., *Informal Science Learning: What the Research Says about Television, Science Museums, and Community-Based Projects* (Dedham, MA: Research Communications, 1994), 18.

12. Based on A. C. Nielsen Media Research data for 1998.

13. Author's analysis of programming on three Washington, DC–area public television stations (WMPT, WETA, WHMM) during one week in August 1992. Of 367.5 hours of content broadcast by all stations, science and science-related (nature, medicine, and public health) programming totaled 59 hours, or 16 percent.

14. Data from Robin Dunbar, *The Trouble with Science* (Cambridge, MA: Harvard University Press, 1996), 147.

15. Richard Seltzer, "New Television Science Series May Draw Both Kudos and Arrows," *Chemical & Engineering News* 73 (April 24, 1995): 52–53.

16. Chen, "Television and Informal Science Education," 23.

17. Quoted in Bruce Gamen, "The Geek Factor," *Chemical & Engineering News* 74 (February 12, 1996): 5.

18. Quoted in Douglas M. Birch, "A Quantum Leap," *Baltimore Sun*, February 5, 1998.

19. Chen, "Television and Informal Science Education," 21–22.

20. Garrett G. Fagan, "Seductions of Pseudoarchaeology: Far Out Television," *Archaeology* 56 (May/June 2003): 46.

21. Chen, "Television and Informal Science Education," 15–56.

22. Ellen Wartella, Aletha C. Huston, Victoria Rideout, and Michael Robb, "Studying Media Effects on Children: Improving Methods and Measures," *American Behavioral Scientist* 52 (April 2009): 1112.

23. J. Michael Oakes, "The Effect of Media on Children: A Methodological Assessment from a Social Epidemiologist," *American Behavioral Scientist* 52 (April 2009): 1136-1151.

24. *Conference on the Interpretation*, 154.

25. Edward D. Miller, quoted in Libby Copeland, "Doctor's Orders: First, Do No Harm, Next, Get No Sleep," *Washington Post*, August 30, 2000.

26. "Hollywood's House Calls," *Brill's Content* 3 (November 2000): 136-137. See also Jeff Stryker, "More Drama Added to the Politics of Transplants," *New York Times*, February 19, 2002.

27. George Gerbner, Larry Gross, Michael Morgan, and Nancy Signorielli, "Health and Medicine on Television," *New England Journal of Medicine* 305 (October 8, 1981): 903; M. Downing, "The World of Daytime Television Serial Drama" (PhD dissertation, 1975), as quoted in Mary Cassata, Thomas Skill, and Samuel Osei Boadu, "Life and Death in the Daytime Television Serial: A Content Analysis," in Mary B. Cassata and Thomas Skill, ed., *Life on Daytime Television: Tuning-in American Serial Drama* (Norwood, NJ: Ablex, 1983), 49; Joseph Turow and Rachel Gans, *As Seen on TV: Health Policy Issues in TV's Medical Dramas* (Menlo Park, CA: Kaiser Family Foundation, 2002); and Mollyann Brodie, Ursula Foehr, Vicky Rideout, Neal Baer, Carolyn Miller, Rebecca Flournoy, and Drew Altman, "Communicating Health Information through the Entertainment Media," *Health Affairs* 20 (January/February 2001): 192-199.

28. For example, Bob Arnott (CBS), Holly Atkinson (NBC), Timothy Johnson (ABC), Ian K. Smith (NBC), Nancy Snyderman (ABC and NBC), and Art Ulene (NBC).

29. Vivian Carol Sobchack, *Screening Space: The American Science Fiction Film*, second edition (New Brunswick, NJ: Rutgers University Press, 1997), 254.

30. Leon M. Lederman, "The Advancement of Science," *Science* 256 (May 22, 1992): 1119-1123. See also William Booth, "Scientists May Not Be Mad, Just Very Unhappy," *Washington Post*, September 10, 1991; Andrew Pollack, "Scientists Seek a New Movie Role: Hero, Not Villain," *New York Times*, December 1, 1998; and Karen Kaplan, "Scientists Say Movie Image Not a Pretty Picture," *Los Angeles Times*, November 18, 1998.

31. Lederman, "Advancement of Science," 1123; James Gorman, "Think of the Laughs a Lab Could Cook Up," *New York Times*, May 19, 1996; and Claudia Dreifus, "Science Is Serious Business to the 'Mel Brooks' of Physics," *New York Times*, July 14, 1998.

32. William W. Warner and Carl Larsen to S. Dillon Ripley, "Television and the Smithsonian," April 18, 1972, SIA (RU416), 20:10.

33. Craig Tomashoff, "The Science of Success: It's All in the Details," *New York Times*, March 3, 2002. See also Michael E. Hill, "High Ratings, Devine Stories," *Washington Post TV Week*, December 30, 2001, 6-7, 52. About twenty-four million people watched the show every week during December 2001.

34. Serena Wade and Wilbur Schramm, "The Mass Media as Sources of Public Affairs, Science, and Health Knowledge," *Public Opinion Quarterly* 33 (Summer 1969): 209.

35. Ien Ang, *Living Room Wars: Rethinking Media Audiences for a Postmodern World* (London: Routledge, 1996), 152.

36. Raymond Bellour, "Double Helix," in Timothy Druckrey, ed., *Electronic Culture: Technology and Visual Representation* (New York: Aperture Books, 1996), 192.

37. Head, *Broadcasting in America*, 420.

38. Roger Silverstone, "Science and the Media: The Case of Television," in S. J. Doorman, ed., *Images of Science: Scientific Practice and the Public* (Aldershot, UK: Gower, 1989), 189.

39. Kurt Lang and Gladys Engel Lang, "The Unique Perspective of Television and Its Effects: A Pilot Study," *American Sociological Review* 18 (1953): 3-12. See also James L. Baughman, *The Republic of Mass Culture: Journalism, Filmmaking, and Broadcasting in America since 1941* (Baltimore: Johns Hopkins University Press, 1992), 57.

40. G. Ray Funkhouser and Eugene F. Shaw, "How Synthetic Experience Shapes Social Reality," *Journal of Communication* 40 (Spring 1990): 83.

41. Ibid., 79, quoting Neil Postman. Daniel Boorstin warned that consumers of television's synthetic reality would find it easier to accept images than to test their authenticity. Daniel J. Boorstin, *The Image: A Guide to Pseudo-Events in America* (New York: Harper and Row, 1961), iii, 185.

42. Constance Penley, *NASA/TREK: Popular Science and Sex in America* (London: Verso, 1997).

43. Bill McKibben, *The Age of Missing Information* (New York: Plume Books, 1993).

44. June Goodfield, *Reflections on Science and the Media* (Washington, DC: AAAS, 1981), 88.

45. Harry Collins, "Certainty and the Public Understanding of Science: Science on Television," *Social Studies of Science* 17 (November 1987): 709.

46. David A. Kirby, "Science Consultants, Fictional Films, and Scientific Practice," *Social Studies of Science* 33 (April 2003): 245.

47. Richard D. Klausner, "Successfully Sharing Our Stories of Science," *The Scientist* 13 (January 18, 1999): 13.

48. Judith S. Trent and Robert V. Friedenberg, *Political Campaign Communication: Principles and Practices* (New York: Praeger, 1983), 146.

49. Shanto Iyengar and Donald R. Kinder, *News That Matters: Television and American Opinion* (Chicago: University of Chicago Press, 1987); and Doris A. Graber, "Failures in News Transmission: Reasons and Remedies," in Philip Gaunt, ed., *Beyond Agendas: New Directions in Communication Research* (Westport, CT: Greenwood Press, 1994), 80.

50. E. E. Schattschneider, quoted in Robert J. Spitzer, ed., *Media and Public Policy* (New York: Praeger, 1993), 8.

51. Frank Baxter, quoted in "Prof. Frank Baxter Dies at 86," *Chicago Tribune*, January 22, 1982.

52. For example, George Gilder, *Life after Television: The Coming Transformation of Media and American Life* (New York: W. W. Norton, 1992); and Peter d'Agostino and David Tafler, eds., *Transmission: Toward a Post-Television Culture*, second edition (Thousand Oaks, CA: Sage Publications, 1995).

53. Robert W. Mason to Carl W. Larsen, June 12, 1972, SIA (RU367), box 8, folder "Telecommunications Committee."

54. William Albig, *Public Opinion* (New York: McGraw-Hill, 1939), 334.

55. Lyman Bryson and Edward R. Murrow, "You and Television," *Hollywood Quarterly* 4 (Winter 1949): 181.

56. Aldo Leopold, "Foreword to Sand County Almanac" (1948), in *A Sand County Almanac, with Essays on Conservation from Round River* (New York: Ballantine Books, 1966), xvii.

Manuscript Sources Consulted

HAGLEY MUSEUM AND LIBRARY

Cavalcade of America Collection.

JOHNS HOPKINS UNIVERSITY ARCHIVES

Office of News and Information Services, Records, 1946–, Record Group 10.020.
Special Collections, Lynn Poole Papers, 1948–1976, MS27.

LIBRARY OF CONGRESS, MANUSCRIPT COLLECTIONS

American Psychological Association Records.

NATIONAL ACADEMIES ARCHIVES

Central Policy Files, 1950–1956.
Central Policy Files, 1957–1961.
International Geophysical Year Records.

SMITHSONIAN INSTITUTION ARCHIVES

Assistant Secretary for External Affairs and Deputy Assistant Secretary for Media, Subject Files, 1986–1990, Accession 91-177.
Assistant Secretary for History and Art, American Revolution Bicentennial Records, 1962–1982, Record Unit 337.
Assistant Secretary for Public Service, Subject Files, 1961–1974, Record Unit 145.
Assistant Secretary for Public Service, Subject Files, circa 1968–1988, Record Unit 367.

Board of Regents, Minutes, 1846–1995, Record Unit 1.

International and Environmental Programs, Records, 1962–1975, Record Unit 218.

National Museum of American History, Office of the Director, Subject Files, circa 1945–1987, and related records, 1920–1930, Record Unit 334.

National Museum of History and Technology, Office of the Director, Records, 1944–1975, Record Unit 276.

National Museum of Natural History, Department of Entomology, Records, circa 1948–1974, Record Unit 247.

National Museum of Natural History, Office of the Director, Records, 1962–1978, Record Unit 309.

National Museum of Natural History, Public Information Officer, Records, circa 1969–1991 and undated, Record Unit 416.

National Zoological Park, Office of the Director, Records, 1928–1986 and undated, Record Unit 380.

National Zoological Park Papers, 1887–1965 and undated, Record Unit 74.

Office of Academic Programs, Records, 1964–1970, Record Unit 102.

Office of Public Affairs, Clippings, 1965–1991, Record Unit 372.

Office of Public Affairs, Director's Records, circa 1965–1976, Record Unit 369.

Office of Telecommunications, Production Records, 1967–1987, 1991–1992, Accession 01-203.

Office of Telecommunications, Production Records, circa 1976–1994, Accession 95-003.

Office of Telecommunications, Production Records, 1979–1991, Accession 03-022.

Office of Telecommunications, Production Records, 1981–1989, Accession 00-114.

Office of Telecommunications, Production Records, circa 1963–1988, Record Unit 586.

Office of Telecommunications, Publication Records, circa 1985–1991, Record Unit 594.

Office of the Secretary, Administrative Records, 1953–1972, Accession T90108.

Office of the Secretary, Administrative Records, 1964–1966, 1972, Accession T90110.

Office of the Secretary, Records, 1949–1964, Record Unit 50.

Office of the Secretary, Records, 1950, 1964–1984, Accession 84-219.

Office of the Secretary, Records, 1964–1971, Record Unit 99.

Office of the Secretary, Records, 1972–1984, Record Unit 613.

Office of the Under Secretary, Records, 1958–1973, Record Unit 137.

Office of the Under Secretary, Records, 1972–1984, and related records, 1971–1978, Record Unit 392.

Oral history interview with Nazaret Cherkezian, 1986, Record Unit 9541.

Science Service Records, 1920s–1970s, Accession 90-105.

Science Service Records, Record Unit 7091.

Smithsonian Institution Press, Records, 1941–1975, Record Unit 535.

Smithsonian Press/Smithsonian Productions, Marketing Records, 1979–1997, Accession 02-096.

Smithsonian Production Records, 1971–1998, Accession 02-118.

Webster P. True Papers, 1914–1972, Accession T91027.

Selected Bibliography

BOOKS

Adler, Richard, ed. *Television as a Social Force: New Approaches to TV Criticism*. New York: Praeger, 1975.

Adler, Richard, and Douglass Cater, eds. *Television as a Cultural Force*. New York: Praeger, 1976.

d'Agostino, Peter, and David Tafler, eds. *Transmission: Toward a Post-Television Culture*. Second edition. Thousand Oaks, CA: Sage Publications, 1995.

Ang, Ien. *Desperately Seeking the Audience*. London: Routledge, 1991.

———. *Living Room Wars: Rethinking Media Audiences for a Postmodern World*. London: Routledge, 1996.

Attenborough, David. *Life on Air: Memoirs of a Broadcaster*. Princeton, NJ: Princeton University Press, 2002.

Barnouw, Erik. *Documentary: A History of the Non-fiction Film*. New York: Oxford University Press, 1974.

———. *The Golden Web: A History of Broadcasting in the United States, 1933-1953*. New York: Oxford University Press, 1968.

———. *The Image Empire: A History of Broadcasting in the United States from 1953*. New York: Oxford University Press, 1970.

———. *The Sponsor: Notes on a Modern Potentate*. New York: Oxford University Press, 1978.

———. *A Tower in Babel: A History of Broadcasting in the United States to 1933*. New York: Oxford University Press, 1966.

———. *Tube of Plenty: The Evolution of American Television*. Revised edition. New York: Oxford University Press, 1982.

Baughman, James L. *The Republic of Mass Culture: Journalism, Filmmaking, and Broadcasting in America since 1941*. Baltimore: Johns Hopkins University Press, 1992.

———. *Same Time, Same Station: Creating American Television, 1948-1961*. Baltimore: Johns Hopkins University Press, 2007.

Berger, Arthur Asa. *Essentials of Mass Communications Theory*. Thousand Oaks, CA: Sage Publications, 1995.

Bianculli, David. *Teleliteracy: Taking Television Seriously*. New York: Continuum, 1992.

Bluem, A. William. *Documentary in American Television: Form, Function, Method*. New York: Hastings House, 1965.

Blumler, Jay G., and Michael Gurevitch. *The Crisis of Public Communication*. London: Routledge, 1995.

Boorstin, Daniel J. *The Image: A Guide to Pseudo-Events in America*. New York: Harper and Row, 1961.

Bowler, Peter J. *Science for All: The Popularization of Science in Early Twentieth-Century Britain*. Chicago: University of Chicago Press, 2009.

Boyer, Paul S. *By the Bomb's Early Light: American Thought and Culture at the Dawn of the Atomic Age*. New York: Pantheon, 1985.

———. *Fallout: A Historian Reflects on America's Half-Century Encounter with Nuclear Weapons*. Columbus: Ohio State University Press, 1998.

Burnham, John C. *How Superstition Won and Science Lost: Popularizing Science and Health in the United States*. New Brunswick, NJ: Rutgers University Press, 1987.

Cassata, Mary B., and Thomas Skill, eds. *Life on Daytime Television: Tuning-in American Serial Drama*. Norwood, NJ: Ablex, 1983.

Castleman, Harry, and Walter J. Podrazik. *Watching TV: Four Decades of American Television*. New York: McGraw-Hill, 1982.

Cater, Douglass, ed. *The Future of Public Broadcasting*. New York: Praeger, 1976.

Commission on Freedom of the Press. *A Free and Responsible Press*. Chicago: University of Chicago Press, 1947.

Comstock, George, et al., eds. *Television and Social Behavior*. Washington, DC: GPO, 1972.

Conrad, Peter. *Television: The Medium and Its Manners*. London: Routledge, 1982.

Crane, Valerie, ed. *Informal Science Learning: What the Research Says about Television, Science Museums, and Community-Based Projects*. Dedham, MA: Research Communications, 1994.

Dates, Jannette L., and William Barlow, eds. *Split Image: African Americans in the Mass Media*. Washington, DC: Howard University Press, 1993.

Davidson, Keay. *Carl Sagan: A Life*. New York: John Wiley, 1999.

Davis, Jeffery. *Children's Television, 1947–1990*. Jefferson, NC: McFarland, 1995.

Day, James. *The Vanishing Vision: The Inside Story of Public Television*. Berkeley: University of California Press, 1995.

Diamond, Edwin. *The Rise and Fall of the Space Age*. Garden City, NY: Doubleday, 1964.

Doorman, S. J., ed. *Images of Science: Scientific Practice and the Public*. Aldershot, UK: Gower, 1989.

Druckrey, Timothy, ed. *Electronic Culture: Technology and Visual Representation*. New York: Aperture Books, 1996.

Dunbar, Robin. *The Trouble with Science*. Cambridge, MA: Harvard University Press, 1996.

DuPuy, Judy. *Television Show Business*. Schenectady, NY: General Electric, 1945.

Einstein, Daniel. *Special Edition: A Guide to Network Television Documentary Series and Special News Reports, 1955–1979*. Metuchen, NJ: Scarecrow Press, 1987.

Erikson, Hal. *Syndicated Television: The First Forty Years, 1947–1987*. Jefferson, NC: McFarland, 1989.

Ettema, James S., and D. Charles Whitney, eds. *Audiencemaking: How the Media Create an Audience*. Thousand Oaks, CA: Sage Publications, 1994.

Fielding, Andrew. *The Lucky Strike Papers: Journeys through My Mother's Television Past*. Duncan, OK: BearManor Media, 2007.

Friedman, Sharon M., Sharon Dunwoody, and Carol L. Rogers, eds. *Communicating Uncertainty: Media Coverage of New and Controversial Science*. Hillsdale, NJ: Lawrence Erlbaum, 1999.

———. *Scientists and Journalists: Reporting Science as News*. New York: Free Press, 1986.

Fyfe, Aileen, and Bernard Lightman, eds. *Science in the Marketplace: Nineteenth-Century Sites and Experiences*. Chicago: University of Chicago Press, 2007.

Gabler, Neal. *Life the Movie: How Entertainment Conquered Reality*. New York: Knopf, 1999.

———. *Walt Disney: The Triumph of the American Imagination*. New York: Knopf, 2006.

Gamson, Joshua. *Claims to Fame: Celebrity in Contemporary America*. Berkeley: University of California Press, 1994.

Gandy, Oscar H. *Beyond Agenda Setting: Information Subsidies and Public Policy*. Norwood, NJ: Ablex, 1982.

Gans, Herbert J. *Deciding What's News: A Study of CBS Evening News, NBC Nightly News, Newsweek, and Time*. New York: Pantheon, 1979.

Gaunt, Philip, ed. *Beyond Agendas: New Directions in Communication Research*. Westport, CT: Greenwood Press, 1994.

Geier, Leo. *Ten Years with Television at Johns Hopkins*. Baltimore: Johns Hopkins University, 1958.

Gianakos, Larry James. *Television Drama Series Programming: A Comprehensive Chronicle, 1947–59*. Metuchen, NJ: Scarecrow Press, 1980.

Gilbert, James Burkhardt. *Redeeming Culture: American Religion in an Age of Science*. Chicago: University of Chicago Press, 1997.

Gilder, George. *Life after Television: The Coming Transformation of Media and American Life*. New York: W. W. Norton, 1992.

Glut, Donald F., and Jim Harmon. *The Great Television Heroes*. New York: Doubleday, 1975.

Goodell, Rae. *The Visible Scientists*. Boston: Little, Brown, 1977.

Goodfield, June. *Reflections on Science and the Media*. Washington, DC: AAAS, 1981.

Hawes, William. *American Television Drama: The Experimental Years*. University: University of Alabama Press, 1986.

Hay, James, Lawrence Grossberg, and Ellen Wartella, eds. *The Audience and Its Landscape*. Boulder, CO: Westview, 1996.

Haynes, Rosalynn D. *From Faust to Strangelove: Representations of the Scientist in Western Literature*. Baltimore: Johns Hopkins University Press, 1994.

Head, Sydney W. *Broadcasting in America: Survey of Television and Radio*. Cambridge, MA: Riverside Press, 1956.

Hester, Harriet H., H. L. Fishel, and Martin Magner. *Television in Health Education*. Chicago: American Medical Association, 1955.

Himmelstein, Hal. *Television Myth and the American Mind*. New York: Praeger, 1984.

Iyengar, Shanto, and Donald R. Kinder. *News That Matters: Television and American Opinion*. Chicago: University of Chicago Press, 1987.

Jacobs, Norman, ed. *Culture for the Millions? Mass Media in Modern Society*. Boston: Beacon Press, 1961.

Jowett, Garth. *Film: The Democratic Art*. Boston: Focal Press, 1985.

Kirby, David A. *Lab Coats in Hollywood: Science, Scientists, and Cinema*. Cambridge, MA: MIT Press, 2011.

Klaidman, Stephen. *Health in the Headlines: The Stories behind the Stories*. New York: Oxford University Press, 1991.

Lambourne, Robert, Michael Shallis, and Michael Shortland. *Close Encounters? Science and Science Fiction*. Bristol, UK: Adam Hilger, 1990.

Lazarsfeld, Paul F., and Patricia L. Kendall. *Radio Listening in America: The People Look at Radio—Again*. New York: Prentice-Hall, 1948.

Lears, Jackson. *Fables of Abundance: A Cultural History of Advertising in America*. New York: Basic Books, 1994.

Lightman, Bernard. *Victorian Popularizers of Science: Designing Nature for New Audiences*. Chicago: University of Chicago Press, 2007.

Livingstone, Sonia, and Peter Lunt. *Talk on Television: Audience Participation and Public Debate*. London: Routledge, 1994.

Luciano, Patrick, and Gary Colville. *American Science Fiction Television Series of the 1950s: Episode Guides for Casts and Credits for Twenty Shows*. Jefferson, NC: McFarland, 1998.

Malmsheimer, Richard. *"Doctors Only": The Evolving Image of the American Physician*. New York: Greenwood Press, 1988.

Mankiewicz, Frank, and Joel Swerdlow. *Remote Control: Television and the Manipulation of American Life*. New York: Times Books, 1978.

McCurdy, Howard E. *Space and the American Imagination*. Washington, DC: Smithsonian Institution Press, 1997.

McKibben, Bill. *The Age of Missing Information*. New York: Plume Books, 1993.

McLuhan, Marshall. *Understanding Media: The Extensions of Man*. New York: McGraw-Hill, 1964.

Melody, William. *Children's Television: The Economics of Exploitation*. New Haven, CT: Yale University Press, 1973.

Mitman, Gregg. *Reel Nature: America's Romance with Wildlife on Film*. Cambridge, MA: Harvard University Press, 1999.

National Association of Science Writers. *Science, the News, and the Public: Who Gets What Science News, Where They Get It, and What They Think about It*. New York: New York University Press, 1958.

Nelkin, Dorothy. *Selling Science*. New York: W. H. Freeman, 1987.

Neuman, W. Russell. *The Future of the Mass Audience*. Cambridge: Cambridge University Press, 1991.

Nye, Russel. *The Unembarrassed Muse: The Popular Arts in America*. New York: Dial Press, 1970.

O'Connor, Alan, ed. *Raymond Williams on Television: Selected Writings*. London: Routledge, 1989.

Palmer, Chris. *Shooting in the Wild: An Insider's Account of Making Movies in the Animal Kingdom*. San Francisco: Sierra Club Books, 2010.

Penley, Constance. *NASA/TREK: Popular Science and Sex in America*. London: Verso, 1997.

Perkins, Marlin. *My Wild Kingdom*. New York: E. P. Dutton, 1982.

———. *Zooparade*. Chicago: Rand McNally, 1954.

Poole, Lynn. *Science via Television*. Baltimore: Johns Hopkins University Press, 1950.

Poole, Robert. *Earthrise: How Man First Saw the Earth*. New Haven, CT: Yale University Press, 2008.

Poundstone, William. *Carl Sagan: A Life in the Cosmos*. New York: Henry Holt, 1999.

Price, Jennifer. *Flight Maps: Adventures with Nature in Modern America*. New York: Basic Books, 1999.

Reagan, Leslie J., Nancy Tomes, and Paula A. Treichler, eds. *Medicine's Moving Pictures: Medicine, Health, and Bodies in American Film and Television*. Rochester, NY: University of Rochester Press, 2007.

Real, Michael R. *Mass-Mediated Culture*. Englewood Cliffs, NJ: Prentice Hall, 1977.

Rose, Brian G., ed. *TV Genres: A Handbook and Reference Guide*. New York: Greenwood Press, 1985.

Rosenberg, Bernard, and David Manning White, eds. *Mass Culture: The Popular Arts in America*. Glencoe, IL: Free Press, 1957.

Sanders, Marlene, and Marcia Rock. *Waiting for Prime Time: The Women of Television News*. New York: Harper and Row, 1990. First published by University of Illinois Press, 1988.

Schickel, Richard. *Intimate Strangers: The Culture of Celebrity in America*. Chicago: Ivan R. Dee, 2000.

Schiller, Herbert I. *Information Inequality: The Deepening Social Crisis in America*. New York: Routledge, 1996.

Schwartz, Tony. *The Responsive Chord*. Garden City, NY: Anchor Books, 1973.

Schwoch, James. *Global TV: New Media and the Cold War, 1946–69*. Urbana: University of Illinois Press, 2009.

Seed, David. *American Science Fiction and the Cold War: Literature and Film*. Edinburgh: Edinburgh University Press, 1999.

Shinn, Terry, and Richard Whitley, ed. *Expository Science: Forms and Functions of Popularisation*. Special issue of *Sociology of the Sciences* 9, 1985.

Smith, Anthony. *The Shadow in the Cave: The Broadcaster, His Audience, and the State*. Urbana: University of Illinois Press, 1973.

——, ed. *Television: An International History*. Oxford: Oxford University Press, 1995.

Sobchack, Vivian Carol. *Screening Space: The American Science Fiction Film*. Second enlarged edition. New Brunswick, NJ: Rutgers University Press, 1997.

Spitzer, Robert J., ed. *Media and Public Policy*. New York: Praeger, 1993.

Spragens, William C. *Electronic Magazines: Soft News Programs on Network Television*. Westport, CT: Praeger, 1995.

Stedman, Raymond. *The Serials: Suspense and Drama by Installment*. Second edition. Norman: University of Oklahoma Press, 1977.

Stettel, Irving, ed. *Top TV Shows of the Year, 1954-1955*. New York: Hastings House, 1955.

Streeter, Thomas. *Selling the Air: A Critique of the Policy of Commercial Broadcasting in the United States*. Chicago: University of Chicago Press, 1996.

Terrace, Vincent. *The Complete Encyclopedia of Television Programs, 1947-1976*. Volumes 1 and 2. South Brunswick, NJ: A. S. Barnes, 1976.

Thomas, Erwin K., and Brown H. Carpenter, eds. *Handbook on Mass Media in the United States: The Industry and Its Audiences*. Westport, CT: Greenwood Press, 1994.

Trent, Judith S., and Robert V. Friedenberg. *Political Campaign Communication: Principles and Practices*. New York: Praeger, 1983.

Triangle Publications, Radio and Television Division. *The University of the Air: Commercial Television's Pioneer Effort in Education*. Philadelphia: Triangle Publications, 1959.

Tuomey, Christopher P. *Conjuring Science: Science as Meaning in American Culture*. New Brunswick, NJ: Rutgers University Press, 1996.

Turney, Jon. *Frankenstein's Footsteps: Science, Genetics, and Popular Culture*. New Haven, CT: Yale University Press, 1998.

Turow, Joseph. *Playing Doctor*. Oxford: Oxford University Press, 1989.

Turow, Joseph, and Rachel Gans. *As Seen on TV: Health Policy Issues in TV's Medical Dramas*. Menlo Park, CA: Kaiser Family Foundation, 2002.

Udelson, Joseph H. *The Great Television Race: A History of the American Television Industry, 1925-1941*. University: University of Alabama Press, 1982.

U.S. Commission on Civil Rights. *Window Dressing on the Set: Women and Minorities in Television*. Washington, DC: GPO, 1977.

von Schilling, James. *The Magic Window: American Television, 1939-1953*. New York: Haworth Press, 2003.

Walker, J. Samuel. *Three Mile Island: A Nuclear Crisis in Historical Perspective*. Berkeley: University of California Press, 2004.

Wallis, Roger, and Stanley J. Baran. *The Known World of Broadcast News: International News and Electronic Media*. London: Routledge, 1990.

Warren, Frank. *Television in Medical Education: An Illustrated Handbook*. Chicago: American Medical Association, 1955.

Watts, Steven. *The Magic Kingdom: Walt Disney and the American Way of Life*. Boston: Houghton Mifflin, 1997.

Weart, Spencer R. *Nuclear Fear: A History of Images*. Cambridge, MA: Harvard University Press, 1988.

Weinstein, David. *The Forgotten Network: DuMont and the Birth of American Television*. Philadelphia: Temple University Press, 2004.

Williams, Raymond. *Television: Technology and Cultural Form*. New York: Schocken Books, 1975.

Wilson, Alexander. *The Culture of Nature: North American Landscape from Disney to the Exxon Valdez*. Cambridge: Blackwell Publishers, 1992.

Winkler, Allan M. *Life under a Cloud: American Anxiety about the Atom*. New York: Oxford University Press, 1993.

Wober, J. M., ed. *Television and Nuclear Power: Making the Public Mind*. Norwood, NJ: Ablex, 1992.

Zaller, John R. *The Nature and Origins of Mass Opinion*. Cambridge: Cambridge University Press, 1992.

JOURNAL ARTICLES AND REPORTS

Ackerman, William C. "The Dimensions of American Broadcasting." *Public Opinion Quarterly* 9 (Spring 1945): 1–18.

———. "U.S. Radio: Record of a Decade." *Public Opinion Quarterly* 12 (October 1948): 440–454.

Adams, Paul C. "Television as Gathering Place." *Annals of the Association of American Geographers* 82 (March 1992): 117–135.

Amrine, Michael. "Psychology in the News." *American Psychologist* 19 (1959): 74–78.

Aucourt, Joan. "Television: A Double Take." *Hollywood Quarterly* 3 (Spring 1948): 258–261.

Badash, Lawrence. "Nuclear Winter: Scientists in the Political Arena." *Physics in Perspective* 3 (2001): 76–105.

Bader, Renate G. "How Science News Sections Influence Newspaper Science Coverage: A Case Study." *Journalism Quarterly* 67 (Spring 1990): 88–96.

Bauer, Martin W., Kristina Petkova, Pepka Boyadjieva, and Galin Gornev. "Long-Term Trends in the Public Representation of Science across the 'Iron Curtain,' 1946–1995." *Social Studies of Science* 36 (February 2006): 99–131.

Becker, Ron. "'Hear-and-See Radio' in the World of Tomorrow: RCA and the Presentation of Television at the World's Fair, 1939–1940." *Historical Journal of Film, Radio, and Television* 21 (October 2001): 361–378.

Beier, Carl, Jr. "A New Way of Looking at Things." *Hollywood Quarterly* 2 (October 1946): 1–10.

Bensaude-Vincent, Bernadette. "A Historical Perspective on Science and Its 'Others.'" *Isis* 100 (2009): 359–368.

Boot, William. "NASA and the Spellbound Press." *Columbia Journalism Review* 25 (July/August 1986): 23–29.

Boyer, Paul S. "From Activism to Apathy: The American People and Nuclear Weapons, 1963–1980." *Journal of American History* 70 (March 1984): 821–844.

Brodie, Mollyann, Ursula Foehr, Vicky Rideout, Neal Baer, Carolyn Miller, Rebecca Flournoy, and Drew Altman. "Communicating Health Information through the Entertainment Media." *Health Affairs* 20 (January/February 2001): 192–199.

Bryson, Lyman, and Edward R. Murrow. "You and Television." *Hollywood Quarterly* 4 (Winter 1949): 178–181.

Bucher, François. "Television (an Address)." *Journal of Visual Culture* 4 (2005): 5–15.

Campbell, Vincent. "The Extinct Animal Show: The Paleoimagery Tradition and Computer Generated Imagery in Factual Television Programs." *Public Understanding of Science* 18 (2009): 199–213.

Cassirer, Henry R. "Educational Television: World-Wide." *Quarterly of Film, Radio, and Television* 8 (Summer 1954): 367–374.

Chambers, David Wade. "History of Science on the Silver Screen." *Isis* 57 (Winter 1966): 494–497.

Chessin, Paul L. "Spies, Electric Chairs, and Housewives." *American Mathematical Monthly* 65 (June/July 1958): 416–421.

Chimba, Mwenya, and Jenny Kitzinger. "Bimbo or Boffin? Women in Science: An Analysis of Media Representations and How Female Scientists Negotiate Cultural Contradictions." *Public Understanding of Science* 19 (2010): 609–624.

Collins, Harry. "Certainty and the Public Understanding of Science: Science on Television." *Social Studies of Science* 17 (November 1987): 689–714.

"Competition and TV Program Content." *University of Chicago Law Review* 19 (Spring 1952): 556-573.

Crichton, Michael. "Ritual Abuse, Hot Air, and Missed Opportunities." *Science* 283 (March 5, 1999): 1461-1463.

Curtin, Michael. "The Discourse of 'Scientific Anti-Communism' in the 'Golden Age' of Documentary." *Cinema Journal* 32 (Autumn 1992): 3-25.

Davies, Gail. "Science, Observation and Entertainment: Competing Visions of Post-War British Natural History Television, 1946-1967." *Ecumene* 7 (2000): 432-459.

Dawson, Max. "Home Video and the 'TV Problem.'" *Technology and Culture* 48 (July 2007): 524-549.

DeFleur, Melvin, and Lois DeFleur. "The Relative Contribution of Television as a Learning Source for Children's Occupational Knowledge." *American Sociological Review* 32 (1967): 777-789.

DelGaudio, Sybil. "If Truth Be Told, Can 'Toons Tell It? Documentary and Animation." *Film History* 9 (1997): 189-199.

Dimmick, John. "The TV Western Program Cycle: Decision Uncertainty and Audience Habituation." *Mass Comm. Review* 4 (Spring 1977): 10-15.

Disney, Walt. "Mickey as Professor." *Public Opinion Quarterly* 9 (Summer 1945): 119-125.

Dudo, Anthony, Dominique Brossard, James Shanahan, Dietram A. Scheufele, Michael Morgan, and Nancy Signorielli. "Science on Television in the 21st Century: Recent Trends in Portrayals and Their Contributions to Public Attitudes toward Science." *Communication Research* 20 (December 2010): 1-24.

Du Mont, Allen B. "Television Now and Tomorrow." *Journal of Marketing* 9 (January 1945): 278-279.

Elena, Alberto. "Skirts in the Lab: Madame Curie and the Image of the Woman Scientist in the Feature Film." *Public Understanding of Science* 6 (1997): 269-278.

Fagan, Garrett G. "Seductions of Pseudoarchaeology: Far Out Television." *Archaeology* 56 (May/June 2003): 46.

Feldman, Stanley, and Lee Sigelman. "The Political Impact of Prime-Time Television: 'The Day After.'" *Journal of Politics* 47 (June 1985): 556-578.

Flicker, Eva. "Between Brains and Breasts—Women Scientists in Fiction Film: On the Marginalization and Sexualization of Scientific Competence." *Public Understanding of Science* 12 (2003): 307-318.

Fly, James Lawrence. "Regulation of Radio Broadcasting in the Public Interest." *Annals of the American Academy of Political and Social Science* 213 (January 1941): 102-105.

Fones-Wolf, Elizabeth. "Creating a Favorable Business Climate: Corporations and Radio Broadcasting." *Business History Review* 73 (Summer 1999): 221-255.

Freeman, Ira M. "Interpreting Science." *Science* 136 (June 8, 1962): 902-903.

Funkhouser, G. Ray, and Eugene F. Shaw. "How Synthetic Experience Shapes Social Reality." *Journal of Communication* 40 (Spring 1990): 75-87.

Gamson, William A., and Andre Modigliani. "Media Discourse and Public Opinion on Nuclear Power: A Constructionist Approach." *American Journal of Sociology* 95 (July 1989): 1-37.

Gerbner, George, Larry Gross, Michael Morgan, and Nancy Signorielli. "Health and Medicine on Television." *New England Journal of Medicine* 305 (October 8, 1981): 901-904.

Gibson, Robert L. "Some Preferences of Television Audiences." *Journal of Marketing* 10 (January 1946): 289-290.

Gilbert, Douglas L. "Television as a Wildlife Education Medium." *Journal of Wildlife Management* 20 (October 1956): 456-458.

Gilbert, James B. "Popular Culture." *American Quarterly* 35 (Summer 1983): 141-154.

Gitlin, Irving J. "Radio and Atomic-Energy Education." *Journal of Educational Sociology* 22 (January 1949): 327–328.

Gitlin, Todd. "Prime Time Ideology: The Hegemonic Process in Television Entertainment." *Social Problems* 26 (February 1979): 251–266.

Gorney, Carole. "Numbers versus Pictures: Did Network Television Sensationalize Chernobyl Coverage?" *Journalism Quarterly* 69 (Summer 1992): 460–463.

Gould, Jack. "Television: Boon or Bane?" *Public Opinion Quarterly* 10 (Autumn 1946): 314–320.

Graber, Doris A. "Seeing Is Remembering: How Visuals Contribute to Learning from Television News." *Journal of Communication* 40 (Summer 1990): 134–155.

Greenberg, Michael R., David B. Sachsman, and Peter M. Sandman. "Risk, Drama, and Geography in Coverage of Environmental Risk by Network TV." *Journalism Quarterly* 66 (Summer 1989): 267–276.

Harrison, Helen A. "Stuart Davis's 'World of Tomorrow.'" *American Art* 9 (Autumn 1995): 97.

Head, Sydney W. "Content Analysis of Television Drama Programs." *Quarterly of Film, Radio, and Television* 9 (1954): 175–194.

Hilgartner, Stephen. "The Dominant View of Popularization: Conceptual Problems, Political Uses." *Social Studies of Science* 20 (1990): 519–539.

Hodgens, Richard. "A Brief, Tragical History of the Science Fiction Film." *Film Quarterly* 13 (Winter 1959): 30–39.

Holbert, R. Lance. "A Typology for the Study of Entertainment Television and Politics." *American Behavioral Scientist* 49 (November 2005): 436–453.

Horak, Jan-Christopher. "Wildlife Documentaries: From Classical Forms to Reality TV." *Film History* 18 (2006): 459–475.

Hoskins, Andrew. "Mediating Time: The Temporal Mix of Television." *Time and Society* 10 (2001): 213–233.

Jones, Ayrlene McGahey. "Television Activity, Department of Mathematics, University of Alabama." *American Mathematical Monthly* 65 (June/July 1958): 421.

Kalisch, Philip A., and Beatrice J. Kalisch. "Nurses on Prime-Time Television." *American Journal of Nursing* 82 (February 1982): 264–270.

Karp, Ivan. "High and Low Revisited." *American Art* 5 (Summer 1991): 14.

Kingson, Walter. "The Second New York Television Survey." *Quarterly of Film, Radio, and Television* 6 (Summer 1952): 317–326.

Kirby, David A. "Science Consultants, Fictional Films, and Scientific Practice." *Social Studies of Science* 33 (April 2003): 231–268.

Korsmo, Fae L. "Shaping Up Planet Earth: The International Geophysical Year (1957–1958) and Communicating Science through Print and Film Media." *Science Communication* 26 (December 2004): 162–187.

Koszarski, Richard. "Coming Next Week: Images of Television in Pre-war Motion Pictures." *Film History* 10 (1998): 128–140.

Lang, Kurt, and Gladys Engel Lang. "In the Plural." *Journal of Communication* 38 (Summer 1988): 130–131.

———. "The Unique Perspective of Television and Its Effects: A Pilot Study." *American Sociological Review* 18 (1953): 3–12.

Lasswell, Harold. "Educational Broadcasters as Social Scientists." *Quarterly of Film, Radio, and Television* 7 (Winter 1952): 150–162.

Lazier, Benjamin. "Earthrise; or, The Globalization of the World Picture." *American Historical Review* 116 (June 2011): 602–630.

Lehr, Marguerite. "An Experiment with Television." *American Mathematical Monthly* 62 (January 1955): 15–21.

Lerner, Irving. "Director's Notes." *Hollywood Quarterly* 1 (January 1946): 183.

Livingstone, Sonia. "Half a Century of Television in the Lives of Our Children." *Annals of the American Academy of Political and Social Science* 625 (September 2009): 151–163.

Long, Marilee, Greg Boiarsky, and Greg Thayer. "Gender and Racial Counter-Stereotypes in Science Education Television: A Content Analysis." *Public Understanding of Science* 10 (2001): 255–269.

Long, Marilee, Jocelyn Steinke, Brooks Applegate, Maria Knight Lapinski, Marne J. Johnson, and Sayani Ghosh. "Portrayals of Male and Female Scientists in Television Programs Popular among Middle School-Age Children." *Science Communication* 32 (2010): 356–382.

MacDonald, Scott. "Up Close and Political: Three Short Ruminations on Ideology in the Nature Film." *Film Quarterly* 59 (Spring 2006): 4–21.

Manvell, Roger. "Experiments in Broadcasting and Television." *Hollywood Quarterly* 2 (July 1947): 388–392.

Marshall, Roy K. "Televising Science." *Physics Today* 2 (January 1949): 26.

Mazur, Allan. "The Journalists and Technology: Reporting about Three Mile Island and Love Canal." *Minerva* 22 (Spring 1984): 45–66.

Mechling, Elizabeth Walker, and Jay Mechling. "The Atom According to Disney." *Quarterly Journal of Speech* 81 (November 1995): 436–457.

Merron, Jeff. "Murrow on TV: *See It Now*, *Person to Person*, and the Making of a 'Masscult Personality.'" *Journalism Monographs* 106 (July 1988).

Metz, Anneke M. "A Fantasy Made Real: The Evolution of the Subjunctive Documentary on U.S. Cable Science Channels." *Television and New Media* 9 (July 2008): 333–348.

Meyerowitz, Joshua. "We Liked to Watch: Television as Progenitor of the Surveillance Society." *Annals of the American Academy of Political and Social Science* 625 (September 2009): 32–48.

Midas, Margaret. "Without TV." *American Quarterly* 3 (Summer 1951): 152–166.

Mitman, Gregg. "Cinematic Nature: Hollywood Technology, Popular Culture, and the American Museum of Natural History." *Isis* 84 (December 1993): 636–661.

Morrisett, Lloyd N. "The Age of Television and the Television Age." *Peabody Journal of Education* 48 (January 1971): 112–121.

Mulkay, Michael. "Frankenstein and the Debate over Embryo Research." *Science, Technology, and Human Values* 21 (Spring 1996): 157–176.

Newsom, C. V. "Radio and Television." *Scientific Monthly* 79 (October 1954): 248–252.

Nisbet, Matthew C., Dietram A. Scheuffle, James Shanahan, Patricia Moy, Dominique Brossard, and Bruce V. Lewenstein. "Knowledge, Reservations, or Promise? A Media Effects Model for Public Perceptions of Science and Technology." *Communication Research* 29 (October 2002): 584–608.

Nucci, Mary L., and Robert Kubey. "'We Begin Tonight with Fruits and Vegetables': Genetically Modified Food on the Evening News, 1980–2003." *Science Communication* 29 (December 2007): 147–176.

Oakes, J. Michael. "The Effect of Media on Children: A Methodological Assessment from a Social Epidemiologist." *American Behavioral Scientist* 52 (April 2009): 1136–1151.

Pandora, Katherine, and Karen A. Rader. "Science in the Everyday World." *Isis* 99 (June 2008): 350–364.

Patterson, Jody. "Modernism and Murals at the 1939 New York World's Fair." *American Art* 24 (Summer 2010): 50–73.

Pigott, Leonard D. "Biology Reaches a New Horizon: Science on Television." *AIBS Bulletin* 1 (July 1951): 7.

Pinckney, Charles E. "'Your Lease on Life' in Denver." *Public Health Reports* 69 (June 1954): 606–608.

Pion, Georgine M., and Mark Lipsey. "Public Attitudes toward Science and Technology: What Have the Surveys Told Us?" *Public Opinion Quarterly* 145 (1981): 303–316.

Poole, Lynn. "The Challenge of Television." *College Art Journal* 8 (Summer 1949): 299–304.

Postman, Neil. "The Las Vegasizing of America." *National Forum* 62 (Summer 1982): 6–9.

Rakow, Lana F., and Kimberlie Kranich. "Woman as Sign in Television News." *Journal of Communication* 41 (Winter 1991): 8–23.

"Report of the Committee on the Role and Opportunities in Broadcasting—To President Edward H. Levi, August 15, 1972." *University of Chicago Record* 7 (April 21, 1973): 156–186.

Robertson, R. S. "Don't Waste the Wasteland." *American Biology Teacher* 27 (April 1965): 283–284.

Rubin, David M. "How the News Media Reported on Three Mile Island and Chernobyl." *Journal of Communication* 37 (Summer 1987): 42–57.

Rubin, David M., and Constance Cummings. "Nuclear War and Its Consequences on Television News." *Journal of Communication* 39 (Winter 1989): 39–58.

Russell, Christine. "The Man behind 'Ascent of Man': Jacob Bronowski." *BioScience* 25 (January 1975): 9–12.

Rydell, Robert W. "The Fan Dance of Science: American World's Fairs in the Great Depression." *Isis* 76 (December 1985): 525–542.

Saltman, Paul. "The Softest Hard Sell: Bronowski's Approach to Communicating Science." *Leonardo* 18 (1985): 243–244.

Sarnoff, David. "Probable Influence of Television on Society." *Journal of Applied Physics* 10 (July 1939): 428.

Schreiber, Flora Rheta. "Television's New Idiom in Public Affairs." *Hollywood Quarterly* 5 (Winter 1950): 144–152.

"Science on Television." *Discovery* 14 (April 1953): 103–105.

Scott, David K., and Robert H. Gobetz. "Hard News/Soft News Content of National Broadcast Networks, 1972–1987." *Journalism Quarterly* 69 (Summer 1992): 406–412.

Seldes, Gilbert. "The Nature of Television Programs." *Annals of the American Academy of Political and Social Science* 213 (January 1941): 138–144.

Shepherd, R. Gordon. "Science News of Controversy: The Case of Marijuana." *Journalism Monographs* 62 (August 1979).

Siepmann, Charles A., and Sidney Reisberg. "'To Secure These Rights': Coverage of a Radio Documentary." *Public Opinion Quarterly* 12 (Winter 1948/1949): 649–658.

Singer, Maxine. "Heroines and Role Models." *Science* 253 (July 19, 1991): 249.

Smith, Anthony. "Technology, Identity, and the Information Machine." *Daedalus* 115 (Summer 1986): 155–170.

Smythe, Dallas W. "A National Policy on Television?" *Public Opinion Quarterly* 14 (Autumn 1950): 461–474.

———. "Reality as Presented by Television." *Public Opinion Quarterly* 18 (Summer 1954): 143–156.

———. "Television in Relation to Other Media and Recreation in American Life." *Hollywood Quarterly* 4 (Spring 1950): 256–261.

Spangler, Romayne Wicks. "We're on TV Every Week." *American Journal of Nursing* 55 (May 1955): 592–593.

Steinke, Jocelyn. "Cultural Representations of Gender and Science: Portrayals of Female Scientists and Engineers in Popular Films." *Science Communication* 27 (September 2005): 27–63.

———. "A Portrait of a Woman as a Scientist: Breaking Down Barriers Created by Gender-Role Stereotypes." *Public Understanding of Science* 6 (1997): 409–428.

———. "Women Scientist Role Models on Screen: A Case Study of *Contact*." *Science Communication* 21 (December 1999): 111–136.

Steinke, Jocelyn, and Marilee Long. "A Lab of Her Own? Portrayals of Female Characters on Children's Educational Science Programs." *Science Communication* 18 (December 1996): 91–115.

Streckfuss, Richard. "Objectivity in Journalism: A Search and a Reassessment." *Journalism Quarterly* 67 (Winter 1990): 973–983.

Terzian, Sevan G., and Andrew L. Grunzke. "Scrambled Eggheads: Ambivalent Representations of Scientists in Six Hollywood Film Comedies from 1961 to 1965." *Public Understanding of Science* 16 (2007): 407–419.

Tsutsui, William M. "Looking Straight at *Them!* Understanding the Big Bug Movies of the 1950s." *Environmental History* 12 (April 2007): 237–253.

University of Chicago Committee on Educational Television. "Television and the University." *School Review* 61 (April 1953): 202–225.

Wade, Serena, and Wilbur Schramm. "The Mass Media as Sources of Public Affairs, Science, and Health Knowledge." *Public Opinion Quarterly* 33 (Summer 1969): 197–209.

Walsh, Peter L. "This Invisible Screen: Television and American Art." *American Art* 18 (Summer 2004): 2–9.

Wang, Zuoyue. "Responding to *Silent Spring*: Scientists, Popular Science Communication, and Environmental Policy in the Kennedy Years." *Science Communication* 19 (December 1997): 141–163.

Warner, Harry P. "Television and the Motion Picture Industry." *Hollywood Quarterly* 2 (October 1946): 11–18.

Wartella, Ellen, Aletha C. Huston, Victoria Rideout, and Michael Robb. "Studying Media Effects on Children: Improving Methods and Measures." *American Behavioral Scientist* 52 (April 2009): 1111–1114.

Wheatley, Parker. "Radio and Television as Instruments of Education." *Bulletin of the American Academy of Arts and Sciences* 3 (October 1949): 2–4.

Williams, Robert J. "The Politics of American Broadcasting: Public Purposes and Private Interests." *Journal of American Studies* 10 (December 1976): 331.

Wilson, Kris. "Television Weathercasters as Science Communicators." *Public Understanding of Science* 17 (2008): 73–87.

Winkler, Allan M. "The 'Atom' and American Life." *History Teacher* 26 (May 1993): 317–337.

UNPUBLISHED SOURCES

Bailey, Robert Lee. "An Examination of Prime Time Network Television Special Programs, 1948 to 1966." Ph.D. dissertation, University of Wisconsin, 1967.

Hogan, Alfred Robert. "Televising the Space Age: A Descriptive Chronology of CBS News Special Coverage of Space Exploration from 1957 to 2003." Master's thesis, University of Maryland, 2005.

Lewenstein, Bruce V. "'Public Understanding of Science' in America, 1945–1965." Ph.D. dissertation, University of Pennsylvania, 1987.

Soranno, Alexander Mark. "A Descriptive Study of Television Network Prime Time Programming, 1958–59 through 1962–63." Master's thesis, San Francisco State College, 1966.

ELECTRONIC RESOURCES

The Johns Hopkins Science Review episodes, 1950s. Internet Archive. http://www.archive.org/.

"RCA Presentation: Television." Internet Archive. http://www.archive.org/details/RCAP rese1939/.

Science in Action episodes, 1950s and 1960s. Internet Archive. http://www.archive.org/.

Science in Action episodes and script lists, 1950s and 1960s. California Academy of Sciences. http://www.research.calacademy.org/library/collections/archives/SIAtelevision/.

Science Reporter episodes, 1960s. MIT Museum. http://museum.mit.edu/150/34.

Tales of Tomorrow episodes, 1950s. Internet Archive. http://www.archive.org/.

What in the World? episodes, 1950s. Internet Archive. http://www.archive.org/.

What in the World? episodes, 1950s. University of Pennsylvania Museum. http://pennmuseu marchives.wordpress.com/2009/02/14/what-in-the-world/.

Illustration Credits

Frontispiece: Smithsonian Institution Archives, RU50; digital image SIA2011-2219.

1a   Smithsonian Institution Archives, Acc. 90-105; digital image SIA2011-2215.

1b   Smithsonian Institution Archives, Acc. 90-105; digital image SIA2011-2216.

2   Chemistry Collections, Division of Medicine and Science, National Museum of American History, Smithsonian Institution.

3   Photographic History Collection, Division of Culture and the Arts, National Museum of American History, Smithsonian Institution; image 1997.3001.111.

4   Emilio Segrè Visual Archives, American Institute of Physics.

5   Smithsonian Institution Archives, RU50; digital image SIA2011-2218.

6   Smithsonian Institution Archives, RU50; digital image SIA2011-2220.

7a   Smithsonian Institution Archives, RU50; digital image SIA2011-2221.

7b   Smithsonian Institution Archives, RU50; digital image SIA2011-2222.

8   Smithsonian Institution Archives, Acc. 90-105; digital image SIA2009-2513.

9a   Library of Congress, U.S. News & World Report Collection, Prints and Photographs Reading Room, contact sheet 7494-5, negative LC-U9-7495, frame 13.

9b   Library of Congress, U.S. News & World Report Collection, Prints and Photographs Reading Room, contact sheet 7495, negative LC-U9-7495, frame 17.

9c   Library of Congress, U.S. News & World Report Collection, Prints and Photographs Reading Room, contact sheet 7495, negative LC-U9-7494, frame 21.

10a   Smithsonian Institution Archives, digital image OPA-1548.23A.

10b   Smithsonian Institution Archives, digital image OPA-1548.25.

11   Photographic History Collection, Division of Culture and the Arts, National Museum of American History, Smithsonian Institution; image Secretary_S_Dillon_Ripley.

12   Smithsonian Institution Archives, digital image OPA-1666.18A.

13   Institutional History Division Collections, Smithsonian Institution, digital image 74-2646-20A.

14   Nuclear Regulatory Commission collections, file photo.

15   Smithsonian Institution Archives, Collection No. 000549; digital image 2010-3055.

16   Smithsonian Institution Archives, Acc. 02-118; digital image SIA2011-2217.

17   Institutional History Division Collections, Smithsonian Institution; digital image SPS1987-000056.

Index

Page numbers in italics refer to figures and captions.

Abbot, Charles Greeley, 37, *38*

Abbott, Berenice, 214

ABC (American Broadcasting Company), 35, 86, 140; children's programs, 172, 175, 178

ABC Evening News, 140–43, 257n16

Abominable Snowman, 114, 206

About Time, 51, 56–57. *See also* Bell Telephone System: specials

"Academy of Detection Arts and Sciences," 54. *See also* Bell Telephone System: specials

accuracy, 44, 62–64, 106–7, 112–18, 199–200. *See also* authenticity

A. C. Nielsen (company), 242n24; program ratings, 47, 95, 114, 116, 160, 180, 250n50, 253n57

Acrobat Ranch, 172

Action for Children's Television, 178

Actor's Studio, 20–21

actualities, 12, 16–19, 80, 142, 217, 225–26. *See also* atomic energy: weapons testing; eclipses, coverage of; expeditions, coverage of

Adams, Jacqueline, 194

Adams, Paul C., 24

Adams, Robert McCormick, 206–7, 211–14. *See also* Smithsonian Institution

Adorno, T. W., 29

Adventure, 49–50. *See also* American Museum of Natural History

"Adventureland," 49. *See also Disneyland*; *Walt Disney's Wonderful World of Color*

Advertising Council, Inc., 9, 17

Age of Uncertainty, The, 156. *See also* Malone, Adrian

Agnew, Spiro T., 264n28

Albert and Mary Lasker Foundation, journalism award, 64

Albig, William, 229

Alda, Alan, 221

Alex Stone (fictional character), 66

Alien Nation, 224

All Bird TV, 262n58

Allen, Fred, 38–39

Allen, Gracie, 76

Alley, Robert, 63

Allied Corporation, 132

Alphabet Conspiracy, The, 51, 55. *See also* Bell Telephone System: specials

Altschull, J. Herbert, 138

Ambrosino, Michael, 124–26, 133–35, 164, 201, 219

America, 110

American Association for the Advancement of Science (AAAS), 27–28, 75–76, 94, 124, 160–61, 203

American Association of Colleges for Teacher Education, 32
American Astronomical Society, 17
American Medical Association (AMA), 59–68; meeting broadcasts, 11, 59; television and radio advisory committee, 63–66, 245n50
American Museum of Natural History, 45, 49–50
American Psychiatric Association, 67–68
American Psychological Association, 67–68
"Americans on Everest," 89. *See also* National Geographic Society
American Television Dealers and Manufacturers Association, 171
American Trust Company, 16. See also *Science in Action*
America Rock. See *Schoolhouse Rock*
Ames, Joseph Sweetman, 15
Amory, Cleveland, 175
Anderson, Carl D., 243n49
Anderson, Warren, 62. See also *Doctor, The*
Andrea Thomas (fictional character), 187. See also *Isis*
Andrews, Donald H., 14
Andromeda Strain, The (novel and film), 147–48
Ang, Ien, 155–56, 225
Angier, John, 125, 160–61, 221
Animal Clinic, 45, 172–73
"Animal of the Week," 16. See also *Science in Action*
Animal Planet (channel), 222, 262n58
Animals, Animals, Animals, 180
Animal Secrets, 49
Animal World, 180
Annenberg Foundation, 163
Ansara, Michael, 63
Antarctica, television coverage of, 74, 76–77, 159, 168
Antarctica—The Third World, 74
Antoinette, Marie, 116
Apollo 7 (space mission), *103. See also* space, television and
Apollo 8 (space mission), 81. *See also* space, television and
Apollo 11 (space mission), 81, 82. *See also* space, television and
Appalachian Community Service Network, 200

Apsell, Paula, 125, 129, 135, 194. See also *NOVA*
Arendt, Hannah, 29
Ark II, 223
Arlen, Michael, 164
Armstrong, Neil, 81
Arnott, Bob, 272n28
Arrowsmith (novel and television dramas), 64
Arthur Vining Davis Foundation, 132
artificial hearts, 142
Arts & Industries Building. *See* Smithsonian Institution
Ascent of Man, The, 118, 122–23, 125, 156, 159, 198, 208, 254n8
Asimov, Isaac, 114
Asteroidal Society (fictional group), 72
Astounding Science Fiction (magazine), 71
Astro Boy, 177
AT&T (American Telephone and Telegraph Company). *See* Bell Telephone System: specials
Atari Corporation, 161
Atkinson, Holly, 272n28
Atlantic Richfield Company, 132, 158
Atom Bomb Boys (fictional characters), 57. *See also* Bell Telephone System: specials
Atomic Attack (drama), 21
atomic energy: discussion of, 12–13, 32, 77; documentaries on, 18, 21; dramas about, 21, 72, 77; weapons testing, 17–18, 217. *See also* Operation Crossroads; Operation Doorstep
Atomic Energy Commission (AEC), 12, 17, 217
Atom Squad, 72
Attenborough, David, 166–68, 226
audience: growth of, 10; measurement of, 264n38, 261n17, 272n33. *See also* A. C. Nielsen (company)
Audubon's Animal Adventures, 262n58
Audubon Society, 111, 167, 203, 262n58
authenticity, 44–45, 49, 169, 273n41; in medical programs, 62, 68; and the Smithsonian, 114–15, 117. *See also* accuracy

Baird, Bil and Cora, 54
Baltimore, David, 148
Barnouw, Erik, 19, 66, 264n29
Barrier Reef, 175
Bateson, Catherine, 98

Battlestar Galactica, 155, 223
Bauer, Martin W., 141
Baughman, James L., 19, 23
Baum, Rudy, 145
Baxter, Frank C., 5, 25, 51–57, 219, 228.
   See also Bell Television System: specials;
   Dr. Research (fictional character)
Bazell, Robert J., 140
BBD&O, 112. *See also* DuPont Company
   (E. I. duPont de Nemours & Company)
Beakman's World, 183, 194. *See also* Zaloom,
   Paul
Beebe, William, 77–78
Bellour, Raymond, 225
Bell Telephone System: specials, 51–57, *53*,
   55, 71, 87, 243n49; subsidy of television,
   51–57, 134, 159. *See also* Baxter, Frank
   C.; Capra, Frank; *About Time*; *Alphabet
   Conspiracy, The*; *Gateways to the Mind:
   The Story of the Human Senses*; *Hemo the
   Magnificent*; *Our Mr. Sun*; *Strange Case of
   the Cosmic Rays*, The; *Thread of Life, The*;
   Unchained Goddess, The
Ben Casey, 65–66
Ben Casey (fictional character), 65–66
Benchley, Peter, 188
Benny, Jack, 108
Benton, William B., 101
Bergman, Jules, 147
Berle, Milton, 13, 27
Best of Cosmos, The, 158. *See also* Sagan, Carl
BET (Black Entertainment Television), 200
Betty Crocker (fictional character), 63
Bhopal (chemical facility), 143
Bicentennial (United States): Advisory
   Committee, 109; and Smithsonian,
   108–9, 118–19
Bigfoot, 114, 205
Bill Nye the Science Guy, 182–83, 185
Bionic Woman, The, 155, 224
Birtles, Alastair, 168
*Blast Masters: The Science of Explosion,
   The*, 221
Bluem, William, 86
Boeing Company, 182–83
Bogart, John B., 138
Bold Ones, The. See *Doctors, The*
Bones, 188, 214
books, companion, to television series,
   158–60, 261n21

Boone, Richard, 62. *See also* Dr. Konrad
   Styner (fictional character); *Medic*
Boorstin, Daniel, 109, 251n30, 273n41
Borlaug, Norman, 133
Bower, Robert T., 6
Bown, Ralph, 52
Boyer, Paul S., 77
Bradley, Sandra Wentworth, 210–14,
   270n78
Bradley, Truman, 73
Brain, The, 163–64
Brave New World (novel and film), 147
Bride of Frankenstein, The (film), 186
Bridges, Lloyd, 99
Bridges, William, 45
Briggs, Asa, 98
British Broadcasting Corporation (BBC),
   121–26, 130–34, 163, 165–67; and *Cosmos*,
   158; history of, 23; and Smithsonian, 118
Brock, Stan, 47
Bronk, Detlev W., 15, 76
Bronowski, Jacob, 23, 122–23, 164, 219
Bronson, Charles, 63
Brown, Bob, 172
Buck Rogers, 72
Buck Rogers (fictional character), 54, 72
Bullard, George, 205–6
"Bulletin 120" (drama), 64. *See also*
   medicine, on television
Burke, James, 134, 155, 166. See also
   Connections[2]
Burnham, John, 145
Burns, George, 76
Burrud, Bill, 99
Burton, Richard, 161
Bush, Vannevar, 240n69
Buzz Conroy (fictional character), 180. See
   also *Frankenstein, Jr.* (television series)
Buzz Corey (fictional character), 72. See also
   Space Patrol
Byrd, Richard, 77–78, 89

Cable Health Network, 165
cable television: and children's
   programming, 182; growth of, 104–5,
   156, 200, 220–22; and science, 164–67
California Academy of Sciences, 15–16.
   See also *Science in Action*
California Institute of Technology, 34, 52
Calvin, Melvin, 16

Calypso, 89–90. *See also* Cousteau,
    Jacques-Yves
Camel News Caravan, 18
Camera Three, 31
Campaign against Nuclear War, 151
Capra, Frank, 51–52, 54, 63, 74
Captain Kangaroo, 179–80
Captain Midnight (fictional character), 72
Captain Midnight (radio program), 173
Captain Video (fictional character), 50, 72
Captain Video and His Video Rangers, 72
Captain Z-Ro, 173
Carmichael, Leonard, 36–39, *38*, 46, 90
Carnegie Corporation, 124
Carpenter, Scott, 99
Carr, Martin, 201–8, 268n6, 269n48,
    269n58
Carson, Johnny, 158, 175, 186
Carson, Rachel, 87
Case, Francis, 37
CATV. *See* cable television
Cavalcade of America, 21. *See also* DuPont
    Company (E. I. duPont de Nemours &
    Company)
CBS (Columbia Broadcasting System), 19,
    159–60; and education, 30–32, 177–78;
    and IGY, 75; news coverage, 86–88,
    140–43; and Smithsonian, 35–40, 102–6,
    242n41, 254n90; space coverage, 80–82.
    See also 60 *Minutes*
CBS Evening News, 140, 257n16
CBS News (network division), 36, 102–3, 110
CBS Reports, 87–88
celebrity, 19: and gender, 191–95; hosts
    and narrators, 51, 134, 160, 168, 181;
    physicians, 223; scientists, 11, 157,
    189–91
censorship, 63–64, 181–82; and funding,
    132
Center for Short-Lived Phenomena, 96, 108.
    See also Smithsonian Institution
Cereal Institute, 173. *See also* Mr. Wizard
    (fictional character)
Challenger (accident), 144
Chamberlain, Richard, 65
Chamberlin, Ward B., 201
Chancellor, John, 152–53
Chedd, Graham, 125, 160, 221
Chedd-Angier Productions, 125, 160, 221
Chen, Milton, 222

Cherkezian, Nazaret, 117–18, 198–201, 204,
    208, 268n6
Chernobyl (nuclear accident), 128, 130,
    142–43
children's programs, 171–83, 230, 264n29,
    264n38; evaluation of, 67; and gender,
    185, 189, 194–95; and Smithsonian,
    95, 202
Children's Television Act of 1990, 183
Children's Television Workshop, 183
Chimba, Mwenya, 193
China Syndrome, The (film), 148
Christian Broadcasting Network, 200
Christian Science Monitor, 219
CIBA-Geigy Ltd., 163
CIBA Pharmaceuticals, 60
Cinemax, 200
City Hospital, 62
Civilization, 110, 198
Clark, Kenneth, 98, 110, 122–23
closed-circuit broadcasting, 25–26, *26*, 59–60
CNN (Cable News Network), 140, 167, 200,
    221
Coalition for a Nuclear Freeze, 151
Collins, Harry, 227
Colville, Gary, 73
*Commander Cody, Sky Marshall of the
    Universe*, 72
Committee for Environmental Information,
    259n67. *See also* Commoner, Barry;
    Scientists' Institute for Public
    Information (SIPI)
Committee on Public Understanding of
    Science. *See* American Association for the
    Advancement of Science (AAAS)
Commoner, Barry, 133, 259n67
Common Sense of Science, The, 122. *See also*
    Bronowski, Jacob
Communications Act of 1934, 23
Connections: An Alternative View of Change,
    134, 158–59. *See also* Burke, James
Connections², 134. *See also* Burke, James
Conquest, 74–76. *See also* National Academy
    of Sciences (NAS)
Conrad, Peter, 189
Conrad, William, 99
Contact (novel and film), 188, 260n20.
    See also Sagan, Carl
Continental Classroom, 31–34, *33*. *See also*
    White, Harvey Elliott

"Conversation with Dr. J. Robert Oppenheimer, A," 19-20. *See also* Murrow, Edward R.

Cooke, Alistair, 110

Cooley, Denton, 129

Coon, Carlton, 21

Cooper, Jackie, 65

Corey, Wendell, 67-68

Corporation for Public Broadcasting (CPB), 123-24, 158, 168, 198-99, 210, 214. *See also* Public Broadcasting System (PBS)

Cosmic Connection, The, 157-58. *See also* Sagan, Carl

Cosmos, 156-59, 198, 261n17. *See also* Sagan, Carl

Cousteau, Jacques-Yves, 89-90, 102, 106, 133, 181, 219

Cousteau/Oasis in Space, 133

Cowen, Robert C., 127

Cox, Wally, 40, 65

Coyote, Peter, 167

Crawford, Broderick, 65

Creature, The (film), 188

Crenna, Richard, 63

Crewdson, Richard, 98-99

Crichton, Michael, 147

Cro, 183

Cronkite, Walter, 17, 144, 149, 159, 161

Crosby, John, 11, 35, 54, 57, 71

Cross, Milton, 22

CSI: Crime Scene Investigation, 5, 188, 224, 272n33

C-SPAN, 200

Culliton, Barbara, 160

Curie, Marie, 44, 188

Dave Garroway Show, The, 36. *See also* *Wide Wide World*

Dave Kelsey (fictional character), 66

David Attenborough's Natural World, 166. *See also* Attenborough, David

Davidson, Keay, 261n17

Davies, Paul L., 251n30

Davis, Helen Miles, 217

Davis, Watson, v, 34, 223

Day After, The (film), 151-52

Day After Tomorrow, The (documentary), 181

Day the Earth Stood Still, The (film), 71

Death Trap, 180-81

de Guzman, Michael, 268n6

de Kruif, Paul, 64-65

DePrato, Mario, 46

Destination Moon, 71

Dickens, Charles, 54

"Did Darwin Get It Wrong?," 130. *See also* NOVA

Digital Equipment Corporation, 163

Dimmick, John, 245n47

Discover (magazine), 160

Discover: The World of Science (television series), 160-61, 203

Discovering Women, 195

Discovery (ABC series), 177

Discovery Channel, 158, 165, 212, 221-22

Disney, Walt, 38-39, 41, 76

Disneyland, 48-49, 71, 125, 206, 242n36; and IGY, 74; and space, 76-77. *See also* *Wonderful World of Disney, The*

Disney Studios, 48, 51, 242n36, 262n58. *See also* *Disneyland*; *Wonderful World of Disney, The*

Doctor, The, 62

Doctor, The (fictional character), 62

Doctors, The, 67

documentaries, 74-75, 85-88, 109-11, 168-69, 213; on environment and conservation, 49, 87, 95, 97-99; funding, 97-99, 102-3; on medicine, 64

Donald Duck (fictional character), 19

Donna Reed Show, The, 66

Dostoyevsky, Fyodor, 54

Dougherty, Jill, 194

Douglas, Kirk, 187

Douglas Edwards with the News, 18

Dow Chemical Company, 63

Dr. Amanda Mayson (fictional character), 188

Dragnet, 62

Dragonfly TV, 265n50

dramatization, 5-6, 20-21, 37, 43-57, 223-28; of medicine, 59-69, 147; of nature, 88-90; of space, 77-83

Draper, Benjamin C., 15

Dr. Barton Crane (fictional character), 62

Dr. Beakman (fictional character), 183

Dr. Helena Russell (fictional character), 187. *See also* *Space: 1999*

Dr. Jekyll and Mr. Hyde (novel and film), 44, 186-87

Dr. Kate Morrow (fictional character), 62
Dr. Kildare (fictional character), 62, 65–67, 147
Dr. Kildare (television series), 65–67, 147
Dr. Konrad Styner (fictional character), 62
"Dr. Leakey and the Dawn of Man" (National Geographic special), 89
Dr. Matt Hooper (fictional character), 188
Dr. Pauli (fictional character), 72
Dr. Research (fictional character), 51, 56
Dr. Simon Chase (fictional character), 188
Dr. Von Meter (fictional character), 72
Dubos, René, 97–98
Dudo, Anthony, 188
Du Mont, Allan B., 1
DuMont Television Network, 3, 9, 34–35, 72
DuPont Company (E. I. duPont de Nemours & Company), 21, 96, 110, 112, 116–19

Eames, Charles and Ray, 160
eclipses, coverage of: lunar, 10–11; solar, 16–17. *See also* actualities
Edison, Thomas, 192
educational television, 25–41, *176*; attitudes toward, *ii*, *26*, *41*
Educational Television Facilities Act of 1962, 123
Edwards, Vince, 65
"Effects of an Atomic Explosion, The" (documentary), 17
Ehrlich, Paul R., 133, 151
Einstein, Albert, 19, 32, 56, 181, 192, 214
Eiseley, Loren C., 49
Eisner, Michael, 178
Eleanor Arroway (fictional character), 188
Electric Company, The, 180
Elena, Alberto, 191
Eleventh Hour, The, 67–68
Encyclopaedia Britannica Films, 9
Enola Gay, 115–16
Enormous Egg, The, 95
ER, 223
ESPN, 200
Euell, Julian T., 105–6, 113, 120
Evans, Bergen, 31
Evans, Clifford, 102
evolution: controversy, 145, 181; discussion of, 31, 54, 102, 122, 130, 138, 210–11
exclusivity, 106–19, 201–3

expeditions, coverage of, 17, 77, 89, 95–96, 103, 144. *See also* actualities
Experiment, 174. *See also* Herbert, Don
Exxon Valdez, 142

Face of Violence, The, 122. *See also* Bronowski, Jacob
Face the Nation, 19
Fairness Doctrine, 144
Farnsworth, Philo T., 1
Fauci, Anthony, 17
Federal Communications Commission (FCC), 3, 18, 23, 26, 35, 104, 111, 144, 178, 183
Federal Radio Commission, 1
Federoff, Nina, 211
Fels Planetarium, 11
Ferguson, Eugene, 109, 119
Feynman, Richard, 56, 192
"Filming Nature's Mysteries" (documentary), 124–25. See also *Disneyland*
First Eden, The, 166. *See also* Attenborough, Richard
Fisher, Craig B., 94–95, 97
Flash Gordon (movie serials), 71
"Flash of Darkness," 63. See also *Medic*
Flatow, Ira, 182
Flicker, Eva, 188
Flight: The Sky's the Limit. See *Smithsonian World*
Ford, Henry, 192
Ford Foundation, 32, 37, 99
Fowler, Jim, 47
Franciscus, James, 99
Frank, Reuven, 137, 140
Frankenstein, Jr. (fictional character), 180, 264n43
Frankenstein, Jr. (television series), 171, 180
Frankenstein (fictional character), 6, 151
Frankenstein (novel), 78, 186–87
Frankenstein Meets the Space Monster (film), 186
Franklin Institute, 11, 36
Fred Waring's orchestra, 3
Freud, Sigmund, 192
Friendly, Fred W., 20, 87
Friendship 7, 79–80, 79–80
From the Face of the Earth, 261n33. *See also* Goodfield, June

Fuller, Buckminster, 133
Funkhouser, Ray, 226
Furness, Betty, 192

Gabler, Neal, 30
Gajdusek, Carleton, 163
Galapagos (novel), 211
Galbraith, John Kenneth, 156
Galison, Peter, 214
Gallo, Robert, 146
game shows, 20–21, 22, 31
Gamson, Joshua, 189
Gandy, Oscar, 132
Garber, Paul, 37, *38*
Garroway, Dave, 36, 75, 81
*Gateways to the Mind: The Story of the Human
   Senses*, 51, 55. *See also* Bell Telephone
   System: specials
Gaylin, Willard, 162
Geier, Leo, 14
Gelb, Leslie, 59
Geller, Margaret, 214
General Electric Company, 9, 34, 167, 175;
   advertisements, 246n4
General Electric Theater, 175. *See also*
   Herbert, Don; Mr. Wizard (fictional
   character)
General Hospital, 67
General Mills Company, 63
General Motors Research Laboratories,
   175. *See also* Herbert, Don; Mr. Wizard
   (fictional character)
General Pershing's horse, 38
genetics, discussion of, 56, 129, 147, 160,
   162, 210–11
Gerbner, George, 152, 187–88
Gesell, Arnold, 27
Geyer, John Charles, 14
"G for Goldberger," 21. *See also Cavalcade
   of America*
Gibson, Peggy, 182
"Gift of Dr. Minot, The," 21. *See also
   Cavalcade of America*
Gilligan's Island, 187
Gilling, Dick, 122
Gitlin, Todd, 134–35
Gladstone, Josephine, 125, 194
Glashow, Sheldon, 214
Gleason, Jackie, 39
Glenn, John, 79–80; voice of, 80

Godfrey, Arthur, 39
Goldberger, Joseph, 21
Goodall, Jane, 5, 89, 186, 190
Goodell, Rae, 190
Goodfield, June, 97–99, 162, 194, 219, 227,
   250n57
Gould, Jack, 4, 17, 20, 67; on educational
   television, 26, 40, 56, 172; on televised
   medicine, 61, 64
Gould, Stephen Jay, 130, 192
Graber, Doris, 143
Grammar Rock. See *Schoolhouse Rock*
Graves, Peter, 161
Grayson, William C., 249n33
Greene, Lorne, 65, 99, 180
Gregg, Alan, 217
Groody, Tom, 15
Gross, Mason, 21
Grosvenor, Gilbert M., 111, 165
Ground Zero (organization), 151
GTE Corporation, 221. *See also* Bell
   Telephone System
Gulf Oil Corporation, 119, 165

Hapgood (play), 212
Haraway, Donna, 166, 190
Hard Choices, 162
Harvard Program on Public Conceptions of
   Science, 127
Harvest of Shame (documentary), 87
Haskins, Caryl, 252n41
Hayden Planetarium, 50, 72–73. *See also*
   American Museum of Natural History
Hazard, Patrick, 30
Healey, Bernadine, 194
Hector Heathcote Show, The, 177
Heil, David, 182
Heldenfels, R. D., 27
Hemo the Magnificent, 51–52, 54. *See also* Bell
   Telephone System: specials
Henderson Monster, The, 150–51
Henry, William, 180, 264n38
Herald, Earl S., 11, 15–16
Herbert, Don, 173–75, *176*, 194, 219, 263n11
   246n4. *See also* Mr. Wizard (fictional
   character); *Mr. Wizard's World*; *Watch
   Mr. Wizard*
Hodge Podge Lodge, 180
Hoffman, Robert, 213
Home, 31, 74

Home Box Office, 200
Hope Diamond, 112–13, *113*, 117, 206.
 See also Smithsonian Institution
Hopkins 24/7, 223
Hopper, Dennis, 63
Horizon (BBC), 121, 124, 131, 156, 255n24
Horizons (ABC) (later *Medical Horizons*), 19,
 60, 71, 237n59
Horkheimer, Jack, 261n32
Howdy Doody, 11
Hubbard, Ruth, 211
Huntley, Chet, 17
Hutchinson, Evelyn, 97
Huxley, Aldous, 147
Huxley, Julian, 23

I, Leonardo, 161
IBM, 161, 163, 199
IGY. *See* International Geophysical Year
 (IGY)
*IGY: A Small Planet Takes a Look at
 Herself*, 75
IMF (Impossible Missions Force) (fictional
 organization), 80
Incredible Hulk, The, 155, 187
Indiana Jones, 131
Infinite Voyage, The, 163
Infinity Factory, 179
Information Please (radio), 172
In Search for the Lost World (documentary),
 107
INTELSAT (satellite network), 81
International Geophysical Year (IGY),
 174–76, 261n35
International Polar Year, 74
Internes Can't Take Money (film), 62. See also
 Dr. Kildare (television series)
In the News, 178
Invaders, The, 78
Invisible Man, The (novel and television
 adaptation), 187
Irons, Jeremy, 168
Is Atomic Testing Endangering Your Life?
 (documentary), 18
Isis, 187
It's Fun to Know, 172

James Kildare (fictional character), 65–66
James S. McDonnell Foundation, 201–10,
 268n24

Jaws (novel and movie), 188
Jaws and Claws, 221
Jenkins, Bill, 38
Jenkins, C. Francis, 1
Jetsons, The, 78, 177
Jiminy Cricket (fictional character),
 18–19
"Joe McSween's Atomic Machine"
 (television drama), 20–21
Johanson, Donald, 133, 160
Jo Harding (fictional character), 188
Johnny Jupiter (fictional character), 72
Johns Hopkins Science Review, 12–15,
 24, 34, 68, 74; women on, 192,
 266n35
Johns Hopkins University, televised medical
 procedures, 59, 223
Johnson, Martin and Osa, 45
Johnson, Timothy, 272n28
Johnson & Johnson Company, 132
Joint Committee on Educational
 Television, 26
*Journal of the American Medical Association
 (JAMA)*, 61
Jungle Jim, 46
Jurassic Park, 131

Kaiser Foundation, 140
Kaleidoscope, 172
Kaplan, Joseph, 74–75
Karp, Ivan, 30
Kemske, Donald Jeffrey Herbert.
 See Herbert, Don
Key to the Universe, The, 181
Kidder, Alfred, 21
Kidsworld, 179
Kier, Porter M., 112–13, 116–17
Kieran, John, 172
King Louis XIV of France, 116
Kirby, Durward, 45
Kistiakowsky, George, 192
Kitzinger, Jenny, 193
Klausner, Richard D., 227
Knapp, Peggy, 182
Korsmo, Fae L., 261n35
Kraft Television Theatre, 64
Krofft Supershow, The, 180
Kruse, Cornelius, 14
Krutch, Joseph Wood, 88
Kübler-Ross, Elisabeth, 192

L. A. Law, 224
Langella, Frank, 161
L. A. Science, 224
Lasker Foundation. *See* Albert and Mary
    Lasker Foundation, journalism award
Last of the Wild. See *Lorne Greene's Last of
    the Wild*
Laurence, William L., 217, 271n3
Law and Order, 224
Lazarsfeld, Paul F., 29
Leacock, Richard, 185
Leakey, Richard, 89, 160
Learning Channel (TLC), 134, 165, 183, 221
Lederman, Leon, 163, 224
Lee, Gentry, 156
Legendary Curse of the Hope Diamond, The,
    112, *113*, 116–17, 206. *See also* Wolper,
    David L.
Lehr, Marguerite, 31
Leonardo da Vinci, 161
Leopold, Aldo, 229
Lepkowski, Wil, 197
Lerner, Irving, 43
Lester Ratman (fictional character), 183
Levine, Joe, 221–22
Lewenstein, Bruce V., 28
Lewis, Sinclair, 64
Lewontin, Richard, 192
Ley, Willy, 72–73, 76
Life (magazine), 38–39, 102
Life Game, The, 156
Lifeline, 147
Life of Birds, The, 166, 226
Life on Earth, 166
Lindbergh, Charles, 37–38, *38*, 81, 82
Lipsey, Mark, 126
Living Desert, The, 48
Living Planet, The, 166
Livingstone, Sonia, 179
Living Wild, 168, 194
Loch Ness monster, 114, 205
Lockheed Corporation, 132
Lohman, Sidney, 18
Lone Ranger (fictional character), 79
Long, Marilee, 194
Looking at Animals (BBC), 166
Loomis, Henry, 198
Lorne Greene's Last of the Wild, 99, 180
Los Angeles County Medical Association
    (LACMA), 62–64, 66

Lost in Space, 180
Love, Iris, 102
Lovejoy, Thomas E., 167
Lowell Institute Cooperative Broadcasting
    Council, 254n14
Luciano, Patrick, 73
Lytle, Mark, 87

Mackenzie, John K., 85
Magic School Bus, The, 183
Magnolia Floating Theater, 44
"Making of a Natural History Film, The,"
    124. See also *NOVA*
Making of a President, The (documentary),
    109
Malone, Adrian, 122, 156, 208–14, *209*, 219,
    221, 224
Man, Beast and the Land, 95, 107
"Man and His Environment," 97–99. *See
    also* Goodfield, June; Toulmin, Stephen
Man and the Challenge, The, 74
Man and the Moon, 76. *See also* Disney
    Studios
Man in Space, 76. *See also* Disney Studios
Mann, William M., 45–46
Mansfield, John, 128, 130–31, 168
Many Loves of Dobie Gillis, The, 187
March of Medicine, The (specials), 64,
    68, 109
Marcus Welby, M.D., 66, 147
Marshall, E. G., 65
Marshall, Roy K., 11–12, 15, 27, 31, 60,
    219, 235n13
Massey, Raymond, 65
Matinee Theater, 186
Matters of Medicine (BBC), 23
Mayr, Ernst, 97
McCann Erickson, 16
McClellan, Shawn, 206
McClintock, Miller, 9–10
McCullough, David, 205, 208, 269n58
McDonnell Aircraft. *See* James S. McDonnell
    Foundation
McGraw-Hill Book Publishing Company,
    249n36
McKibben, Bill, 90, 226–27
McLaughlin, Marya, 194
McLean, Evalyn Walsh, 116
McLuhan, Marshall, 95
Mead, Margaret, 19, 50, 98, 133, 186, 190

Media Resource Service. *See* Scientists' Institute for Public Information (SIPI)

Medic, 62–63, 65–66, 71, 223, 244n41

Medical Center, 66

Medical Horizons (initially *Horizons*), 19, 60, 71, 237n59

Medicare, 65

medicine, on television, 13, 19, 59–69, 223; documentaries, 87–88, 128–29, 162–63; news, 140–41, 143–47. *See also* American Medical Association (AMA); *Ben Casey*; *Dr. Kildare* (television series); *Marcus Welby, M.D.*; *Medic*

Medicine Man (film), 188

Medix, 66

Meet Me at the Zoo, 46

Meet the Press, 19

Men into Space, 74

Menzel, Donald, 243n49

Merck & Company, 132, 163

Mesthene, Emmanuel, 127

Meteora (fictional character), 54. *See also* Bell Telephone System: specials

Mickelson, Sig, 50

Microbes and Men, 181

Midas, Margaret, 171

Miles, Vera, 63

Miller, Robert C., 15

Mindel, Joseph, 25

Minneapolis Aluminum Club. *See* Mr. Wizard (fictional character)

Minot, George, 21

Mission: Impossible, 80–81, 161

Mister Rogers' Neighborhood, 179

Mitman, Gregg, 48

Mobil Oil Corporation, 166, 199

Monsanto Company, 75–76

Monsters! Mysteries or Myths?, 112, 114–15, 120, 205

Monster Squad, 264n43

Montagu, M.F. Ashley, 31

Moon landings, coverage of, 77, 80–83. *See also* Apollo 11 (space mission)

Morgan, Henry, 50

Morrison, Philip, 144, 160, 164, 219

Morse, Samuel, 21

Moser, James E., 62–65, 223

Mr. I. Magination, 172

"Mr. Peale's Dinosaur," 21. See also *Cavalcade of America*

Mr. Peepers (fictional character), 40

Mr. Sun (fictional character), 5, 51–52, 53

Mr. Wizard (fictional character), 5, 68, 150, 173–75, 180, 182–83, 226, 246n4. *See also* Herbert, Don

Mr. Wizard Science Clubs, 174

Mr. Wizard's World, 175, 176, 194

Ms. Frizzle (fictional character), 183

Multiplication Rock. See *Schoolhouse Rock*

Muppets, The (fictional characters), 181

Murrow, Edward R., 19–20, 88. See also *Person to Person*; *See It Now*

Museum of History and Technology. *See* Smithsonian Institution

Museum of Science and Industry, 45

museums. *See* American Museum of Natural History; California Academy of Sciences; Franklin Institute; Smithsonian Institution; University of Pennsylvania Museum

Mutual Broadcasting Company, 9

Mutual of Omaha Insurance Company, 47. See also *Wild Kingdom*

My Favorite Martian, 78

Nader, Laura, 133

Nathe, Louis, 257n21

National Academy of Sciences (NAS), 74–76, 106, 163–64, 254n8

National Aeronautics and Space Administration (NASA), 74, 78–80, 144, 165, 226

National Association of Broadcasters (NAB), 20, 178–79

National Association of Radio and Television Broadcasters (NARTB), 61

National Audubon Society, 111, 167, 203, 262n58

National Educational Television (NET), 174

National Endowment for the Humanities (NEH), 133, 159, 162

National Film Board of Canada, 133

National Geographic Society: and cable, 222; and the Smithsonian, 103–4, 111; specials, 88–90, 106, 108–9, 165, 220

National Hurricane Research Project, 54

National Institutes of Health (NIH), 194

National Museum of Natural History. *See* Smithsonian Institution

National Organization for Women (NOW), 193

National Public Radio (NPR), 123

National Rifle Association, 111

National Science Foundation (NSF), 108, 118, 124, 163–64, 181, 183, 199, 224; and Mr. Wizard, 174–75; and *NOVA*, 132, 221

National Wildlife Federation, 111

National Zoological Park. *See* Smithsonian Institution

Nature (series), 129, 167, 262n54

Nature of Things, The (Canadian series), 165. *See also* Suzuki, David

Nature of Things, The (U.S. series), 11–12, 36, 173. *See also* Marshall, Roy K.

Nautilus (fictional submarine), 77

NBC (National Broadcasting Company), 33; children's programming, 172–75; educational programming, 25, 31–35; and IGY, 74; lunar landing broadcast, 82; and *Medic*, 63–64; News Division, 74, 93, 96, 140; and Smithsonian, 37, 93–94, 96, 249n36; World's Fair broadcast, 1, 2, 3

NBC Nightly News, 140, 146, 152, 257n16

NCIS, 188

networks, broadcasting, development of, 3–4, 16–18, 68–69, 85–86. *See also* ABC (American Broadcasting Company); CBS (Columbia Broadcasting System); DuMont Television Network; NBC (National Broadcasting Company); PBS (Public Broadcasting Service)

Newman, Marshall T., 14

"New Range Wars, The" (documentary), 167. *See also* *World of Audubon, The*

New York Academy of Medicine, 59

New York City Board of Education, 25

New Yorker, 21

news, television, 18, 71, 138–53, 267n44; defined, 138; experts on, 267n44; and gender, 191–94; norms and standards, 138–39, 258n41

News Nob, 217. *See also* atomic energy: weapons testing; Operation Tumbler-Snapper

newspapers, science in, 138, 161–62, 257n17, 257n21, 258n41

newsreels, 18, 74, 76

Newton's Apple, 182, 194, 265n50

New York Film Council, 9–10

New York State Board of Education, 31

New York State Medical Society, 61

New York Times, 15–17, 115, 130, 155, 205, 214

New York Zoological Park, 45–46

Nickelodeon, 165, 175, 182, 200, 203

Nielsen ratings. *See* A. C. Nielsen (company)

Nilsson, Lennart, 129

1960? Jiminy Cricket! (documentary), 18–19

"99 Days to Survival," 103. *See also* *Smithsonian Adventure with Walter Schirra*

Nixon, Richard S., 123

Nobel Legacy, The, 221

North American Rockwell, 110

Northwestern Mutual Life Insurance Company, 132

NOVA, 124–26, 128–35, 221–23, 255n20, 255n21, 267n37; and gender, 192; marketing, 159

nuclear energy, television coverage of, 14, 20, 77, 142–43, 148–52, 149; weapons, 57, 116, 122. *See also* *Day After, The* (film); Three Mile Island (nuclear plant)

Nuclear Threat to You, The (documentary), 116

Nye, Bill, 182–83, 185

O'Brien, Miles, 221

Odishaw, Hugh, 74, 118

Odyssey, 133

Ogburn, W. F., 28

"Olympics of the Mind," 161. *See also* *I, Leonardo*

Omni (magazine), 155, 160

Omnibus, 37, 49–50, 68, 87, 90, 201

Omni: The New Frontier, 160

"Operation Big Shot," 17

Operation Crossroads, 17

Operation Deepfreeze, 74

Operation Doorstep, 17

Operation Neptune, 173

Operation Tumbler-Snapper, 17

Oppenheimer, J. Robert, 19–20, 160

Orphans of the Wild, 222

Osgood, Charles, 159

Our Friend the Atom, 77. *See also* Disney Studios

Our Mr. Sun, 51–52, 53, 109. *See also* Bell Telephone System: specials

Outer Limits, The, 77–78

Page, Clarence, 150
Page, George, 262n54
Palance, Jack, 187
Palfreman, Jon, 135
Pauling, Linus, 20, 189
Peale, Charles Willson, 21
Penley, Constance, 226
Perkins, Marlin, 5, 46–48
Perl, Martin, 214
Person to Person, 19
Pfizer Inc., 163
Philco Television Playhouse, The, 64
Physicians Advisory Committee on Radio, Television, and Motion Pictures. *See* American Medical Association (AMA)
"Physics for the Atomic Age," 32. See also *Continental Classroom*
Pion, Georgine M., 126
Plague on Your Children, A (BBC), 156
Planet Earth, 163, 261n35
Planet for the Taking, A, 165
Playtime, 46
"Plutonium Connection, The," 130. See also *NOVA*
Poe, Edgar Allen, 54
Polaroid Corporation, 124, 133, 164
Polytel International, 158
Poole, Lynn, 9, 12–16, 27, 34–35, 40, 131, 165, 198, 219
Population Explosion, The (documentary), 87
Postman, Neil, 164, 169, 273n41
Powers of Ten, 160
President's Commission on the Accident at Three Mile Island, 149
Price, Vincent, 180–81
Prime Computer, Inc., 132
Procter & Gamble Company, 199
Producer's Showcase, 64
Professor Anatomy (fictional character), 52
Professor Ludwig von Drake (fictional character), 49
Program on Technology & Society, 127
Project for Excellence in Journalism, 140, 257n17
Public Broadcasting Act of 1967, 123
Public Broadcasting Service (PBS), 123, 134, 158, 163, 168, 177; and science, 182, 221
Public Understanding of Science Program. *See* National Science Foundation (NSF)
Pyle, Denver, 63

"Quantum Universe, The," 212–14. See also *Smithsonian World*
Quest for the Killers, 162–63, 261n33. *See also* Goodfield, June
Quincy, M.E., 155

radio: and gender, 191; and medicine, 59; as model for television, 3–4, 18, 27, 35, 40, 225
Rae Crane (fictional character), 188
"Rage over Trees," 167
Raging Planet, 221
Rainey, Froelich, 21
Randolph, Jennings, 38
Ranger Hal, 46
Rathbone, Basil, 187
Rathbone, Perry, 21
ratings. *See* A. C. Nielsen (company); audience
Raytheon Corporation, 132
RCA (Radio Corporation of America), 1, 2, 3, 14, 34
Reed, Theodore, 46
Reisberg, Sidney, 18
Restless Earth, The, 181
Restless Sea, The, 51, 242n45. *See also* Bell Telephone System: specials
Rhodes, Mick, 125
Richter, Burton, 214
Ring of Truth, The, 164
Ripley, S. Dillon, 90–117, 91, 197–206, 250n49
Rise and Fall of the Third Reich, The, 109
Ritterbush, Philip, 94
Robert Montgomery Presents Your Lucky Strike Theatre, 64
Roberts, Walter Orr, 94, 243n49
Robinson J. Peepers (fictional character), 40
Rocket Ship X-M, 71
Rocky Jones (fictional character), 72
Rod Brown (fictional character), 72
Rod Brown of the Rocket Rangers, 72
Romine, Charles, 39
Roosevelt, Eleanor, 19
Roosevelt, Franklin D., 3
Roots, 119, 253n57
Rossetti, Franco, 14
Rossi, Bruno, 55, 243n49

Safari to Adventure, 99

Sagan, Carl, 5, 94, 133, 151, *157*, 219, 261n20; attitudes to popularization, 144, 152-53, 155; celebrity, 189-91; and *Cosmos*, 156-60

Saint, Eva Marie, 65

Salk, Jonas, 19

"Salute to Mickey Mouse," 48. See also *Disneyland*

San Diego Zoo, 49

Sanger, Margaret, 192

Sarnoff, David, 1, 2, 43

Saudek, Robert, 37-39, 201

Saving Wild Animals—What's It All About?, 181

Sawyer, Diane, 159

Say Goodbye, 111

Schickel, Richard, 189

Schirra, Walter, 102, *103*

Schoolhouse Rock, 178

Schramm, Wilbur, 225

Schreiber, Flora Rheta, 191

Science All-Stars, 177

Science and Human Values (book), 122

science and technology studies (STS), 127

Science and Technology Week, 140-41

Science Calling, 34

Science Circus, 172

science fiction, on television, 71-74, 76-77, 81, 172, 180, 189, 245n47

Science Fiction Theater, 73

Science '58, 33-34

Science in Action, 15-16, 71

Science Review (BBC), 23

Science Rock. See *Schoolhouse Rock*

Sciences, The (magazine), 155

Science Service, 34

Science, Technology & Human Values (journal), 127

"Science Times" (newspaper section), 155

Scientific American (magazine) 155

Scientific American Frontiers, 221, 227

Scientists and Engineers for Social and Political Action, 127

Scientists' Institute for Public Information (SIPI), 150, 259n67

Scooby-Doo, 180

Sea around Us, The, 87

Seal Island, 48

Search, The, 39

Search for the Goddess of Love, 102

secrecy. *See* censorship

Secrets of the Deep, 99

See It Now, 19

Seldes, Gilbert, 16, 189

Serling, Rod, 73, 114, 117

Serving through Science, 9-10, 234n3

Sesame Street, 181

Sevareid, Eric, 39, 87

Seven Dwarfs (fictional characters), 19

Shales, Tom, 206, 208, 217

Shamos, Morris H., 174

Shapiro, Irwin, 213-14

Shapley, Harlow, 27

Shaw, Eugene, 226

Shayon, Robert Lewis, 24, 78, 171

Shelley, Mary, 186, 264n43

Shelly, Freeman M., 46

Shepard, Alan B., 77

Shepherd, R. Gordon, 258n41

Sherr, Lynn, 194

Shimony, Abner, 214

Shirer, William, 109

Siepmann, Charles, 18

Silent Spring (book), 87

"*Silent Spring* of Rachel Carson, The" (documentary), 87

Silent World, The (Le Monde du Silence), 90

Silverstone, Roger, 44, 225

Simmons, Robert Hilton, 206

Sinclair Oil Company, 95

Singer, Maxine, 190-91, 210-11

Six Million Dollar Man, The, 155

60 Minutes, 159

"Skeletons in the Closet," 14. See also *Johns Hopkins Science Review*

Skinner, B. F., 192

Sloan-Kettering Institute, 64

Smith, Anthony, 30, 137-38

Smith, Ian K., 272n28

Smith Kline and French Laboratories, 60, 87; Medicine Television Unit, 64

Smithsonian, The (television program), 95-96, 249n36

Smithsonian Adventure with Walter Schirra, *103*

Smithsonian Discovery Theater, 202

Smithsonian Institution: Arts & Industries Building, 81-83, *82*; Board of Regents, 37, 111-13; Museum of History and

Smithsonian Institution (cont.)
Technology (National Museum of
American History), 108-11, 118-19;
National Museum of Natural History,
14-15, *113*; National Zoological Park,
45-46, 95, 205; Project Discovery,
212; revenues from television, 249n36;
television projects, 35-39, 45-46,
90-99, 101-20. See also *Smithsonian,
The* (television program); *Smithsonian
Adventure with Walter Schirra*;
Smithsonian Discovery Theater;
Smithsonian World
Smithsonian World, 129, 197-215, 209,
268n6. *See also* Bradley, Sandra
Wentworth; Carr, Martin; Cherkezian,
Nazaret; Malone, Adrian; Ripley, S. Dillon
Snowbird, Albert, 232
Snyderman, Nancy, 272n28
Social Studies of Science (journal), 127
Soranno, Alexander Mark, 245n47
Southwestern Bell Telephone Company,
210, 214, 270n89
space, television and, 73-74, 76-83, 103;
coverage of space missions, 79-80, 82;
news, 140-44, 194. *See also* Apollo 11
(space mission); *Friendship 7*; National
Aeronautics and Space Administration
(NASA)
Space Ghost, 180
Space: 1999, 187, 223
Space Patrol, 72. *See also* Buzz Corey
(fictional character)
Spaceship Earth, 98
Spanish Television Network, 200
Spirit of St. Louis, 37, *38*, 81, 82
Sputnik, 28, 32, 73
Squibb Corporation, 163
Stahl, Lesley, 194
Standard Oil of Great Britain, 98
Stanley, Wendell, 75
Stanton, Frank, 30
Stapleton, Maureen, 168
Star Gazer, 261n32. *See also* Horkheimer,
Jack
Star Trek, 5, 80, 180, 223, 226
Star Wars (defense initiative), 129
Star Wars (film), 156, 160
State of the Planet, 166
Steadman, David, 210

Steffens, Roy, 173
Steiger, Rod, 65
Stein, Sonia, 17-18
Steinke, Jocelyn, 194
Stephenson, William, 127
Stevenson, Robert Louis, 186
Stewart, T. Dale, 14-15
Stoppard, Tom, 212
Strange Case of the Cosmic Rays, The, 51,
54, *55*. *See also* Bell Telephone System:
specials
Strong, John, 14
Sunday at the Bronx Zoo, 45
Sunrise Semester, 32-33, 198
Super Friends, 180
Superman, 79
Survival Anglia Ltd., 181
Suspense, 187
Suzuki, David, 165, 167, 219
Swayze, John Cameron, 18, 21, 172
synthetic reality, 225-26, 273n41

Talbot, Lee M., 95-96, 250n49
Talbot, Marty, 95-96, 250n49
Tales of Tomorrow, 72
Taylor, Deems, 138
Taylor, Telford, 26-27
Teichner, Martha, 194
television stations, Boston, in 1977, 264n39
television stations (individual): KCET-TV
(Los Angeles), 158; KCTS-TV (Seattle),
162, 183; KQED-TV (San Francisco),
20; KTLA-TV (Los Angeles), 217;
WETA-TV (Washington, D.C.), 199-214,
268n29, 269n48, 271n13; WFIL-TV
(Philadelphia), 22; WGBH-TV (Boston),
124-26, 130-33, 179, 239n44, 254n14;
WHMM-TV (Washington, D.C.), 271n13;
WMAL-TV (Washington, D.C.), 10, 12;
WMAR-TV (Baltimore), 12; WMPT-TV
(Baltimore), 271n13; WNET-TV (New
York), 262n54; WPTZ-TV (Philadelphia),
11; WQED-TV (Pittsburgh), 32;
WTOP-TV, *38*
Teller, Edward, 20, 130, 151, 189, 192
Think Fast, 21
Thomas, Elizabeth Marshall, 211
Thomas, Lewis, 129
Thomas Debs Henderson (fictional
character), 150-51

Thoreau, Henry David, 1
Thread of Life, The, 51, 56. *See also* Bell
    Telephone System: specials
Three Mile Island (nuclear plant), 130, 143,
    148-50, *149*
3-2-1 Contact, 181-82. *See also* children's
    programs
Three Two One Zero (documentary), 18
Time-Life, Inc., 96, 118, 121-22, 134, 159
Today, 17, 157
Today with Mrs. Roosevelt, 19
Tom Corbett (fictional character), 72
Tom Corbett, Space Cadet, 72, 173
Tomlin, Lily, 183
Tom Mix, 173
"Tomorrowland" (segments), 49, 76-77.
    See also *Disneyland*
Tomorrow's World (BBC), 121, 134
Tomorrow the Moon, 76
Tonight Show, The, 31, 158
Tors, Ivan, 73
Toulmin, Stephen, 97-99, 250n57
Trembley, Francis J., 31
Trials of Life, 167
Triangle Television, 31
Tripp, Paul, 172
"Trip to the Moon, A," 11. *See also* Marshall,
    Roy K.
Trout, Robert, 21
TRSH ("transparent ratings seeking hype"),
    132
"True-Life Adventures." See *Disneyland*
TRW Inc., 132
Turner, Ted, 167
Turner Broadcasting System, 167, 200
Turow, Joseph, 63, 244n41, 245n50
TV Guide, 114, 151, 155-56, 183
Twain, Mark, 39
20,000 Leagues Under the Sea (film), 77.
    See also Disney Studios
Twilight Zone, 73
Twister (film), 188
230,000 Will Die (documentary), 18. *See also*
    medicine, on television

Ulene, Art, 272n28
Unchained Goddess, The, 51, 54, 74. *See also*
    Bell Telephone System: specials
"Undersea Archeology," 49. *See also*
    Cousteau, Jacques-Yves

*Undersea World of Jacques Cousteau,
    The,* 90
United Technologies, 181
Universe (CBS), 159-61, 202. *See also*
    Cronkite, Walter
University of Alabama, 33
University of Chicago, 33-34; Committee on
    Educational Television, 219; Committee
    on the Role and Opportunities in
    Broadcasting, 101, 104-5, 119; University
    Broadcasting Council, 104
University of Michigan, 33
University of Pennsylvania Museum, 21
University of the Air, 31
University of Utah, 33
University of Wisconsin, 35
U.S. Air Force, 16, 74
U.S. Department of Defense, 74
U.S. Department of Energy, 224
U.S. Department of Health, Education,
    and Welfare. *See* Learning Channel
    (TLC)
U.S. Naval Observatory, 10
U.S. Naval Task Force, 74
U.S. News & World Report, 71, 79-80
U.S. Office of Education, 181
U.S. Rubber Company, 9
Ustinov, Peter, 160

V, 224
Valens, Evans G. ("Red"), 75
Vanderbilt University News Archive, 140,
    257n16
Vanishing Prairie, The, 48
Vassar, David, 268n6
*Verdict on the Silent Spring of Rachel Carson,
    The* (documentary), 87
Verne, Jules, 71, 83
Victor Frankenstein (fictional character),
    6, 78, 151, 186-87
Vidal, Gore, 187
Vietnam war, 127, 140
violence, on television, 48, 99, 177, 180,
    222, 264n29
Violent Earth, The, 108
Violent Universe, The, 181
Visit to Lincoln Park Zoo, 47. *See also* Perkins,
    Marlin
von Braun, Wernher, 76
Vonnegut, Kurt, 211

Wade, Serena, 225
Wald, George, 243n49
Walt Disney Company, 183, 222
Walt Disney Studios, 48–49, 51, 76–77, 178
Walt Disney's Wonderful World of Color, 93
Walter Cronkite's Universe, 159–61, 202
Ward, Jonathan, 159
Warner, Jr., Jack, 91
Warner, William W., 96, 98–99, 102, 105
Warner Brothers Studios, 51–52, 55–57, 91, 166
War of the Worlds (radio broadcast), 152
War of the Worlds (television series), 224
Wartella, Ellen, 222
Watch Mr. Wizard, 68, 71, 173–75, 177. *See also* Herbert, Don
Watch the World, 172
Watergate (political crisis), 123, 140
Water World, 99
Watson, Thomas J., 199
Watts, Steven, 48
Weather Machine, The (BBC), 181
Weaver, Warren, 28, 52, 139, 174
Webb, James E., 252n41
"Web of Life: Exploring the Genetic Landscape," 210–11. *See also Smithsonian World*
Weiner, Jonathan, 164, 181–82
Weiss, Paul, 97
Welles, Orson, 152, 168
Wells, H.G., 1, 71, 187
Wentworth Films, 270n78. *See also* Bradley, Sandra Wentworth; *Smithsonian World*
westerns, on television, 35, 172, 245n47
Wetmore, Alexander, 36
WGBH Educational Foundation, 254n14
"What Hath God Wrought," 21. See also *Cavalcade of America*
What in the World?, 21, 237n72
What Is the Future of Television?, 24
What's New, 22
Wheldon, Huw, 121, 135
When Worlds Collide, 71
White, Harvey Elliott, 31–32, 33, 40
White, Theodore H., 109
Whitehead, Clay, 264n26
White Sands Test Facility, 75

Who Said That?, 21
Wide Wide World, 75, 81. See also *Dave Garroway Show, The*
Wild, Wild World of Animals, 99
Wild Cargo, 49
Wild Kingdom, 46–48, 179
Wildlife International, 222
Wildlife Journeys, 222
Wildlife Tales, 222
Wild Sanctuaries, 222
Williams, Raymond, 43, 123
Williams, Robert J., 24
Winkler, Allan M., 151
Winston, Harry, 116
Wolf, Thomas H., 203–7, 209, 269n58
Wolfle, Dael, 28
Wolman, Abel, 14
Wolper, David L., 202, 219, 251n30, 253n57; and Smithsonian, 97, 109–20, 113, 122, 198, 249n33, 254n90
women, and television, 129, 185–95, 267n37
Wonderful World of Disney, The, 81, 179. *See also* Disney Studios
World of Audubon, The, 167
World of Survival, The, 180
World's Fair, New York (1939), 1, 2, 3
World War II, effect on television, 3
World Wildlife Fund, 203
Wright Brothers (Orville and Wilbur), 36, 81, 82
Wylie, Philip, 10

X Files, The, 224

"Yellow Jack" (drama), 64–65
Young, Fred, 192
Young, Wesley A., 45
Young Dr. Malone, 66–67
Your World Tomorrow, 25

Zaloom, Paul, 183. See also *Beakman's World*
Zoom!, 179
Zoo Parade, 47. *See also* Perkins, Marlin
Zoo Quest (BBC), 166
Zoorama, 49
zoos, and television, 44–49, 95, 142, 173